Michael Werner Tobias Krayer

A hydrodynamical perspective on the turbulent transport of bacteria in rivers

Dissertationsreihe am Institut für Hydromechanik
Karlsruher Institut für Technologie (KIT)

Heft 2023/1

Eine Übersicht aller bisher in dieser Schriftenreihe erschienenen Bände finden Sie am Ende des Buches.

A hydrodynamical perspective on the turbulent transport of bacteria in rivers

by
Michael Werner Tobias Krayer

KIT Scientific Publishing

Karlsruher Institut für Technologie
Institut für Hydromechanik

A hydrodynamical perspective on the turbulent transport of bacteria in rivers

Zur Erlangung des akademischen Grades eines Doktors der Ingenieurswissenschaften
von der KIT-Fakultät für Bauingenieur-, Geo- und Umweltwissenschaften
des Karlsruher Instituts für Technologie (KIT) genehmigte Dissertation

von Michael Werner Tobias Krayer, M.Sc.

Tag der mündlichen Prüfung: 10. November 2022
Referent: Prof. Dr. Markus Uhlmann
Korreferent: Prof. Dr.-Ing. habil. Michael Manhart

Impressum

KIT Scientific Publishing

Karlsruher Institut für Technologie (KIT)
KIT Scientific Publishing
Straße am Forum 2
D-76131 Karlsruhe

KIT Scientific Publishing is a registered trademark
of Karlsruhe Institute of Technology.
Reprint using the book cover is not allowed.

www.ksp.kit.edu

Print on Demand 2023 – Gedruckt auf FSC-zertifiziertem Papier

ISSN 1439-4111
ISBN 978-3-7315-1310-0
DOI 10.5445/KSP/1000160619

Dedicated to Sophie & Lennard.

Abstract

The transport of bacteria in river systems is a phenomenon which occurs on a multitude of length scales ranging from the size of individual microbes up to the size of an entire estuary. At the same time the understanding of the spreading of microbial populations after a localised contamination event such as a combined sewer overflow is crucial for the prediction of the water quality downstream of the source, which is in turn essential to managing public health. It is well-established that microbial populations in fluvial systems may preferably be found on the surface of small particles rather than solely freely suspended in the water body. The attachment to particles provides an environment beneficial to the survival of bacteria due to the improved access to nutrients and the shielding from environmental stressors, but also alters their dispersion characteristics as the transport of bacteria is then coupled to the trajectories of heavy particles.

The importance in the distinction between the particle-attached and the freely-suspended mode of transport has been recognised in the mechanistic modelling of bacteria fate and transport. However, due to the multiscale nature of the problem, the mechanisms which govern the transport of particles in river-like flows are never resolved explicitly, and hence, the models profoundly rely upon the availability of accurate descriptions thereof. The associated problem of particles settling in a turbulent carrier flow is an active topic of research by itself, and is rich in emerging phenomena such as the emergence of spatial inhomogeneities or non-trivial modifications of the settling characteristics compared to quiescent environments. In particular, the transient settling of particles in horizontal open channels, which serves as an abstraction of particle-attached bacteria transport in rivers, has hitherto received only little attention in the literature. As a consequence, the knowledge on the impact of its defining features such as boundedness, anisotropy and vertical inhomogeneity on the settling characteristics is limited and needs to be addressed to enable the formulation of reliable models thereof.

The aim of this thesis is to fill the knowledge gap on the transport characteristics of heavy particles in turbulent horizontal open channel flows, and to identify phenomena which may be of importance in the context of bacteria transport modelling. For this purpose, the incompressible Navier–Stokes equations and the momentum balance equations for dispersed particles are solved using direct numerical simulations and the immersed boundary method. This approach resolves all relevant scales of turbulence and the microscopic flow around each particle explicitly, and thus, describes the particle-fluid interaction from fundamental principles of physics without the need of additional modelling. Apart from the contaminated particles, which are introduced near the free surface of the flow, the simulation domain includes approximately 100,000 fully resolved particles at the bottom of the domain, which form a realistic sediment bed, and enable the examination of the interaction between contaminated particles and mobile sediments.

Concerning the parameter space, the value of the friction Reynolds number is varied within the range $Re_\tau \in [241, 838]$, while the contaminant parameter space is chosen such that the resulting relative turbulence intensities—defined as the ratio between the friction velocity and the undisturbed terminal velocity—lie within the range $I_T \in [0.47, 2.88]$. Moreover, two types of sediment bedforms are investigated in order to assess their effect on contaminant transport, namely a macroscopically flat bed and a bed featuring ripples.

The analysis of the simulation data shows that the settling velocity of the contaminant particles is enhanced in the ensemble-averaged sense, yet, the time from beginning of the settling until the initial deposition is prolonged when compared to the ratio between the channel height and the terminal velocity. The enhancement is demonstrated to be a result of the preferential sampling of turbulent sweep events, which also implies that the streamwise component of the particle velocity is increased compared to the mean fluid velocity at the same position. A closer examination of the spatial organisation of contaminated particles reveals that they tend to accumulate in large-scale high-speed velocity streaks in the outer region of turbulence. Due to this focusing mechanism, the mean-squared lateral displacement of the settling particles stagnates in the lower half of the channel such that contaminants are not further dispersed in cross-stream direction until shortly before deposition. The same behaviour could be reproduced using a time-invariant exact coherent flow state resembling a hairpin vortex as a proxy for turbulence, and an extended parameter sweep in this setup suggests that this transport barrier effect persists even at high relative turbulence intensities. It is speculated that this phenomenon might confine contaminated particles to a region close to the river bank over a considerable downstream distance in the aftermath of a combined sewer overflow event, which might seriously impact decisions regarding public health measures.

Near the sediment bed, the barrier effect of the large-scale motions is inactive and contaminants are found to disperse laterally at a rate which presumably depends on the Shields parameter. The interaction between the sediment and the contaminants is distinct for the two bed topologies under investigation. In the case of macroscopically flat beds, the contaminated particles are transported towards sediment ridges which are in turn known to be a result of the action of large-scale fluid motions, and the mixing of contaminants and sediment particles is restricted to the thin layer of sediment near the interface. In contrast, the presence of ripples leads to a capturing effect where contaminated particles are preferentially deposited in the trough of the ripple, and subsequently buried by a thick layer of sediment due to the propagation of the bed feature. This mechanism temporarily immobilises a large share of all contaminated particles until the displacement of the ripple has sufficiently progressed for them to be eroded on the windward side. During the immobilisation, the associated bacteria are shielded from solar radiation to a substantial degree, which likely has a significant impact on their inactivation, especially in shallow waters. Moreover, the cyclic nature of this phenomenon may provide one of many explanations for bacteria storages which are known to exist in river sediments and may cause bursts in fecal bacteria indicator levels even in absence of immediate contamination events.

It is concluded that direct numerical simulation can be a valuable tool for the analysis of bacteria transport, and recommendations are made on how the conjectures compiled in this thesis can be targeted in laboratory experiments to examine their relevance.

Kurzfassung

Der Transport von Bakterien in Flusssystemen ist ein Phänomen welches auf einer Vielzahl von Längenskalen auftritt, die von der Größe einzelner Mikroben bis hin zur Größe eines ganzen Ästuars reichen. Zugleich ist die Ausbreitung mikrobieller Populationen nach einem lokal begrenztem Verschmutzungsereignis wie es beispielsweise nach dem Überlauf eines Mischwasserkanalisationssystems auftritt entscheidend für die Vorhersage der Wasserqualität flussabwärts der Eintrittsstelle, welche wiederum für das Management der öffentlichen Gesundheit unerlässlich ist. Es ist allgemein bekannt, dass mikrobielle Populationen in Fließgewässern vorzugsweise auf der Oberfläche kleiner Partikel aufzufinden sind und nicht ausschließlich frei schwebend im Gewässer. Die Anhaftung an Partikel bietet ein günstiges Umfeld für das Überleben von Bakterien aufgrund des verbesserten Zugangs zu Nährstoffen und der Abschirmung vor Umweltstressfaktoren, verändert aber auch ihre Ausbreitungseigenschaften, da der Bakterientransport dann an die Trajektorie schwerer Partikel gekoppelt ist.

Die Bedeutsamkeit der Unterscheidung zwischen dem an Partikel gebundenen und dem frei schwebenden Transportmodus ist bei der mechanistischen Modellierung des Schicksals und Transports von Bakterien bereits erkannt worden. Allerdings werden aufgrund des mehrskaligen Charakters des Problems die Mechanismen, die den Transport von Partikeln in flussähnlichen Strömungen steuern nie explizit dargestellt und daher sind die Modelle in hohem Maße von der Verfügbarkeit von genauen Beschreibungen abhängig. Das damit verbundene Problem der Ablagerung von Partikeln in einer turbulenten Trägerströmung ist selbst ein aktives Forschungsthema und ist reich an emergenten Phänomenen wie z.B. dem Auftreten von räumlichen Inhomogenitäten oder nicht-triviale Veränderungen der Absetzcharakteristiken im Vergleich zu einer ruhenden Umgebung. Insbesondere die unstetige Ablagerung von Partikeln in horizontalen offenen Kanälen, die als Abstraktion des partikelgebundenen Bakterientransports in Flüssen dient fand bisher in der Literatur nur wenig Beachtung. Infolgedessen ist das verfügbare Wissen über die Auswirkungen der bestimmenden Merkmale wie Begrenztheit, Anisotropie und vertikale Inhomogenität auf das Absetzungsverhalten limitiert und muss adressiert werden um die Formulierung von zuverlässigen Modellen zu ermöglichen.

Das Ziel dieser Arbeit ist es die Wissenslücke über die Transporteigenschaften von schweren Partikeln in turbulenten offenen Kanälen zu füllen und Phänomene zu identifizieren, die im Kontext der Bakterientransportmodellierung von Bedeutung sein könnten. Zu diesem Zwecke werden die inkompressiblen Navier–Stokes-Gleichungen und die Impulsbilanzgleichungen für dispergierte Partikel mittels direkter numerischer Simulation und der Immersed Boundary-Methode gelöst. Dieser Ansatz löst alle relevanten Skalen der Turbulenz und die mikroskopische Strömung um jeden Partikel explizit auf und beschreibt so die Partikel-Fluid-Wechselwirkung auf der Grundlage fundamentaler physikalischer Prinzipien ohne dass eine zusätzliche Modellierung erforderlich ist. Abgesehen von den kontaminierten Partikeln, die in der Nähe der freien Oberfläche der Strömung eingebracht werden, umfasst der Simulationsbereich etwa 100.000 vollständig aufgelöste Partikel am Boden des Strömungsgebiets, die ein realistisches Geschiebe bilden und eine Untersuchung der Wechselwirkung zwischen kontaminierten Partikeln und mobilen Sedimentpartikeln ermöglichen.

Was den Parameterraum betrifft, so wird der Wert der Reynoldszahl basierend auf der Wandschubspannungsgeschwindigkeit im Bereich von $Re_\tau \in [241, 838]$ variiert während der Parameterraum für die Verunreinigungen so gewählt wird, dass die resultierenden relativen Turbulenzintensitäten definiert als das Verhältnis zwischen der Wandschubspannungsgeschwindigkeit und der ungestörten Endgeschwindigkeit innerhalb des Bereichs $I_\tau \in [0.47, 2.88]$ liegen. Darüber hinaus werden zwei Arten von Sedimentstrukturen untersucht um deren Auswirkungen auf den Schadstofftransport zu bewerten, nämlich ein makroskopisch flaches Bett und ein Bett mit Rippeln.

Die Analyse der Simulationsdaten zeigt, dass der Ensemble-Mittelwert der Absetzgeschwindigkeit der Schadstoffpartikel erhöht ist, aber die Zeit vom Beginn des Absetzens bis zur ersten Ablagerung ist im Vergleich zum Verhältnis zwischen der Kanalhöhe und der Endgeschwindigkeit verlängert. Die Erhöhung ist das Ergebnis des bevorzugten Aufsuchens von turbulenten Sweep-Ereignissen. Dies beinhaltet auch, dass die Hauptströmungskomponente der Partikelgeschwindigkeit im Vergleich zur mittleren Strömungsgeschwindigkeit an der gleichen Stelle erhöht ist. Eine genauere Untersuchung der räumlichen Organisation der kontaminierten Partikel zeigt, dass sie dazu neigen sich in großskaligen Hochgeschwindigkeitsstreifen in der äußeren Region der Turbulenz zu sammeln. Aufgrund dieses Fokussierungsmechanismus stagniert die mittlere quadratische Seitenverschiebung der sich absetzenden Partikel in der unteren Hälfte des Kanals, sodass die Verunreinigungen erst kurz vor der Ablagerung weiter in Querstromrichtung dispergiert werden. Das gleiche Verhalten konnte reproduziert werden indem ein zeitinvarianter, exakt kohärenter Strömungszustand, der einem Haarnadelwirbel ähnelt, als Ersatz für Turbulenz verwendet wird. Eine erweiterte Parametervariation in dieser Anordnung deutet darauf hin, dass dieser Transportbarriereeffekt auch bei hohen relativen Turbulenzintensitäten bestehen bleibt. Es wird vermutet, dass dieses Phänomen feine kontaminierte Partikel über eine beträchtliche Strecke flussabwärts in eine Region nahe des Flussufers eingrenzen kann, was Entscheidungen über Maßnahmen für die öffentliche Gesundheit ernsthaft beeinflussen könnte.

In der Nähe des Sedimentbettes ist die Barrierewirkung der großskaligen Bewegungen inaktiv und die Schadstoffe breiten sich seitlich mit einer Geschwindigkeit aus, die vermutlich vom Shields-Parameter abhängt. Die Wechselwirkung zwischen dem Sediment und den Schadstoffen ist für die beiden untersuchten Topologien des Geschiebes unterschiedlich. Bei makroskopisch flachen Sedimentbetten werden die verunreinigten Partikel in Richtung der Sedimentgräte transportiert, die selbst ein Ergebnis großskaliger Fluidbewegungen sind und die Vermischung von Schadstoffen und Sedimentpartikeln ist auf die dünne Sedimentschicht nahe der Grenzfläche beschränkt. Im Gegensatz dazu führt das Vorhandensein von Rippeln zu einem Einfang-Effekt bei dem kontaminierte Partikel bevorzugt im Tal der Rippel abgelagert und anschließend von einer dicken Sedimentschicht verdeckt werden durch die Bewegung des Geschiebes. Durch diesen Mechanismus wird ein großer Anteil an kontaminierten Partikel vorübergehend immobilisiert bis die Bewegung der Rippel so weit vorangeschritten ist, dass sie auf der Luvseite erodiert werden. Während der Immobilisierung werden die anhaftenden Bakterien von der Sonneneinstrahlung abgeschirmt was wahrscheinlich einen erheblichen Einfluss auf ihre Inaktivierung hat, insbesondere in flachen Gewässern. Darüber hinaus könnte die zyklische Eigenschaft dieses Phänomens eine von mehreren Erklärungen für Bakterienspeicher sein, die bekanntermaßen in Flusssedimenten vorhanden sind und zu Ausbrüchen der Indikatorwerte für fäkale Bakterien führen können auch wenn keine unmittelbaren Verschmutzungsereignisse vorliegen.

Die Schlussfolgerung lautet, dass die direkte numerische Simulation ein wertvolles Instrument für die Analyse des Bakterientransports sein kann, und es werden Empfehlungen gegeben, wie die in dieser Arbeit aufgestellten Vermutungen in Laborexperimenten auf ihre Relevanz hin überprüft werden können.

Acknowledgements

Although the purpose of a doctoral thesis is to demonstrate the ability to conduct novel research independently, the final product is certainly not independent from the human interactions which took place over the research period. This section is therefore dedicated to the colleagues, friends, family and institutions, who directly and indirectly contributed to this thesis.

By far the strongest impact on my work has been exerted by my supervisor, Markus Uhlmann, who I highly regard as an exceptional scientist and mentor. They say, that a researcher's style of thinking and approaching problems is shaped by their supervisor, and if there is a truth to this, I am certainly very privileged. I cannot express enough how much I appreciate your supportiveness, availability, and willingness to discuss even the tiniest details of my research. The conversations I had during my time as a board member of the doctoral committee showed me that this is not the norm, and that I was very lucky to enjoy such an outstanding supervision.

It should also be acknowledged that many ideas have been initiated, refined or discarded within valueable discussions with colleagues and friends.[1] Honourable mentions in this regard are certainly Kay Schäfer, with whom I spent a large share of my break time and who helped me establish a respectable collection of mate lemonade bottles,[2] and Tiago Pestana, who was always willing to discuss—even when our children did not want us to. I appreciate the true friendships which have formed over time and created many memories to keep, be it badly planned motorcycle trips with Yoshiyuki Sakai—who somehow surpassed my wife as the person that I crossed the most international borders with—or enjoying expensive drinks with Antoine Morente in Paris to comply with the Canadian customs regulations. On a more serious note, the regular chats with Helmut Oppmann, Heidi Lauinger and Markus Scherer are greatly acknowledged as they (successfully) aided me through various difficult times. Unfortunately, the limited space of this section does not allow me to reminisce as much as I would like to, nor to give all my colleagues the credit they deserve. Nonetheless, my time at the Institute of Hydromechanics would not have been the same without my wonderful co-workers Agathe Chouippe, Aman Kidanemariam, Angelika Fels, Christian Bauer, Christof Gromke, Herlina, Manuel Moriche, Michele Pinelli, Michele Trevisson, Swagat Nayak and Yulia Akutina, among others, who I very much appreciate.

Science lives from the exchange with researchers outside one's direct sphere of influence who can provide unbiased feedback. In this regard, I first and foremost thank Michael Manhart for reviewing my thesis, for being a member of my committee and for the kind words in his evaluation of my work. Likewise, I greatly acknowledge Harald Horn for being a member of my committee and for sharing his perspective and expertise from another field of research. Moreover, I thank Olivier Eiff for chairing the committee and his support through the years. I very much enjoyed the vivid discussion during my defense with all of you.

[1] There certainly does not exist a sharp distinction between those two groups.

[2] A valueable asset which still needs to be cashed out.

Some of the more advanced ideas in this thesis originated from a research exchange programme with Osaka University, which was funded by the German Academic Exchange Service (DAAD) and the Japan Society for the Promotion of Science (JSPS). The discussions with and guidance of Genta Kawahara and Susumu Goto are deeply appreciated. Among all other members of the programe, I especially thank Julius Lustro for patiently explaining many concepts of the theory of nonlinear systems to me.

The work of the students I have supervised during my time as research assistant at KIT is appreciated and I wish Mehrdad Ordoobadi, Markus Müller and Shaobin Zhuo all the best for their future careers.

All of this research would not have been possible without various sources of fundings. I acknowledge the Baden-Württemberg Stiftung, which funded the majority of my employment through the project "Microorganisms and Turbulence" within the call "High-performance computing II", as well as KIT and the Karlsruhe House of Young Scientists for funding the remaining parts. Regarding the computational resources, the funding of the large-scale project "Finite-size Particle Dynamics and Scalar Transport in Turbulent Open Channel Flow Over a Mobile Sediment Bed" by the Gauss Centre for Supercomputing is greatly acknowledged, as well as the support by the state of Baden-Württemberg through bwHPC. The work of the High-performance Computing Center Stuttgart and the Steinbuch Centre for Computing is appreciated, especially with respect to the resources they provided (Hazel Hen, Hawk, ForHLR II, bwUniCluster 2.0, LSDF, bwDataArchive).

Contents

Nomenclature

Acronyms

BC	boundary condition
CSO	combined sewer overflow
DEM	discrete element method
ECS	exact coherent state / exact coherent structure
EQ	equilibrium solution
HBG	Homann et al. (2013)
HR	high Reynolds number (case acronym)
IBM	immersed boundary method
KIT	Karlsruhe Institute for Technology
LR	low Reynolds number (case acronym)
LSM	large-scale motion
PB	particle-attached bacteria
PO	periodic orbit
RP	ripple (case acronym)
SB	suspended bacteria
SN	Schiller and Naumann (1933)
TW	travelling wave
UV	ultraviolet radiation
VLSM	very-large-scale motion

Averaging operators and conditioned sets

$\langle \bullet \rangle_{\mathbb{T}}$	ensemble-average at a given instance in time (eq. (4.1))
$\langle \bullet \rangle_{\mathbb{X}}$	ensemble-average at the time of first crossing of a wall-parallel plane (eq. (4.5))
$\langle \bullet \rangle_{\Gamma}$	average of the Gamma distribution
$\langle \bullet \rangle_{\Delta}$	binned average
$\langle \bullet \rangle_{\mathbb{P}}$	particle-conditioned average
$\langle \bullet \rangle_{\mathbb{S}}$	average over virtual sphere
$\langle \bullet \rangle_{\mathbb{S}\mathbb{X}}$	average over virtual sphere at the time of first crossing of a wall-parallel plane
$\langle \bullet \rangle_{x}$	spatial average over streamwise direction
$\langle \bullet \rangle_{z}$	spatial average over spanwise direction
$\langle \bullet \rangle_{t}$	temporal average

$\langle \bullet \rangle_{xt}$	streamwise-temporal average
$\langle \bullet \rangle_{xz}$	streamwise-spanwise average
$\langle \bullet \rangle_{xzt}$	streamwise-spanwise-temporal average
$\langle \bullet \rangle_{xyzt}$	average over all spatial directions and in time
$[\bullet]_{\mathbb{S}}$	set of samples on virtual sphere
$[\bullet]_{\mathbb{S}\mathbb{X}}$	set of samples on virtual sphere at the time of first crossing of a wall-parallel plane
$\uparrow, \downarrow, \uparrow\downarrow$	indicates evaluation of at the time of upward, downward or all crossings

Latin symbols and variables

$\mathbf{a}_f$	fluid acceleration with components $(a_{p,x}, a_{p,y}, a_{p,z})^T$
A_V	Voronoï cell area
$\mathbf{c}$	propagation velocity
c_{RP}	propagation velocity of ripple
c_{TW}	propagation velocity of travelling wave
$c_{d,norm}$	collision dissipation coefficient (normal)
$c_{d,tang}$	collision dissipation coefficient (tangential)
C_d	drag coefficient
C_{SB}, C_{PB}	concentration of suspended/particle-attached bacteria
$\mathcal{D}_{SB}, \mathcal{D}_{PB}$	dispersion coefficient of suspended/particle-attached bacteria
$\mathcal{D}_{c,z}$	lateral particle dispersion coefficient
d_p, d_c, d_s	particle diameter
$\mathbf{F}_{p,col}$	particle collision force
Fr	Froude number
$\mathbf{g}, g$	gravitational acceleration (vector, magnitude)
Ga	Galileo number
h_b	instantaneous and spatially-resolved bed height
$h_{p,0}$	spacing of particle for initial condition of settling in travelling wave
H	channel height in case of open channel, centreline position for closed channel
H_f	mean clear fluid height
H_b	mean bed height
$\mathbf{I}_p, I_p$	particle moment of inertia (vector, magnitude)
I_η	relative turbulence intensity based on u_η
I_τ	relative turbulence intensity based on u_τ
I_v	relative turbulence intensity based on $\langle v_f^2 \rangle_{xzt}^{1/2}$
I_{UV}	intensity of ultraviolet irradiance
$j_{c,\mathbb{X}}$	number flux of particles through wall-parallel plane
k_{att}	attenuation coefficient of Beer's law
k_s	sand roughness
k_{elas}	collision force stiffness coefficient
ℓ	filter aspect ratio

L_x, L_y, L_z	domain size
L_w	wake length
m_p	particle mass
N_p	number of particles per realisation
$N_{p,bed}$	number of particles classified as bed
$N_{p,if}$	number of particles classified as interface
$N_{p\Sigma}$	number of particles over all realisations
N_{ref}	refinement factor
N_{run}	number of realisations
N_{sub}	number of sub-steps of collision model
N_x, N_y, N_z	number of grid points
N_σ	number of filter levels
$N_{c,\mathbb{X}}$	number of crossings
$\mathcal{P}_{\mathbb{X}}$	probability density of crossing
Pe	Péclet number
Pr	Prandtl number
q_{PB}	deposition flux of particle-attached bacteria in the stream model of Jamieson et al. (2005)
$q_{p,x}$	local streamwise particle flux
$Q_{p,x}$	streamwise particle flux
R_{h_b}	two-point, two-time correlation of bed interface fluctuation
R_{NN}	distance to nearest neighbouring particle
Re	Reynolds number
Re_b	bulk Reynolds number
Re_c	Couette Reynolds number
Re_τ	friction Reynolds number
Re_p	particle Reynolds number
$Re_{p,\infty}$	particle terminal Reynolds number
$\mathcal{S}$	shear parameter
S_p	particle surface
S_v	surface of virtual sphere
S_{xy}	fluid mean shear
Sc	Schmidt number
St	Stokes number
St^+	Stokes number based on viscous scales
$t_{c,d}$	deposition time
t_p	particle response time
$t_{p,s}$	characteristic settling time
$t_{c,\mathbb{X}}$	time at contaminant particle crossing a given wall-parallel plane
t_f	characteristic fluid time scale
t_b	bulk time unit
t_v	viscous time scale

$\mathbf{T}_{p,col}$	particle collision torque
u_b	bulk velocity
u_H	centreline velocity
u_τ	friction velocity
u_η	Kolmogorov velocity scale
$\mathbf{u}_f$	fluid velocity with components $(u_f, v_f, w_f)^T$
$\hat{\mathbf{u}}_f$	Fourier-transformed fluid velocity
$\hat{\mathbf{u}}_f^{(k,l)}$	Fourier mode of fluid velocity field
$v_{f,def}$	velocity deficit
$\mathbf{u}_p$, $\mathbf{u}_c$, $\mathbf{u}_s$	particle velocity with components $(u_p, v_p, w_p)^T$
$\mathbf{u}_w$	velocity of wall
$\bar{v}_{c,\mathbb{X}}$	group velocity of contaminant particles
$v_{c,\mathbb{X}}$	velocity of contaminant particle while crossing a given wall-parallel plane
$v_{p,g}$	gravitational velocity
$v_{p,\infty}$	terminal velocity under quiescent and isolated conditions
V_p	particle volume
$\mathbf{x}_p$, $\mathbf{x}_c$	particle position with components $(x_p, y_p, z_p)^T$
$\overrightarrow{\mathbf{x}}_c$	contaminant trajectory extended across periodic boundary
$x_{p,0}$, $y_{p,0}$, $z_{p,0}$	initial position of settling particle
$x_{c,\mathbb{X}}$, $z_{c,\mathbb{X}}$	horizontal particle position at time of crossing
y_{SB}	lower boundary of layer of suspended bacteria

Greek symbols and variables

β	fraction of surface area which is exposed to sunlight
δ_v	viscous length scale
σ_{box}	width of box filter
Δ_{col}	collision force range
$\delta_{col}^{(i,j)}$	collision overlap
η	Kolmogorov length scale
$\overrightarrow{\delta x_c}, \overrightarrow{\delta z_c}$	displacement from initial position
$\overrightarrow{\delta x_{c,d}}, \overrightarrow{\delta z_{c,d}}$	displacement from initial position at deposition
$\Delta x, \Delta y, \Delta z$	mesh spacing $(\Delta x = \Delta y = \Delta z)$
Δt	time step
ε_{rest}	coefficient of restitution
Γ_p	boundary of solid subdomain
ϱ	Pearson correlation coefficient
ρ_f	fluid density
ρ_p, ρ_c, ρ_s	particle density

Ψ	gravitational potential
ς_{SB}	smoothing length of layer of suspended bacteria
μ_{fric}	Coulomb friction coefficient
κ_x, κ_z	wavenumber
λ_x, λ_z	wavelength
θ	Shields parameter
τ_f	fluid shear stress
Φ_p	solid volume fraction
ϕ_p	local solid volume fraction
χ_p	solid indicator function
ν_f	fluid viscosity
$\boldsymbol{\omega}_f$	fluid vorticity with components $\left(\omega_{f,x}, \omega_{f,y}, \omega_{f,z}\right)^T$
$\boldsymbol{\omega}_p$	particle angular velocity
$\boldsymbol{\Omega}_p$	particle angular position
Ω	simulation domain

1 Introduction

To this day, it has been around 200 years since Claude-Louis Navier published his seminal work on the laws of viscous fluids (Navier 1823) in which he was the first to derive the equations nowadays referred to as the Navier–Stokes equations. Perhaps it is delineative for the complexity which is comprised by these equations that the existence and smoothness of its solutions in three dimensions is yet to be proven and that such a proof constitutes one of the seven "Millennium Prize Problems" selected by the Clay Mathematics Institute in 2000. The most prominent manifestation of this complexity is turbulence—the chaotic state of motion where the fluid flow is characterized by the presence of eddies which emerge on a wide continuous range of spatial and temporal scales. Turbulence is the norm rather than the exception in natural and technical environments. In fact, it is safe to say that life as we know it would not be able to exist without it, as turbulence is essential to virtually all ecosystems, for instance in the form of large-scale phenomena such as the weather, the hydrological cycle and the carbon cycle, or on a smaller scale in form of crop pollination and the locomotion of the majority of fish, birds and flying insects to name but a few. This essentiality becomes immediately perceptible whenever the flow activity is reduced, with prominent examples being the undesired growth of blue-green-algae in water bodies (Thomas and Gibson 1990) or the lack of pollutant dispersion during smog events (Neiburger 1957). Presumably it is this combination of complexity, ubiquitousness and relevance which inveigles scientists from various academic disciplines such as meteorology, oceanography, engineering, applied mathematics, or perhaps less evidently, cosmology or philosophy, to entrain themselves into the study of fluid motion, and which at the same time serves as an inspiration for various famous works of fine art such as Leonardo Da Vinci's "Studies of water", Hokusai's "The Great Wave off Kanagawa" or Vincent van Gogh's "The Starry Night" which all share a fascinating passion for detail regarding the inherent features of turbulence (Colagrossi et al. 2021, Marusic and Broomhall 2021, Cartwright and Nakamura 2009, Aragón et al. 2008).

It may not have gone unnoticed from the previously provided examples that turbulence rarely arises isolated in natural environments. In fact, more often than not, it is the interaction of a fluid flow with its environment which is appreciated by the observer—think of the flickering of distance stars due to chaotic variations in the density of the atmosphere or the rustling of leaves in the wind. While Navier's research was, inter alia, actuated by the search for a more accurate description of the interaction between a solid object and its surrounding fluid, it was an unfortunate stroke of fate that he was misguided in the choice of appropriate boundary conditions for his equations, which left him ultimately unsuccessful (Darrigol 2002). Motivated by the accurate description of pendulums used in geodesic measurements, it was George Gabriel Stokes in 1845 who came to the conclusion that a vanishing velocity near the solid-fluid interface is the most natural (Stokes 1845), introduced the concept of flow instabilities and subsequently formulated a law for the flow resistance experienced by a sphere under creeping flow conditions (Stokes 1851). With these findings, Stokes laid much of the groundwork for today's mathematical analysis of particle-laden flows—the study of the combined motion of particles and their embedding fluid. Particle-laden flows are similarly to turbulence omnipresent in nature, and frequently both go hand in hand. Phenomena such as precipitation in clouds, sand dunes in a desert or the bed forms in rivers are all results of the multifarious interplay between a solid and a predominantly turbulent fluid phase, and besides their ability to inspire scientists and artists alike, many of these phenomena remain far from being fully understood down to the present day.

1.1 Motivation

In a world rich of flow phenomena, researches kin to the field of fluid dynamics of the last two centuries have been able to tackle increasingly complex problems involving fluid flows and their interactions. While many studies are dedicated to the improvement of the fundamental understanding of turbulence and particle-laden flows in the matter, the field has advanced sufficiently to address a multitude of practical problems, and on many occasions, progress in one aspect comes along with progress in the other as well, just like with the discoveries of Navier and Stokes. The motivation of this thesis is rooted in the numerical prediction of the propagation of solid contaminants in fluvial systems, and as will be elaborated in the following, the transport processes relevant to this problem are not yet comprehensively understood neither with respect to their inferences on the practical application, nor in terms of the underlying hydrodynamical phenomena.

The guiding theme of this thesis has its origin in the issue of the anthropogenic contamination of streaming freshwater ecosystems which unfortunately still negatively affects the health and quality of life of many people today. One substantial source of hazardous river pollution are *combined sewer overflows* (CSOs), which are discharge events of sewage water into a nearby water body. As the name suggests, CSOs predominantly occur in a combined sewer system—a unified waste water system for municipal/industrial sewage and surface runoff. The load on such a system may fluctuate immensely due to the amount of precipitation which is received, and it is at times of heavy rainfall that the capacity of the infrastructure may be exceeded and the storm sewage is partially discharged without adequate treatment.

Combined sewer systems are by no means rare. In the United Kingdom, France and Germany an estimate of around 70% of the existing sewerage infrastructure by length are combined sewers (Butler and Davies 2004) and this is especially the case for Europe's major historic cultural centres such as London, Madrid or Paris (Gasperi et al. 2008).[1] The situation is similar in the Northeastern and in the Great Lakes region of the United States of America where more than 40 million people are served by such systems (USEPA 1995). In Canada, 890 million cubic meters of untreated sewage water has been discharged during CSOs into 844 different water bodies between 2013 and 2017 (ECCC 2019),[2] which highlights the fact that this mode of pollution is still a common occurrence today even in industrialized countries. Combined sewer overflows are also in no way restricted to the Western hemisphere, as some of the largest metropolitan regions of East Asia—and thus the world—suffer from the very same problem with examples including Shanghai (Li et al. 2010), Shenzhen (Talamini et al. 2016), Tokyo (Ham et al. 2008) and Osaka (Wada et al. 2002). Also, it should not go unnoticed that similar discharge events can take place in separated sewer systems if the dedicated sewage system is allowed to be considerably infiltrated by rainwater, which has been reported e.g. in the USA (USEPA 1970, 1971) or Brazil (Reda et al. 2014).

1 Although perhaps not of the same cultural importance, it can be noted that the city of Karlsruhe, where most of this thesis was compiled, has both combined and separated sewer systems and is planning to convert the existing infrastructure into a unified combined sewer in the future (City of Karlsruhe 2010).

2 To put this number into perspective, this corresponds to a mean yearly wastewater discharge of nearly four times the freshwater discharge of the small river Pfinz (LUBW 2009) located just outside the writer's window.

The adverse effects of CSOs on general freshwater quality and human health have been known for more than half a century (Palmer 1950, USPHS 1964). Although the effluent in CSOs is to a large portion composed of storm water, its ability to pollute is substantial nonetheless as surface runoff itself is known to be a non-negligible source of pollutants such as heavy metals, mineral oils, pesticides, but also coliform bacteria (Ellis 1986),[3] and the intermittent nature of heavy storms may trigger a sudden release of depositions held back in the sewers (Butler and Davies 2004). The composition of contaminants found in the effluent varies depending on the origin of the wastewater and the runoff, but typically includes pathogens, oxygen-demanding substances, nutrients, toxics and sanitary debris (House et al. 1993). The contamination may therefore impair the viability of aquatic habitats (eutrophication due to nutrients; oxygen depletion; toxics), pose a potential threat to public health (pathogens) or lead to a general degradation in aesthetics of the receiving waters (eutrophication; sanitary debris), cf. Butler and Davies (2004). More recent studies suggest that the discharge of CSOs may also constitute an environmental source to pollutants whose long-term consequences are not yet well understood, such as hormones and micropollutants (Phillips et al. 2012, Dittmer et al. 2020) or antibiotic resistance genes (Brown et al. 2019, 2020).

Naturally, actions have been taken already in many countries to improve the environmental situation. For instance, the city of Osaka was able to reduce their CSO loads by 75% by means of a storm water reservoir which dampens the instantaneous peak loads on water treatment plants during heavy rainfall events (Wada et al. 2002). While this certainly is an effective method to solve the problem and is employed worldwide at various scales, it is at the same time cost-intensive, and hence, not always a viable solution. Moreover, the expanding urbanization and the associated sealing of soils increases the surface runoff in many places, which leads to increasing peak loads on the sewer system, and hence, the requirements on reservoir capacities are not static (Semadeni-Davies et al. 2008). There are also indications that climate change may exacerbate this situation in the future in regions where rainfall intensity increases (Semadeni-Davies et al. 2008, Kleidorfer et al. 2009, Fortier and Mailhot 2015).

The severity of the adverse impacts of CSOs also depends on the engineering design of the outlet. The major risk factor for public health are pathogens such as bacteria, which exhibit the tendency to attach onto solids present in the wastewater (Gannon et al. 1983, Auer and Niehaus 1993, Characklis et al. 2005). For this reason, it is often attempted to separate these solids from the storm sewage before it is discharged by the use of stilling ponds or other types of separators. The effectiveness of this separation step depends heavily on the settling velocity of the individual solid particles. Particles which are large or whose density differs considerably from that of water separate more rapidly from the effluent than small or nearly neutrally-buoyant particles, and are thus more likely to be removed under high throughput conditions. On the other hand, the concentration of fine suspended solids and dissolved material in the spill flow is known not to differ considerably from that of the continuation flow to the treatment plant (Butler and Davies 2004), while at the same time it is known that most enteric bacteria are attached to small particles (Walters et al. 2014). The attachment of bacteria onto suspended particles also obstructs disinfection measures which are at times applied to the effluent (Madge and Jensen 2006, Dickenson and Sansalone 2012, Gibson et al. 2017).

The environmental impacts caused by CSOs are severe, highly relevant worldwide and difficult to fully avoid in the near future. It is thus worthwhile to investigate the processes which occur when untreated wastewater enters a natural fluvial system in order to assess the fate of the entering pollutants. Generally, these processes can be divided into two types—biotic and abiotic processes. Biotic processes are

[3] The hazardousness of surface runoff is one of the reasons why separated sewer systems are not always the preferred solution, as in these systems runoff is typically released with only little treatment (De Toffol et al. 2007).

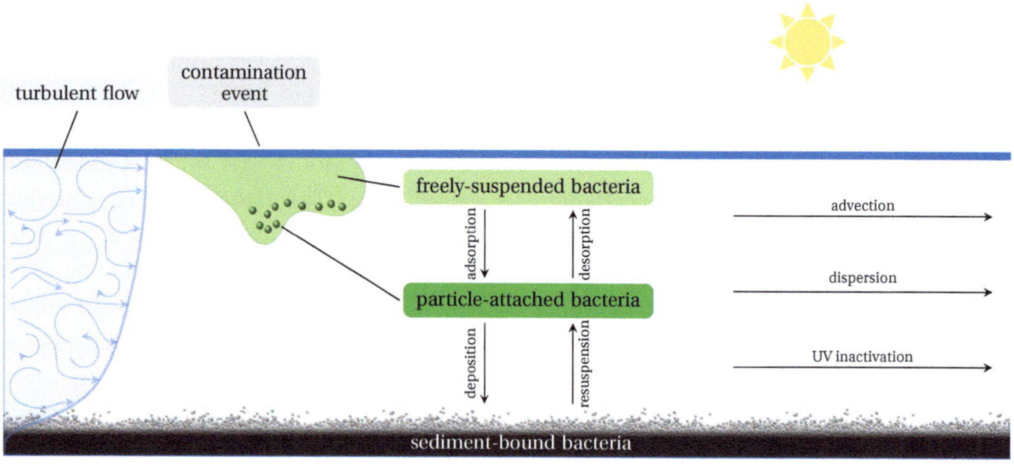

Figure 1.1: Schematic portrayal of abiotic processes associated to a localized contamination event in a fluvial environment. Image adapted from Walters (2013).

directly related to the living nature of microorganisms and examples may include the growth of bacterial populations, the natural inactivation thereof, predation by other organisms or competition with other microbial populations. As pathogens are one of the major threats to public health, a comprehensive understanding of biotic processes is crucial for the assessment and prediction of water quality. However, due to their living nature, these processes usually cannot be solely understood from first principles, and thus, their mathematical description commonly relies on empirical findings. On the contrary, abiotic processes are immediately subject to the laws of physics, and in many cases, to the laws of (hydro-)mechanics. Prominent examples may include transport and mixing processes, the deposition of solids, but also the attachment/detachment of bacteria onto/from solids. Moreover, both types of processes regularly dependent on each other. For instance, the evolution of microbial populations is largely influenced by the local availability of oxygen and nutrient which is a result of physical transport mechanisms. Conversely, the presence of microorganisms modifies the concentration field of these substances, which in turn affects the transport behaviour.

For the prediction of water quality after a spill event, it is crucial to know how the contamination propagates downstream. In the case of dissolved pollutants, the spreading is in its essence a mixing problem between freshwater and sewer effluent, which is rather well understood from the point of view of fluid mechanics. However, the exclusive consideration of dissolved contamination is known to be insufficient to adequately describe the characteristics of pollutant propagation, especially in regard to pathogens (Gannon et al. 1983, Auer and Niehaus 1993, Wilkinson et al. 1995, Jamieson et al. 2005). This is for the most part a result of the tendency of microorganisms to attach onto solid matter, which is thought to be a major mode of bacteria transport in fluvial systems (Jamieson et al. 2004, Droppo et al. 2009, Walters et al. 2014). For this reason, it is inevitable to also consider the transport of suspended solids within the receiving water body in any detailed analysis or model of this phenomenon, which implies that a turbulent particle-laden flow problem has to be studied.

The interactions of microbiologically contaminated particles with their surroundings are manifold and not well understood. Compared to freely suspended bacteria, particle-associated bacteria are known to survive over longer periods of time (Garcia-Armisen and Servais 2009), which is thought to be a combined result of protection against detrimental environmental factors such as UV irradiation or predation, a better availability of nutrients, and, for some materials, an enhanced protection by biofilms (Roper and Marshall 1974, Erkenbrecher 1981, Wright et al. 1995, Craig et al. 2004, Schultz-Fademrecht et al. 2008). There are believed to be two mechanisms of bacterial adsorption—weak adsorption where van der Waals forces overcome repulsive forces, and strong adsorption due to cellular appendages or extracellular polymers excreted from the cell (Jamieson et al. 2005). Whereas the latter is considered to be irreversible, the former allows for a transition between the suspended and the particle-attached state which constitutes a mass transfer problem in the framework of fluid mechanics.

Figure 1.1 shows an abstraction of the major abiotic processes which occur downstream of a contamination event according to Walters (2013). Apart from the processes which were already elaborated on, the graphic also indicates that contaminated particles may interact with the natural river sediment bed present in a river. The ability of particles to be deposited and resuspended has striking implications on water quality prediction, as it provides a way for pathogens to be stored and released at a later time, possibly long after the primordial event occurred (Sherer et al. 1992). The release may be triggered by an increase in discharge of the contaminated river, e.g. due to precipitation, which promotes the transport of sediment, and increases in bacterial levels by a factor of 10 to 100 have been reported in literature (McDonald and Kay 1981, McDonald et al. 1982, Sherer et al. 1988) which demonstrates the relevance of this effect.

1.2 State of research in bacteria transport modelling

The prediction of water quality using models serves two major purposes—the issuance of public warnings in the case of an acute contamination event, and the promotion of the general understanding of pollutant spreading, e.g. to make informed engineering decisions. For the former purpose, real-time forecasts ("nowcasts") are typically obtained using empirical statistical models which traditionally rely on ordinary least squares regression, but more recently also employ more advanced regression methods from the field of machine learning, see Brooks et al. (2016) for an overview. For the latter purpose, a physics-driven approach is more expedient, and hence, so-called deterministic or mechanistic models which attempt to explicitly model the abiotic processes displayed in fig. 1.1 as well as biotic processes are typically employed. In the following, the characteristics of these deterministic models will be summarized largely based on the instructive review provided by de Brauwere et al. (2014).

The variable to be predicted by the vast majority of water-quality models is the total concentration of *fecal indicator bacteria* (FIB). FIB concentrations are used as a proxy for the straightforward quantification of all pathogens present in a natural water body in order to simplify regulations and analyses. In order for FIB to be tracers of external contaminations, these indicators are selected such that they occur e.g. in the faeces of humans or animals, do not grow naturally in water bodies, but are subject to similar processes as other pathogens. Nowadays, Escherichia coli is a commonly used FIB for water quality regulations.

The fundamental principle behind all mechanistic models is a mass balance, which implies that the concentration of bacteria is influenced by sources (the contamination event), sinks (internal processes and reactions which lead to inactivation) and transport. The level of detail which is employed in the mathematical modelling of these terms determines the complexity of the model.

Source terms may be either incorporated explicitly, which means that the drainage network is included in the modelling domain, or implicitly by imposing, possibly varying, boundary conditions of the domain. The choice of the method greatly depends on what processes are to be subject of investigation. Generally speaking, if the interest lies in an investigation of how pollutants enter the system, it would be preferred to model the source explicitly, whereas for studies primarily concerned with the downstream spreading, it is often sufficient to simply implement a boundary condition.

Sink terms describe processes of bacteria decay. Experimental data usually suggest a first-order kinetic of decay (Chick 1910), although an initial lag in response is often observed before exponential decay is achieved (Darakas 2002). The decay rate may be a function of various external factors such as temperature, sunlight intensity, salinity or predation, however, the constitutive equations describing their influence are predominantly local, and thus, the level of detail at which the decay is described depends on the resolution of the transport model. This is particularly the case for temperature and salinity whose distribution is mostly determined by hydromechanics.

Hydromechanics is also the key factor to all transport processes. As outlined in the introductory words, the flow inside a fluvial environment is eminently complex and the degrees of freedom a transport model was to require in order to resolve all relevant scales exceeds the computational resources available today and in the foreseeable future by far. For this reason, compromises have to be made in the mathematical description of the advection and dispersion of FIB concentration, and regularly, low-dimensional models are employed.

Spatially-unresolved models

The simplest mechanistic models from a hydromechanics point of view do not consider any spatially-resolved transport processes at all. As an example, Auer and Niehaus (1993) investigated the time-evolution of bacteria concentration in lakes solely in terms of first-order loss kinetics with a decay rate depending on temperature, UV irradiation levels, but also the sedimentation of particle-attached bacteria. Such spatially homogeneous models have also been applied to river flows, where typically a second balance equation is evaluated for bacteria deposited in the sediment. Resuspension and deposition then act as a source or sink to the bacteria concentration in the stream, respectively, and the coupling between the two balance equations is achieved by an ordinary differential equation relating the concentration in the sediment to that in the flow (Jenkins et al. 1984, Wilkinson et al. 1995).

It should be noted that spatially-unresolved water-quality studies are not uncommon, as many of them are conducted in the framework of hydrologic models, especially if they aim at simulating large geographical scales or a high level of geographical details. A widely used model is the "Soil & Water Assessment Tool" (SWAT) which is a public domain hydrological transport model capable of modelling virtually any water or soil based process within a river basin. A module for bacteria transport was developed by Sadeghi and Arnold (2002) which essentially implements various sources for bacteria runoff into receiving streams. However, the model does not consider any transport within the receiving water body itself, but solely bacteria decay and resuspension of bacteria deposited in the sediment depending on the local flow rate (Neitsch et al. 2011), which is calculated for each river section using Manning's equation for open-channel flows (Manning 1890).

1D transport models

Numerous studies use an approach where only the downstream coordinate is resolved. An early example is Boehm (2003) who studied the spreading of a point-source contamination along the shoreline of an ocean coast. In the study, the hydrodynamic reasoning of Inman et al. (1971) is followed to arrive at a one-dimensional advection-reaction equation which governs the bacterial transport. A similar approach was employed by Steets and Holden (2003) who studied the transport along the shoreline of a coastal lagoon which, however, also required the recognition of dispersion yielding an advection-diffusion-reaction (ADR) equation for their model. Moreover, their contamination originated from the sediment of the lagoon, and consequently, a second balance equation and a coupling term were employed similar to the spatially-unresolved models. The first study to explicitly distinguish between the transport of particle-attached and freely-suspended FIB was Jamieson et al. (2005) who formulated an ADR equation for each mode of transport, respectively. Moreover, their study was motivated by a contamination event which is localized in space and time and subsequently propagates in a stream, and hence, their boundary conditions were compatible to the investigation of CSOs in streaming water bodies. However, the process of adsorption and desorption is not represented by their model which leaves the two modes of transport uncoupled and is justified by the supposition that the attachment of bacteria is predominantly irreversible.

Another class of one-dimensional transport models exclusively resolves the vertical direction. Vergeynst et al. (2012) investigated decay processes in a storm water basin for a combination of particle-attached and freely-suspended bacteria using a discretisation in vertically-stacked layers. The settling of particles was modelled for multiple size classes using a class-specific settling velocity. Their model explicitly treated adsorption and desorption in each layer, and decay rates were set to be height-dependent, e.g. due to the variation of light intensity with respect to the water depth. A methodologically similar study was conducted by Qian et al. (2016) who reproduced the observations of the large-scale flume experiment by Walters (2013) where turbidity was found to be a key factor. For this reason, the transport of the concentration of all suspended solid matter was modelled using a one-dimensional advection-diffusion equation, where the effect of vertical dispersion of particles by turbulence was the motivation of the diffusive term. The study distinguished between particle-attached and freely-suspended bacteria, where the former were subject to an ADR equation and the latter only to diffusion and decay, but no coupling was implemented as irreversible attachment was assumed. Using this method, they were able to account for light attenuation by suspended solids at varying vertical position in the water body, which in turn was deemed crucial in order to explain the decay rates observed in the flume experiments at high levels of turbidity.

2D transport models

Two-dimensional transport models are typically based on the spatially-averaged Navier-Stokes equations under the justification that the flow is well mixed in the direction of averaging. Dewey and Palmer (1984) solved the depth-averaged Navier-Stokes equations, commonly denoted as the *shallow water equations*, using the method of Leendertse and Nelson (1978) in order to simulate the advection and dispersion of untreated storm water and CSOs within a St. John's harbour in Newfoundland. In a subsequent study for an urban beach on the Ottawa River, they extended their model to account for bacteria decay (Palmer and Dewey 1984). Kashefipour et al. (2002) used a similar model, but included a time-variant decay rate in order to accommodate radiative inactivation of FIB a bay. Schnauder et al. (2007) evaluated the depth-averaged Navier-Stokes equations with a turbulence model based on Prandtl's mixing length hypothesis to study FIB transport in the Bristol Channel. Their hydrodynamic model was sufficient to resolve some rather complex large-scale flow structures induced by the tides which were shown to have a considerable impact on the distribution of FIB. In all of the aforementioned studies, the tangible geographic occurrences and the dominance of transport processes were the main source of motivation for the choice of a transport model with higher dimensionality. A more methodologically motivated study was conducted by Gao et al. (2011) who extended the idea of Jamieson et al. (2005) to distinguish between particle-attached and freely-suspended bacteria to two dimensions and modelled the deposition and resuspension of particle-attached bacteria using a continuous Eulerian approach.

Laterally-averaged variants of the Navier-Stokes equations are significantly less common in the prediction of water-quality, with an early example being Kim et al. (1983) who applied the hydrodynamical model described in Edinger and Buchak (1980) to the Cherokee reservoir in order to explain the observed high FIB concentrations observed by means of circulation patterns. Regarding river flow, Liu and Huang (2012) studied the concentration of salinity and FIB in a tidal estuarine system inter alia under discharge conditions. Their streamwise-vertically resolved model used a mixing-length based turbulence model and a single ADR equation for FIB concentration which takes into account both suspended and particle-attached bacteria. The authors also remark the capability of vertically-resolved models to incorporate the interaction between the sediment and FIB concentrations more explicitly.

3D transport models

Turbulence is an inherently three-dimensional phenomenon, and as such, lower dimensional transport models inevitably neglect some of its defining features. Similar to their two-dimensional counterpart, the emergence of 3D bacteria fate and transport models is widely expedited by the availability of appropriate hydrodynamic models which are most commonly borrowed from coastal ocean modelling. In fact, the bulk of published 3D models fall into two main categories.

The models of the first category use the internal/external mode decomposition described e.g. in Blumberg and Mellor (1987). In this method, the deformation of the free surface of the water body is computed using the two-dimensional shallow water equations, whereas the internal flow is a solution to the three-dimensional Navier-Stokes equations. Both sets of equations are coupled, as the latter depends on the water surface computed by the former, but also provides bottom friction and vertical dispersion terms to it (Cugier and Le Hir 2002). Furthermore, the equations are typically Reynolds-averaged, and hence, a turbulence model is used in order to limit the range of scales which needs to be resolved. The advantage of the decomposition lies in less stringent restrictions on the time-step as the stability criterion of the 3D model becomes independent of the free surface waves (Cugier and Le Hir 2002). An early study on water quality using this method was conducted by Connolly et al. (1999) who studied the transport of three indicator organisms in Mamala Bay, Oahu. Their circulation model included the entire island in order to account for the particular macroscopic flow conditions, was coupled to contaminant transport using ADR equations, and also included the transport of suspended solids using an ADR equation where the reaction term corresponds to the settling of the solids. However, particle-attached FIB were not considered in this study. Special care was taken in the modelling of the decay rates which were depth-dependent in terms of light attenuation and scattering due to suspended solids, thus justifying the necessity of a three-dimensional model. Garcia-Armisen et al. (2006) conducted a similar investigation in the Seine estuary in France within the framework of a multidisciplinary research programme in which a comprehensive hydrodynamical model of this region has been developed by Cugier and Le Hir (2002) using the external/internal mode decomposition. The study distinguished explicitly between particle-attached and freely-suspended FIB where both concentrations were described by ADR equations and deposition/resuspension was modelled for the former. It should, however, be noted that the river Seine itself was represented one-dimensionally in their model and thus served the purpose of a variable boundary condition to the three-dimensional discretisation of the bay (Cugier and Le Hir 2002).

The second category of models utilises the approach of Zhang and Baptista (2008) where the shallow water equations are extended to three dimensions using the continuity equation. Therefore, in contrast to the approach previously described, no momentum equation is solved for the vertical velocity component, and thus, stricter assumptions on the flow conditions are made. The advantage is, however, that the set of equations to be solved is less complex, as no decomposition between internal and external waves has to be employed. Again, the equations are solved in their Reynolds-averaged form and the implementation of Zhang and Baptista (2008), which is employed by all studies mentioned in the following, uses the turbulence model of Umlauf and Burchard (2003). Rodrigues et al. (2011) applied this hydrodynamic model to the Aljezur coastal stream in Portugal, and in opposition to the previously described studies, the stream itself was discretised in three dimensions with the smallest resolved length scale being around 2 meter. Their bacteria model included particle-attached and freely-suspended bacteria in a single ADR equation, but resuspension was not considered. Liu et al. (2015) employed an essentially similar model to the Tamsui estuarine in Taiwan which also included the river system in the discretisation. Subsequently, Chen and Liu (2017) extended this study to include a continuous sediment transport model where the rate of erosion was determined from the bottom shear provided by the hydrodynamical model, thus accounting for deposition and resuspension of particle-attached FIB.

1.3 Research objectives

Bacteria fate and transport models have advanced significantly over the last decades. However, the range of length scales which are typically relevant for the prediction of transport in water quality is immense, reaching from the size of entire river systems down to the size of small suspended particles or less, i.e. from several hundreds of kilometre down to several micrometres. For this reason, any hydrodynamical model which aims at investigating tangible geographic situations is ought to rely on models to represent the processes on the smallest scales, and will be in the foreseeable future. These small-scale processes are at the same time by no means insignificant as becomes apparent from the efforts which have been made in the past to parameterize processes such as turbulent dispersion, sediment resuspension or the settling rate of particle-bound bacteria in transport models.

At the same time, the hydrodynamic processes at the smallest scales are far from being fully understood. In fact, the sedimentation of solid particles in turbulent flows constitutes an active field of research with a rich variety of emergent phenomena such as the formation of patterns, clustering and the interaction with coherent fluid motions whose implications are not yet explicitly captured by simplified models. As will become evident in the subsequent literature review on particle-laden flows, the transient process of particles settling in wall-bounded turbulence has received only little attention in literature yet despite its relevance. The goal of this thesis to contribute to the hydromechanical understanding of this process, and to make inferences on the significance for bacteria transport.

It should be emphasized that the aspiration of this thesis is not to reproduce a specific contamination problem, nor to provide a full bacteria fate and transport model. In fact, bacteria fate will only play a subordinate role, and the emphasis is laid for the greater part on turbulent transport processes. In contrast to previous works in the field of deterministic bacteria transport models, the approach employed here should be regarded as "bottom-up" in the sense that the goal is to resolve the smallest relevant scales with as few assumptions as possible, with the endeavour to make inferences towards emergent phenomena of relevance at larger scales. The drawback of this approach is that the domain of investigation has to be kept general, and that the largest length scales which can be described using this method are still smaller than the smallest length scales in more traditional "top-down" approaches. This inevitability can be seen at a glance in fig. 1.2 where the disparity in the scales described by classical bacteria transport models and the current work is outlined.

More specifically, the following key parameters and phenomena will be subject of investigation in this thesis:

Figure 1.2: Schematic depiction of scales involved in water-quality models. The majority of three-dimensional water-quality fate and transport models have been applied on a mesoscale with the intention to describe entire estuaries, thus regions with length scales of the order of 100 km. For this purpose, it is inevitable to model all processes which occur on a microscale. The research objective of this thesis is to directly assess the effect of turbulence and sediment dynamics on the transport of contaminations from first principles and on the smallest scales. As a consequence, the size of the system which can be represented is many orders of magnitudes smaller than in previous studies. Illustrations: Top row—Geba River estuary in Guinea-Bissau (USGS/NASA Landsat, public domain). Bottom right—Hunter River in New South Wales, Australia (Rygel, M.C., CC BY-SA 3.0). Bottom left—Direct Numerical Simulation of ripple patterns.

Settling velocity during the transient The settling velocity of suspended particles is a paramount parameter for the modelling of particle-attached bacteria. In the bacteria transport models, this quantity is frequently set to the value of the terminal velocity in a quiescent environment despite hydromechanical studies suggesting that turbulence has a non-trivial significant effect on it.

Spatio-temporal inhomogeneity The dynamics of particles in turbulent flows are distinctly different from the behaviour of solubles due to inertial effects, and this distinction is far too momentous to be reduced solely to settling effects. In many cases particles exhibit the tendency to cluster and de-mix, and the effect of such inhomogeneities on the environmental conditions encountered by particle-attached bacteria lacks qualitative and quantitative analysis.

Sediment interaction It is well established that the interaction between bacteria and river sediment plays a significant role in the spreading and storage of contaminations. A detailed analysis of the interplay between natural river bedforms and entering solid contaminants may be beneficial for its understanding.

Identification of emergent phenomena Complex systems are often more than merely a combination of their individual parts. Due to the large disparity in length scales, some emergent phenomena of relevance might have been overlooked in the classical "top-down" approaches, and might be more accessible by an investigation starting from the smallest scales.

1.4 Outline of this thesis

The content of this thesis is organized as follows: Chapter §2 provides an introduction to the relevant principles of hydromechanics and particle-laden flows, and contains a literature review on the state of research concerning the transient settling of particles in wall-bounded turbulent flows. In chapter §3 the exact configuration to be simulated and investigated is introduced along with the mathematical framework and numerical method which is employed. It further contains a summary of the major parameter points which are discussed throughout the thesis and details on the determination of reference quantities. Chapter §4 focuses on the velocity statistics of the solid particles during the phase of transient settling, and chapter §5 aims at explaining these observations in terms of the interaction with the turbulent carrier flow. In chapter §6 the spatial organization of the solid phase into clusters and voids, and the interaction with sediment bedforms is examined, as well as the dispersion and displacement of the contaminants from their initial position. A coarse-graining method is employed in chapter §7 in order to determine the length scales of turbulence which are most relevant for the phenomena previously described. Chapter §8 introduces a novel approach to the investigation of turbulent particle-laden flows by assessing the interaction of settling particles with exact coherent structures, which are non-linear flow states of significantly reduced complexity compared to turbulence, yet are thought to be closely related to coherent fluid motions observed in turbulent flows. From these interactions, scaling laws are derived with reduced computational effort. Chapter §9 transfers the knowledge gained from the hydrodynamical investigations towards bacteria transport models by summarizing and contextualising the recognitions made in the preceding chapters. Finally, the thesis is concluded in chapter §10 by stating the main findings.

2 Wall-bounded turbulence and particle dynamics

2.1 Fundamental principles of hydromechanics

2.1.1 Navier–Stokes and transport equations

The fundamental equations of hydromechanics are the Navier–Stokes equations (NSEs) for incompressible flows which are deduced from the conservation of mass and the balance of linear momentum for a fluid parcel with constant mass and constant volume, viz.

$$\nabla \cdot \mathbf{u}_f = 0 \tag{2.1}$$

$$\frac{\partial \mathbf{u}_f}{\partial t} + \nabla \cdot \left(\mathbf{u}_f \otimes \mathbf{u}_f\right) = -\frac{1}{\rho_f}\nabla P + \nu_f \nabla^2 \mathbf{u}_f + \frac{1}{\rho_f}\sum \mathbf{f}, \tag{2.2}$$

where $\mathbf{u}_f(\mathbf{x},t) = \left(u_f, v_f, w_f\right)^T$ denotes the fluid velocity field, $P(\mathbf{x},t)$ the associated pressure field, ρ_f the density of the fluid, ν_f the kinematic viscosity and $\mathbf{f}(\mathbf{x},t)$ is a placeholder for the acceleration induced by any external forces, or the sum thereof. Moreover, the three-dimensional spatial coordinates are referred to by $\mathbf{x} = \left(x, y, z\right)^T$, the temporal coordinate by t, the unary spatial differential operator by $\nabla := \left(\partial/\partial x, \partial/\partial y, \partial/\partial z\right)^T$, the binary tensor product operator by $\otimes$ and the binary inner product operator by $\cdot$.

The Navier–Stokes equations model the underlying fluid as a continuum rather than discrete molecules and are therefore subject to the continuum hypothesis which starts from the premise that the length scales of macroscopic motion are sufficiently separated from the length scales of molecular motion such that an ensemble of molecules may be regarded as a fluid parcel of infinitesimal size. In fact, the momentum equations shown in eq. (2.2) are a particular instance of the general Cauchy momentum equations for any non-relativistic continuum, and are obtained by specifying the stress tensor for a Newtonian fluid, viz.

$$\sigma_{ij} := -\delta_{ij} P + \rho_f \nu_f \left(\frac{\partial u_{f,i}}{\partial x_j} + \frac{\partial u_{f,j}}{\partial x_i}\right) \tag{2.3}$$

where δ_{ij} denotes the Kronecker delta. As a consequence, tangential stresses are proportional to the rate of strain with a scalar constant of proportionality—the viscosity.

Generally, the suppositions of a continuous medium, an incompressible flow and Newtonian fluid properties are non-restrictive for the flows investigated in the context of water-quality prediction models.[1]

[1] The flow of dense suspensions, which occurs for example in sediment transport problems, is sometimes described using a non-Newtonian stress tensor if both the suspended solids and the carrying fluid are modelled as a single continuous phase. In the context of this thesis, this is not necessary as both phases will be resolved properly.

Therefore, eq. (2.1) and (2.2) may be regarded as the most fundamental description of the relevant hydrodynamic processes, and indeed, all higher-dimensional hydrodynamic models discussed in §1.2 are simplifications of these equations.

The NSEs are sufficient in order to describe the evolution of the fluid velocity and pressure fields if appropriate boundary and initial conditions are provided. If the transport of another continuous quantity by the flow is to be described, the NSEs are oftentimes coupled to an advection-diffusion-reaction equation (ADR) for said quantity, viz.

$$\frac{\partial \varphi}{\partial t} + \nabla \cdot \left(\mathbf{u}_f \varphi\right) = \nabla \cdot \left(\mathcal{D}_\varphi \nabla \varphi\right) + \sum q, \tag{2.4}$$

where $\varphi(\mathbf{x}, t)$ is a placeholder for any transported quantity, $\mathcal{D}_\varphi$ its diffusivity and $q(\mathbf{x}, t)$ a placeholder for its sources or sinks.[2] Hereby, this coupling is denoted as passive if the influence of $\varphi(\mathbf{x}, t)$ is insignificant for the fluid momentum balance, or as active if the coupling is reciprocal and an external force $\mathbf{f}_\varphi(\mathbf{x}, t)$ which depends on φ is introduced in eq. (2.2). Common examples of transported quantities in the context of water-quality prediction are salinity or temperature—both of which are typically modelled as active—or the concentration of FIB which is typically regarded as passive.

Regarding the application of an ADR equation for the transport of bacteria, two clarifying remarks should be made. Firstly, in the bacteria fate and transport models described in §1.2, ADR equations are sometimes used for both freely-suspended and particle-attached bacteria, although the latter is discrete at the smallest length scales of the flow. Here, the treatment of particle-attached bacteria as a continuum is a result of the truncation of length scales, and is inadequate if the fluid movement at the length scales of the suspended particles is resolved. Secondly, it should be kept in mind that bacteria are capable of locomotion and should therefore be regarded as active matter (Ben-Jacob et al. 1994, Vicsek et al. 1995). Therefore, an ADR equation is not the most fundamental way to describe bacteria transport and a more appropriate model is to be employed whenever this effect is expected to be relevant.

2.1.2 Dimensional analysis

In the absence of external forces, the NSEs comprise three fundamental units—length, time and mass— and four independent dimensional parameters—a characteristic velocity, a characteristic length, the density and the viscosity. According to the Buckingham π theorem (Bertrand 1878, Buckingham 1914) the equations can hence be parametrised by a single non-dimensional parameter, which is conventionally chosen to be the Reynolds number defined as

$$Re := \frac{\mathcal{U}\mathcal{L}}{\nu_f}, \tag{2.5}$$

where $\mathcal{U}$ denotes a characteristic scale of velocity and $\mathcal{L}$ a characteristic length scale of the problem to be investigated. The name of this non-dimensional number was given in honour of Osborne Reynolds who famously studied the transition between laminar and turbulent fluid flow in a pipe (Reynolds 1883) and later discovered that the criterion for this transition to occur is solely provided by this parameter (Reynolds 1895). When non-dimensionalising the NSEs, the inverse of the Reynolds number emerges as a prefactor to the diffusive term, and hence, its value is decisive for the relative importance of viscous

[2] Indeed, the ADR equation also comprises the transport of fluid momentum as stated in eq. (2.2) for the particular case that φ corresponds to a fluid velocity component, $\mathcal{D}_\varphi$ to the viscosity and q includes the pressure-gradient.

effects. The specification of the characteristic velocity and length scales are problem dependent and generally not unique. In fact, as will be established when the phenomenon of turbulence is discussed, the tangible definition of the Reynolds number may dependent on the aspect of the problem which is to be assessed.

The amount of non-dimensional parameters which are required to fully describe a flow problem may increase when external forces are considered. In the context of geophysical flows, one external force which is ever-present is induced by gravity and enters eq. (2.2) by the force term $\mathbf{f}_g := \rho_f \mathbf{g}$ with $\mathbf{g}$ being the gravitational acceleration. The non-dimensional number which determines the relevance of this force is termed the Froude number in honour of William Froude[3] and its definition is given by[4]

$$Fr := \frac{\mathcal{U}^2}{g\mathcal{L}} \tag{2.6}$$

where $g := |\mathbf{g}|$ is the magnitude of the gravitational acceleration. The Froude number has its primary relevance for free surface phenomena such as external waves.

Regarding transported quantities, a non-dimensionalisation of the ADR equation yields two dimensionless numbers, namely the Péclet number, $Pe := \mathcal{U}\mathcal{L}/D_\varphi$, which similarly to the Reynolds number describes the relative importance of diffusion, and a dimensionless number related to the reaction term which is commonly referred to as the Damköhler number and whose definition depends on the details of the reaction term. Due to the similarity of the Péclet and the Reynolds number, a different parametrisation is employed occasionally by defining the Schmidt number as their ratio, viz.

$$Sc := \frac{Pe}{Re} = \frac{\nu_f}{D_\varphi}. \tag{2.7}$$

The Schmidt number is an important indicator for mixing processes in turbulent flows and can be interpreted as the separation in scales at which fluid momentum and variations in the transported field are dissipated. In the case of active transport, the Schmidt number is also referred to as the Prandtl number and an external force representing buoyancy effects is added to the NSEs. This buoyant force is commonly formulated in terms of the Boussinesq approximation (Boussinesq 1897) which presumes that variations in density sufficiently small such that they are negligible in terms of inertia, and yields $\mathbf{f}_\varphi := \rho_f \beta(\varphi - \varphi_0)\mathbf{g}$ where β is the linear expansion coefficient and φ_0 is a characteristic value of the transported quantity. A non-dimensionalisation of this body force gives rise to the Richardson number which is a measure for the relative importance of accelerations due to inhomogeneities in the specific weight and reads

$$Ri := \frac{\beta \Delta\varphi_0 g \mathcal{L}}{\mathcal{U}^2} \tag{2.8}$$

where $\Delta\varphi_0$ is a characteristic range in which φ may vary within the flow. The Richardson number may be understood as a variant of the inverse Froude number for internal waves rather than external ones.

Further external forces may need to be considered in bacteria transport models if they aim to resolve transport on a geographic mesoscale. For instance, the ocean circulation model of Zhang and Baptista

[3] The attribution of the discovery of this non-dimensional parameter is disputed and arguments can be made that it is a misnomer (Hager and Castro-Orgaz 2017), as Belanger (1828) formulated this non-dimensional group and de Saint-Venant (1871) recognised its relevance for external wave regimes prior to Froude (1874).

[4] The definition of the Froude number varies in the literature, and the current definition is chosen such that the prefactor of the non-dimensionalised force term is equal to the inverse of the Froude number. A definition which is equivalent to the square-root of the current definition is frequently encountered for historical reasons.

(2008) also includes the Coriolis force and tidal forces, which introduce the Rossby number and Love's number to the parametrisation, respectively. As this thesis is concerned with transport at a microscale, these numbers are only mentioned for completeness.

2.1.3 Scales of turbulence

As mentioned in the introductory words, turbulence is a phenomenon which is characterised by the presence of eddies which occur over a wide continuous range of scales. This understanding of turbulence is has been strongly influenced by Lewis Fry Richardson who introduced the idea of a turbulent cascade where energy is injected into the flow at the largest scales, transferred to subsequently smaller scales in a quasi-inviscid process until viscosity is significant enough to dissipate the mechanical energy into heat (Richardson 1922). Richardson understood this process as a continuous breakdown of eddies from large scales downwards, where eddy is a vague term for coherent packets of turbulent motion which exhibit an intrinsic length and time scale. From today's understanding, this so-called energy cascade is far more complex than what Richardson imagined, albeit the notion of the directional energy transfer still withstands on average.

Nearly two decades later, Andrey Nikolaevich Kolmogorov elaborated on Richardson's idea and formulated hypotheses which allowed him to arrive at quantitative statement on the range of scales over which turbulent motion occurs (Kolmogorov 1941). He postulated that turbulence at sufficiently small scales and sufficiently far away from the boundaries is locally isotropic and homogeneous, which entails that small-scale eddies are universal and independent from the particular geometry and nature of the large-scale flow. Moreover, Kolmogorov argued that the statistics of these smallest scales are ought to be fully determined by the rate of dissipation, in the following denoted by ε_f, and the viscosity. Dimensional considerations then yield the so-called Kolmogorov microscales,

$$\eta := \left(\nu_f^3/\varepsilon_f\right)^{1/4}, \quad u_\eta := \left(\nu_f\varepsilon_f\right)^{1/4}, \tag{2.9}$$

where η is the Kolmogorov length scale and u_η the corresponding velocity scale. The intrinsic Reynolds number for eddies at this scale is unity by definition, which highlights their importance for dissipation in the cascade.

In contrast, the fluid motion at the largest scale is not universal, as the corresponding eddies are typically thought to be the result of inertial instabilities which are affected by the geometry and flow characteristics. The kinetic energy carried by these eddies is of the order of u_ℓ^2, and the time scale at which this energy enters the cascade is similar to the eddy turnover time, which is of the order of ℓ/u_ℓ with ℓ and u_ℓ denoting the length and velocity scales of the largest eddies, respectively. In order for energy cascade to be statistically steady, the rate of energy input at the largest scales, u_ℓ^3/ℓ, must match the rate of dissipation at the smallest scales, $\nu_f u_\eta^2/\eta^2$. As a consequence, the range of scales which is encountered in a turbulent flow is quantified by

$$\ell/\eta \propto Re_\ell^{3/4}, \quad u_\ell/u_\eta \propto Re_\ell^{1/4}, \tag{2.10}$$

and depends on the value of the intrinsic Reynolds number of the largest eddies, $Re_\ell := u_\ell \ell/\nu_f$. Since ℓ and u_ℓ are essentially predetermined by the boundary conditions of the flow, eq. (2.10) suggests that the smallest structures which can be observed in a turbulent flow decrease in size with increasing value of the (integral) Reynolds number.

2.1.4 Reynolds decomposition

Apart from his contributions to the study of the onset of turbulent motion, Osborne Reynolds introduced another crucial concept for the examination of turbulence—the decomposition of an arbitrary field φ into its average value $\langle\varphi\rangle$ and a fluctuation φ' about it, such that

$$\varphi(\mathbf{x},t) = \big\langle\varphi(\mathbf{x},t)\big\rangle + \varphi'(\mathbf{x},t). \tag{2.11}$$

The definition of the averaging operator varies depending on the problem to be investigated, however, it is required to be linear and to satisfy the identity $\langle\langle\varphi\rangle\psi\rangle = \langle\varphi\rangle\langle\psi\rangle$ for all φ and ψ where the latter denotes a second arbitrary field. Common examples for averaging operations include temporal averaging of statistically stationary flows, ensemble averaging over multiple realisations, spatial averaging over statistically homogeneous directions, or a combination thereof.

The Reynolds decomposition can be applied to the velocity and pressure field and substituted into the NSEs in order to derive a set of equations which describes the evolution of the average quantities. These so-called Reynolds equations are given by

$$\nabla\cdot\langle\mathbf{u}_f\rangle = 0, \tag{2.12}$$

$$\frac{\partial\langle\mathbf{u}_f\rangle}{\partial t} + \nabla\cdot\big(\langle\mathbf{u}_f\rangle\otimes\langle\mathbf{u}_f\rangle\big) = -\frac{1}{\rho_f}\nabla\langle P\rangle + \nu_f\nabla^2\langle\mathbf{u}_f\rangle - \nabla\cdot\big\langle\mathbf{u}_f'\otimes\mathbf{u}_f'\big\rangle + \frac{1}{\rho_f}\sum\langle\mathbf{f}\rangle, \tag{2.13}$$

and a comparison to eq. (2.2) reveals an additional term in the momentum equations—commonly denoted as the Reynolds stresses—which depends on the variances and covariances of the velocity fluctuations. The appearance of this additional term has severe implications on the solvability of the system of equations, as it causes it to be underdetermined, and hence further constitutive equations are required in order to solve them. This constitutes the closure problem of turbulence which is inevitable as deriving the required constitutive equations from first principles merely leads to the appearance of further terms of higher order.

Despite this complication, it is desirable for various applications to search for solutions to the Reynolds-averaged equations instead of the unmodified NSEs, especially if the primary subject of interest is the mean flow. In fact, as was brought up in §1.2, most, if not all, higher-dimensional bacteria transport models are based on variations of the Reynolds equations. The advantage in doing so lies in the considerably lesser demand on the resolution of the flow problem—whereas solutions to the NSEs require the model to fully resolve even the smallest scales of fluid motion in order to accurately capture the energy cascade, the resolution requirement on the Reynolds equations is solely dictated by the spatio-temporal variations in lower order statistics which generally occur at vastly larger scales. The disadvantage of this approach, however, is that the Reynolds stresses have to be supplied by a model. A broad class of turbulence models is based on the concept of a turbulent eddy viscosity which presumes that the Reynolds stresses exhibit a similar net effect as an increased molecular viscosity, and indeed, all of the discussed bacteria transport models seize on this idea, although the eddy viscosity is commonly chosen to be anisotropic. However, generally speaking the Reynolds stresses are problem-dependent, as the truncated turbulent motions which cause them are typically not sufficiently small to be considered universal, and hence, the applicability of turbulence models routinely require adequate tuning and introduced a substantial uncertainty for the prediction of the mean flow.

2.1.5 Vorticity and locality

One universal feature of turbulence is the ubiquitousness of vorticity, $\boldsymbol{\omega}_f = \left(\omega_{f,x}, \omega_{f,y}, \omega_{f,z}\right)^T$ which is defined as the curl of the fluid velocity, viz.

$$\boldsymbol{\omega}_f := \nabla \times \mathbf{u}_f, \tag{2.14}$$

where $\times$ is the binary cross product operator. Vorticity is a measure for the local rotation of a fluid parcel, and while it is indeed related to the general concept of a vortex, its presence does not demand a global vortical motion.[5] Taking the curl of eq. (2.2) in the absence of non-conservative body forces yields a transport equation for the vorticity which reads

$$\frac{\partial \boldsymbol{\omega}_f}{\partial t} + \left(\mathbf{u}_f \cdot \nabla\right) \boldsymbol{\omega}_f = \nu_f \nabla^2 \boldsymbol{\omega}_f + \left(\boldsymbol{\omega}_f \cdot \nabla\right) \mathbf{u}_f, \tag{2.15}$$

and constitutes a classical advection-diffusion equation with an additional source term which depends on the vorticity itself, as well as on the local velocity gradient. This so-called vortex stretching term describes the change in moment of inertia of a fluid parcel due to stretching, and, on average, provides the coupling which transfers energy from the mean shear into turbulent vorticity as previously discussed.

An important feature of vorticity is that localised distributions of it can only spread locally. In fact, by reverting eq. (2.14) using the law of Biot-Savart[6],

$$\mathbf{u}_f(\mathbf{x}, t) = \frac{1}{4\pi} \int \frac{\boldsymbol{\omega}_f(\mathbf{x}', t) \times \left(\mathbf{x} - \mathbf{x}'\right)}{\left|\mathbf{x} - \mathbf{x}'\right|^3} \mathrm{d}\mathbf{x}', \tag{2.16}$$

it becomes apparent that the velocity field is induced by any vorticity within the domain, and while this in turn will affect the evolution of existing vorticity globally, eq. (2.15) does not provide a way to generate vorticity from a distance.

In contrast, the velocity field in incompressible flows is inherently non-local due to the role of pressure. By taking the divergence of eq. (2.2), a Poisson equation for the pressure is obtained, viz.

$$\nabla^2 P = \frac{1}{\rho_f} \left(\sum (\nabla \cdot \mathbf{f}) - \nabla \cdot (\mathbf{u}_f \otimes \mathbf{u}_f) \right). \tag{2.17}$$

Analogously to eq. (2.16), the solution of the Poisson problem depends on the velocity distribution everywhere in the domain. However, as the gradient of pressure directly enters eq. (2.2), linear momentum can be redistributed instantaneously over any distance contrary to vorticity. For this reason, the vorticity field may be considered as more fundamental to turbulence than the velocity field, and as the concept of a turbulent eddy is intrinsically local, it should rather be understood in terms of vorticity.

Vorticity is inhomogeneously distributed within a turbulent flow and appears intermittent rather than in an uncorrelated random manner (Batchelor et al. 1949). When the spatial distribution of intense vorticity in turbulence is investigated, it is found to be organised in coherent vortices, which are termed "worms" due to their shape often being cylindrical or ribbon-like (Siggia 1981, Jiménez et al. 1993). In

[5] A prime example from the current context for a flow which contains vorticity, but no global vortical motion, is laminar channel flow. For this reasons, some definitions of turbulence further specify the occurrence of vorticity to be random and at small scales, see e.g. Corrsin (1961) or Mathieu and Scott (2000) for such a definition.

[6] Equation (2.16) is formulated for an unbounded domain for the sake of simplicity, as formulations for bounded domains are considerably more complex (Enciso et al. 2018). However, the inferences remain the same.

general, coherency is an essential aspect of turbulence, and its significance will be discussed further for the specific case of wall-bounded turbulence.

2.2 Wall-bounded turbulence

Hitherto, the discussion on turbulence was predominantly focused on its universal features. In this section, the particular phenomena which are inherent to turbulent flows bounded by walls are reviewed. The presence of a solid boundary is reflected in the hydromechanic equations by two boundary conditions for the fluid velocity field on the wall—the impermeability condition and the no-slip condition—which are given by

$$(\mathbf{u}_f - \mathbf{u}_w) \cdot \mathbf{n} = 0, \tag{2.18}$$

$$\mathbf{u}_f - \big((\mathbf{u}_f - \mathbf{u}_w) \cdot \mathbf{n}\big)\,\mathbf{n} = \mathbf{u}_w, \tag{2.19}$$

respectively, with $\mathbf{u}_w$ being the velocity of the wall and $\mathbf{n}$ denoting its normal vector. In essence, the combination of these boundary conditions implies that any relative motion between the fluid and the solid phase vanishes exactly at the wall, which is an excellent supposition for most flows.[7] The implication of these boundary conditions is that any bulk fluid motion relative to a wall forms a boundary layer—a region where the mean relative fluid velocity decays from its undisturbed value to zero—where the fluid is sheared on average.

It can be demonstrated that in the presence of a mean shear, the Reynolds stresses induce a global transfer of energy from the mean flow towards turbulent eddies by means of stretching and intensification of turbulent vorticity, see e.g. Davidson (2015). As a consequence, the presence of a mean gradient in fluid velocity is one mechanism which generates and sustains turbulence. Vice versa, the mean flow itself is modified by turbulent fluctuations as is suggested by the additional stresses introduced in the Reynolds equations. Therefore, the overarching characteristic of all wall-bounded turbulent flows is the persistent complex interplay between turbulent fluctuations and wall-induced shear.

The varieties in which walls may be arranged in a flow are sheer endless, as boundary layers may occur at all scales and shapes—think of a rather flat canal lining versus the severely curved surface of a pebble. For this reason, the study of wall-bounded turbulence is generally conducted in so-called canonical flows, which denote a canon of well-defined, and often idealised, flow configurations which guarantee the replicability and comparability of experiments. Examples of internal canonical flows include, but are not limited to, the pressure-driven flows through a circular pipe or through a channel, or the flow observed in the gap between two parallel moving plates, which all share certain similarities, but may differ in their particularities.

[7] The no-slip boundary condition may be an inaccurate supposition for flows with low densities or in domains with exceptionally small dimensions, i.e. whenever the molecular density is sufficiently small. Hereby, the deviation commonly occurs before the continuum hypothesis fails, and more adequate results may be achieved by imposing a slip velocity.

2.2.1 Open-channel flow

The canonical flow most relevant for fluvial systems is open-channel flow—a body-force-driven flow bounded by a solid wall and a free surface in one direction, and unbounded in the two remaining perpendicular directions. Open-channel flow is typically horizontal, which means that the bounded direction is coaligned with the direction in which gravity is acting, and this direction will be referred to as the vertical or wall-normal direction and denoted by y in the following. Moreover, without loss of generality, the streamwise coordinate x is chosen to be coaligned with the primary flow direction and the spanwise coordinate z is set to be perpendicular to both x and y.

The unboundedness of the streamwise and spanwise directions is an idealised concept which, in the most cases, can only be achieved approximately, and therefore should rather be understood in the sense that there is no influence exerted by any walls normal to these directions. Hence, unboundedness serves as an approximation to the core of a river where the region of interest is within a reasonable distance to the river shore such that its effect can be neglected. In terms of numerical modelling, unboundedness can be achieved by enforcing a periodicity of the flow with periods of L_x and L_z in streamwise and spanwise direction, respectively. However, this comes at the cost of introducing nonphysical self-interactions if these periods are not chosen adequately.

Concerning the vertical direction, the bottom boundary[8] of the domain is considered to be a stationary solid wall which is subject to the no-slip boundary condition given in eq. (2.19). In numerical models, the upper boundary is often considered to be non-deformable for the sake of simplicity, which corresponds to the limiting case of a vanishing Froude number, and thus prohibits the existence of any external waves. As a result, the free surface may be treated as an impermeable wall which is not subject to the no-slip boundary condition, but rather to the free-slip boundary condition,

$$\frac{\partial}{\partial \mathbf{n}}\left(\mathbf{u}_f - (\mathbf{u}_f \cdot \mathbf{n})\,\mathbf{n}\right) = \mathbf{0}, \tag{2.20}$$

which is of Neumann-type rather than of Dirichlet-type. For the specified coordinate system, the boundary condition reads $\partial u_f/\partial y = 0$, $v_f = 0$ and $\partial w_f/\partial y = 0$, and hence, allows for finite streamwise and spanwise fluid velocities at the upper boundary.

Despite the difference in the upper boundary condition, open-channel flow is closely related to the flow in closed channels, i.e. the configuration where no-slip is imposed at both walls. In fact, the free-slip boundary condition can be interpreted as a mirror symmetry which is imposed on the centre line of a closed channel, and hence, the structure of turbulence found in these configurations is generally comparable apart from some discrepancies which will be discussed at a later point (Calmet and Magnaudet 2003, Bauer et al. 2022).

[8] To be precise, terms such as low/high and bottom/top are used with respect to the direction of gravity, i.e. the bottom is the location an object would be moving towards, if it was solely subject to gravitational acceleration.

2.2.2 Turbulence in channels

The flow in channels is statistically homogeneous in the two unbounded directions, and also in time if the Reynolds number is kept constant and the flow is fully established. Therefore, the Reynolds-averaging operator for this flow is conventionally defined as the streamwise-spanwise-temporal average,

$$\langle \varphi \rangle_{xzt}(y) := \frac{1}{L_x L_z t_{obs}} \int_0^{L_x} \int_0^{L_z} \int_0^{t_{obs}} \varphi(\mathbf{x},t)\, \mathrm{d}t\, \mathrm{d}z\, \mathrm{d}x, \tag{2.21}$$

and as a consequence, the flow statistics only vary in the wall-normal direction. Here, t_{obs} denotes the observation time, and for the remainder of the discussion, it is assumed that the sample space $\{\varphi(\mathbf{x},t) : x \in [0,L_x] \wedge z \in [0,L_z] \wedge t \in [0,t_{obs}]\}$ is sufficiently large to convey the converged statistics.

Due to the boundedness of the wall-normal coordinate and the definition of the spanwise coordinate, the mean flow velocity in these two directions vanishes, i.e. $\langle v_f \rangle_{xzt} = \langle w_f \rangle_{xzt} = 0$, where the former still holds if solely the instantaneous spatial-average is considered. A finite value of the mean velocity therefore only exists in the streamwise direction, and an integral flow velocity scale can be obtained by averaging in wall-normal direction, viz.

$$u_b := \frac{1}{H} \int_0^H \langle u_f \rangle_{xzt}\, \mathrm{d}y. \tag{2.22}$$

The velocity u_b is termed the bulk velocity, and H corresponds to the position of the free-slip boundary, or to the position of the centreline if closed channel flow is considered, while the bottom boundary is set to be located at $y = 0$.[9] The bulk velocity and the channel height are integral scales and can be utilised to define an integral Reynolds number—the bulk Reynolds number,

$$Re_b := \frac{u_b H}{\nu_f}, \tag{2.23}$$

which is a principal parameter for the transition from the laminar to the turbulent flow regime. Channel flow is linearly unstable for $Re_b \gtrapprox 3848$ which implies that any infinitesimal perturbation will trigger transition if this value is exceeded (Orszag 1971). However, in practice the transition oftentimes occurs subcritically by instability mechanisms which involve finite-amplitude perturbations (Chapman 2002). In fact, turbulence may be sustained for $Re_b \gtrapprox 1350$, although values of $Re_b \gtrapprox 3000$ are required to observe universal scaling laws for the shear stress on the wall or the mean streamwise velocity (Patel and Head 1969).

Using the averaging operator defined in eq. (2.21), the Reynolds-averaged streamwise momentum equation for channel flow is given by

$$\frac{\mathrm{d}}{\mathrm{d}y} \underbrace{\left(\rho_f \nu_f \frac{\mathrm{d}\langle u_f \rangle_{xzt}}{\mathrm{d}y} - \rho_f \langle u_f' v_f' \rangle_{xzt} \right)}_{:=\tau_f(y)} = \frac{\mathrm{d}\langle P \rangle}{\mathrm{d}x} \tag{2.24}$$

where $\tau_f(y)$ is the total fluid shear consisting of a viscous and a turbulent contribution. From the averaged lateral momentum equation it can be deduced that the mean pressure gradient is uniform across

9 The distinction between open and closed channel flow is of little importance for the statistical analysis, as closed-channel flow is statistically symmetric about the centreline, which is enforced instantaneously in the open-channel.

the flow (Pope 2000), and as a consequence, the fluid shear is required to vary linearly with the vertical coordinate. At $y = H$, the fluid is ought to be stress-free, and hence, the fluid shear can be expressed in terms of the tangential stress exerted on the no-slip wall, viz.

$$\tau_f(y) = \left(1 - \frac{y}{H}\right)\tau_f(0). \tag{2.25}$$

The wall shear can then be used to define another velocity scale—the friction velocity defined as

$$u_\tau := \sqrt{\tau_f(0)/\rho_f}. \tag{2.26}$$

In the vicinity of the wall, the intrinsic Reynolds number is necessarily small as the velocities must decay to zero, and therefore, a viscous length scale,

$$\delta_\nu := \nu_f/u_\tau, \tag{2.27}$$

can be established by a combination of the characteristic shear velocity and viscosity. The quantities u_τ and δ_ν are commonly denoted as the inner scales, as opposed to the outer scales u_b and H. From the definition of the Kolmogorov scales, it can be demonstrated that

$$\delta_\nu/\eta = (\varepsilon_f^+)^{1/4} \tag{2.28}$$

with ε_f^+ denoting the dissipation rate normalised by the inner scales (Pope 2000, Stumpf 2017).[10] Although the dissipation rate, and hence likewise the Kolmogorov scales, vary with the wall-normal position, empirical data for the normalised dissipation rate suggests that the ratio of viscous and Kolmogorov length scales is of the order of unity at all positions (Kim et al. 1987, Pope 2000). Therefore, the inner scales may be regarded as a proxy for the smallest turbulent scales which appear in channel flow. The Reynolds number based on the friction velocity is called the friction Reynolds number,

$$Re_\tau := \frac{u_\tau H}{\nu_f} = \frac{H}{\delta_\nu}, \tag{2.29}$$

and its value directly indicates the separation in length scales between the largest turbulent motions of $\mathcal{O}(H)$ and the smallest eddies of $\mathcal{O}(\delta_\nu)$.

Turbulent channel flow is statistically inhomogeneous in the wall-normal direction, and as a result, various regions with distinct scaling laws for the mean streamwise velocity can be observed. Traditionally, the wall-normal coordinate is classified into three regions—the inner region where $y \ll H$, the outer region where $y \gg \delta_\nu$, and the overlap region where both conditions are satisfied. A sketch of this classification is provided in fig. 2.1(a), and it becomes immediately apparent that the overlap region can

[10] The superscript "+" notation will henceforth be used to denote any non-dimensionalization by a proper combination of the inner units.

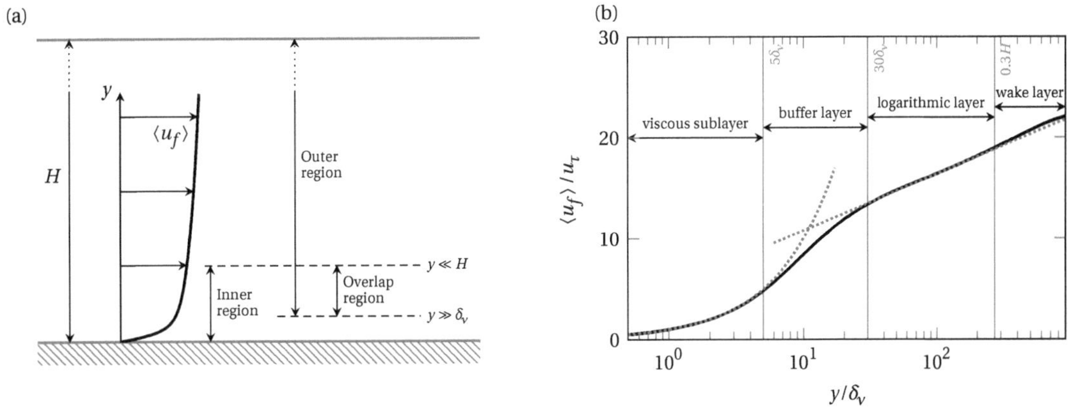

Figure 2.1: (a) The regions of turbulent channel flow. (b) The layers within which different scaling laws for the mean streamwise velocity can be applied. The dashed lines indicate the law of the wall in the viscous sublayer and the log-law respectively. The solid lines shows data obtained by Bauer et al. (2022) for open-channel flow at $Re_\tau = 900$. The illustration is inspired by Davidson (2015).

only exist at high friction Reynolds numbers where the inner and outer scales are sufficiently separated. Using this regions, dimensional analysis suggests the following scaling laws,

$$\frac{\langle u_f \rangle_{xzt}}{u_\tau} = \Phi_{IR}\left(\frac{y}{\delta_v}\right) \qquad \text{“law of the wall” of the inner region,} \tag{2.30}$$

$$\frac{\langle u_f \rangle_{xzt}}{u_\tau} = \frac{1}{\kappa}\ln\left(\frac{y}{\delta_v}\right) + A \qquad \text{“log-law” of the overlap region,} \tag{2.31}$$

$$\frac{u_H - \langle u_f \rangle_{xzt}}{u_\tau} = \Phi_{OR}\left(\frac{y}{H}\right) \qquad \text{“velocity defect law” of the outer region,} \tag{2.32}$$

where Φ_{IR} and Φ_{OR} are functions which require further specification, $\kappa \approx 0.41$ is Kármán's constant, A is a constant of integration and $u_H := \langle u_f \rangle_{xzt}$ is the centreline velocity, see e.g. (Davidson 2015) for derivations. As is suggested by Kolmogorov's theory, the velocity defect law is generally not universal among the canonical wall-bounded flows, and consequently, the scaling behaviour of the outer region may constitute a difference between open- and closed-channel flow. In contrast, the law of the wall can be further specified irrespective of the flow geometry as very close to the wall the flow is laminar to a large degree, which yields (Prandtl 1925)

$$\frac{\langle u_f \rangle_{xzt}}{u_\tau} = \frac{y}{\delta_v} \qquad \text{for } y \lessapprox 5\delta_v. \tag{2.33}$$

Although the log-law has been derived from the premises of the overlap region, its validity extends considerably into what is usually considered to be the outer layer. For instance, in the turbulent open-channel flow at $Re_\tau = 900$ shown in fig. 2.1(b), the log-law accurately describes the mean streamwise velocity in the region $30\delta_v \lessapprox y \lessapprox 0.3H$. Therefore, the regions of channel flow may alternatively be classified into layers within which a associated scaling law holds—the viscous sublayer for which eq. 2.33 holds, the logarithmic layer within which eq. 2.31 is observed, the buffer layer which separates the previous two, and the wake layer which is subject to the non-universal velocity defect law.

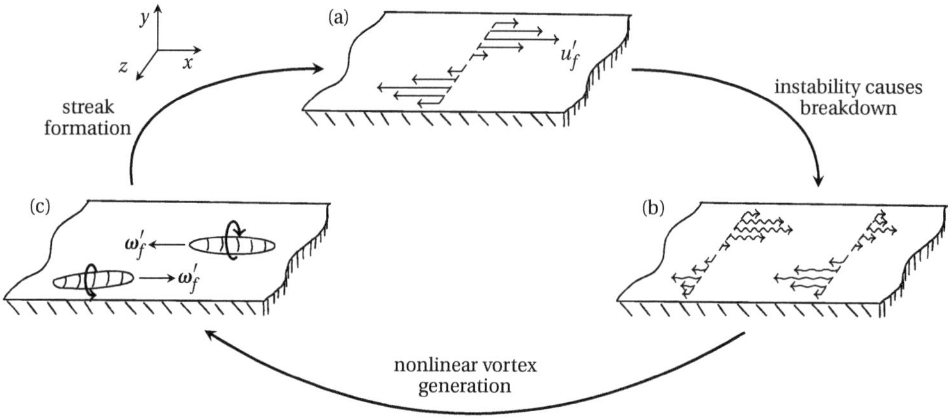

Figure 2.2: Schematic depiction of the self-sustained near-wall cycle as described by Hamilton et al. (1995). (a) The flow is composed of high- and low-speed streaks of streamwise velocity fluctuations. (b) Unstable streamwise modes of the velocity streaks emerge. (c) Quasi-streamwise vortices are generated, which eventually create velocity streaks.

2.2.3 Coherency and structure of turbulence

The previously introduced layers of turbulence have been defined in terms of their statistics. As briefly mentioned in §2.1.5, an essential feature of turbulence is that it comprises coherent packets of vorticity, which may be localised, and in turn may induce coherent patches of flow velocities. As a consequence, turbulent flows are by no means stochastic, but have an intrinsic structure which has interesting implications on transport processes.

In the vicinity of the wall, that is in the buffer layer, the most prominent coherent flow features as so-called velocity streaks, which denote streamwise-aligned, elongated regions of streamwise velocity fluctuations, and have first been described comprehensively by Kline et al. (1967). It is commonly distinguished between low-speed ($u_f' < 0$) and high-speed ($u_f' > 0$) streaks which are most often observed in spanwise-alternating pairs. The dimensions of these streaks scale in inner units in agreement with the arguments made on the inner region, and a spacing of approximately $100\delta_v$ in spanwise-direction between two pairs and streamwise lengths of $\mathcal{O}(1000\delta_v)$ are established values found in literature (Kline et al. 1967, Smith and Metzler 1983, Kim et al. 1987, Waleffe 1991). The buffer-layer streaks contribute significantly to the streamwise share of the turbulent kinetic energy which is reflected by a characteristic peak in $\langle u_f'^2 \rangle_{xzt}$ at $y^+ \approx 12$. Moreover, the streaks are linked to so-called Reynolds-stress events, namely ejections ($u_f' < 0$, $v_f' > 0$), which transport low-speed fluid away from the wall, and sweeps ($u_f' > 0$, $v_f' < 0$), which bring high-speed fluid towards the wall, and hence also contribute significantly towards $\langle u_f' v_f' \rangle_{xzt}$. Blackwelder and Eckelmann (1979) recognised that these "bursting events" are a result of the act of coherent vortices which are quasi-aligned in streamwise direction. In fact, these quasi-streamwise vortices are an instance of the "worms" described in §2.1.5 (Jiménez 2018), and form and breakdown continuously in the near-wall region (Swearingen and Blackwelder 1987). By isolating a single velocity streak in a minimal domain, Waleffe et al. (1993) observed that the constraint flow field exhibits a quasi-periodic behaviour during which the streak undergoes a wavy instability, which dampens its intensity, and subsequently reforms. This recognition lead to the formulation of a self-sustaining process—a sketch of which is provided in fig. 2.2—which describes the alternating formation of velocity streaks and streamwise vortices (Jiménez and Moin 1991, Hamilton et al. 1995). By artificially suppressing velocity fluctuations from the outer region, Jiménez and Pinelli (1999) demonstrated that this near-wall cycle can operate autonomously, and thus, constitutes one of the most fundamental processes in wall-bounded turbulence.

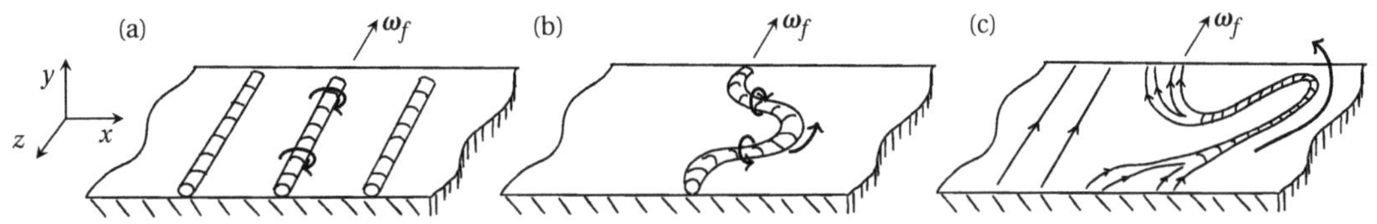

Figure 2.3: Conceptual cartoon of the formation of a hairpin vortex. (a) In undisturbed wall-bounded flow, or in the mean, vorticity is aligned with the spanwise direction. (b) A perturbation causes vorticity to be transported away from the wall which leads to a deformation. (c) Vortex stretching amplifies the intensity of the hairpin vortex. The illustration is inspired by Davidson (2015).

At greater distances from the wall in the logarithmic layer, the occurrence of velocity streaks is often linked to another class of coherent vortices—the hairpin or horseshoe vortex. This arch-shaped, loop-like structure consists of a pair of counter-rotating streamwise-aligned "legs" which are inclined from the neighbouring and connected by an approximately spanwise-aligned "head", and was first described by Theodorsen (1952). Its relevance was demonstrated experimentally by Head and Bandyopadhyay (1981) and numerically by Wu and Moin (2009), who revealed that these structures are ubiquitous and densely populate boundary layers. A potential mechanism for its formation is given by the advection of spanwise vorticity—which is abundant in wall-bounded flow as it is the sole component of vorticity which does not vanish in the mean—away from the wall by means of a velocity fluctuation (Davidson 2015). As the vortex filament is lifted from the wall, it is deformed by the gradient in streamwise velocity and reorientates itself towards an streamwise inclination of approximately 45°, which causes the vortex to be stretched and amplified due to its interaction with the mean shear. A sketch of this process is depicted in fig. 2.3. There are various potential causes for the initial perturbation. While some of them involve the quasi-streamwise vortices of the buffer layer, others involve turbulent eddies originating from the outer layer, though ultimately, turbulence is a chaotic organization of entangled vorticity and a clear causality remains more often than not difficult to establish. A common denominator of all conceptual ideas of hairpin vortices is, however, that the mean shear is responsible for the stretching and intensification of vorticity in order to form elongated vortex structures which are inclined with respect to the wall (Davidson 2015).

While it is certainly possible to observe isolated occurrences, hairpin vortices more frequently occur in the form of packets which contain multiple instances of them (Adrian 2007). Indeed, Zhou et al. (1996) and Zhou et al. (1999) exposed an autogeneration mechanism which may lead to the formation of a vortex packet of streamwise-staggered hairpin vortices, and hence, it is worthwhile to assess the implications of superimposed vortex structures. An important basis for most descriptions of multi-vortex interactions is the attached eddy model introduced by Townsend (1976) based on his study of equilibrium layers in wall turbulence (Townsend 1961). The attached eddy model distinguishes between wall-attached motions whose length scale is consequently proportional to the wall-normal distance over which they span, and wall-detached motions. While the latter is thought to be rather isotropic and to contribute little to the Reynolds stress budget, the former is inevitably anisotropic due to its exposure to wall shear, and is therefore regarded to be more active in terms of velocity scaling laws. Perry and Chong (1982) utilised Townsend's concept and the flow visualizations of Head and Bandyopadhyay (1981) to conceptualise a model in which the wall-attached eddies are considered to be hairpin-like vortices. In a related study, Perry et al. (1986) were able to derive the logarithmic law for the mean streamwise velocity, as well as scaling laws for the spectral density of the fluctuating velocity components, from a continuous random arrangement of such vortices, which emphasises the relevance of coherent motions, and in particular hairpin vortices, to the understanding of turbulent motion.

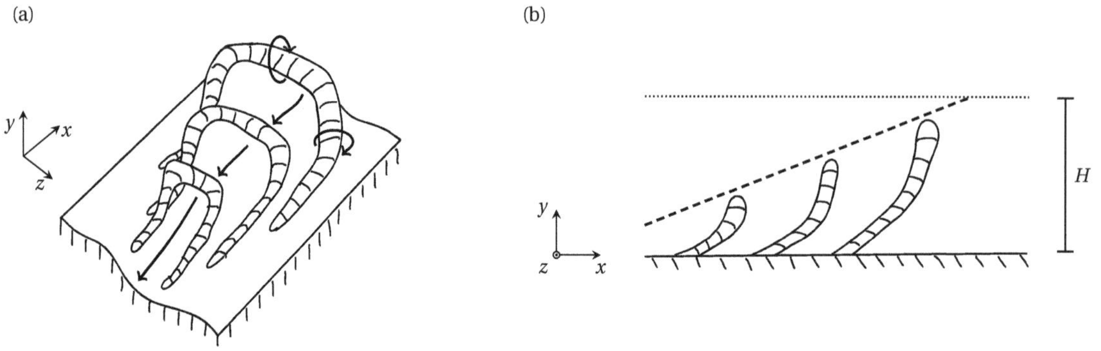

Figure 2.4: Conceptual sketch of hairpin vortex packets as described by Adrian et al. (2000). (a) Hairpin vortices organise in streamwise-staggered packets and induce a coherent low-speed fluctuation in streamwise velocity. (b) Packets may grow away from the wall and attain dimensions of $\mathcal{O}(H)$, and hence, constitute one potential explanation of large-scale motion.

It has been established mainly through the work of Adrian et al. (2000) that vortex packets are common in and essential for wall-bounded turbulence (Smits et al. 2011). The packets are relatively stable implying that they persists for the duration of several turnover times, and whenever the propagation velocity of the individual vortices is similar they may be treated as a larger single coherent structure. The size and propagation speed of this structure increases with its age, and it is not uncommon for them to protrude into the wake layer of the flow in the later stages of their development causing their size and velocity to be dictated by the outer units (Adrian 2007). Vortex packets are associated to the turbulent bulges frequently perceived in flow visualization of the turbulent boundary layer, cf. Head and Bandyopadhyay (1981), and are therefore also denoted as large-scale motions (LSMs). When examining the local flow field it is immediately apparent that a low-speed region is collectively induced by the ensemble of hairpins Adrian et al. (2000), Ganapathisubramani et al. (2003), see fig. 2.4. In turbulent channel flow, these features emerge as large-scale streaks of the streamwise velocity whose maximum dimensions are 2–3 H and 1–1.5 H in streamwise and spanwise direction, respectively, and thus scale in terms of outer units (Brown and Thomas 1977, Kim and Adrian 1999, Zhou et al. 1999, Guala et al. 2006, Balakumar and Adrian 2007), which clearly distinguishes them from the buffer-layer streaks discussed earlier. It should be noted that not all LSMs fulfil Townsend's criteria for attached eddies. In fact, experiments by Bailey and Smits (2010) indicate that hairpin packets in the outer layer are frequently detached from the wall, while those in the logarithmic layer are more likely to be attached.

Vortex packets are not the largest structures of wall-bounded turbulence as coherent streamwise velocity fluctuations with length scales of $\mathcal{O}(3\text{–}100H)$ have been reported for both internal and external wall-bounded flows (Kim and Adrian 1999, Guala et al. 2006, Monty et al. 2007, Balakumar and Adrian 2007, Lozano-Durán and Jiménez 2014a, Bauer 2020). These very large scale motions (VLSMs), which are also known as superstructures, manifest themselves in the form of very long, but laterally confined, meandering streamwise velocity streaks, and can be found in the logarithmic layer as well as the wake layer, although indications exist that for external flows these motions are restricted to the former (Monty et al. 2009). A popular theory on their origin interprets VLSMs as the streamwise organization of vortex packets (Kim and Adrian 1999), though the similarities in lateral scaling suggests that exclusively detached LSMs of the wake layer are responsible for their creation, whereas attached LSMs are located too close to the wall, propagate too slowly and have too narrow lateral scales to be involved (Bailey and Smits 2010, Smits et al. 2011).

Although the interpretation of LSMs and VLSMs as the (pen-)ultimate realization of a self-similar hairpin vortex organization process is alluring, it should not go unnoticed that concurring theories on an underlying mechanism exist. An alternative model to vortex packets is the self-similar vortex cluster proposed by del Álamo et al. (2006), whose primary distinction is its reduced internal organization compared to hairpin packets. Self-similar vortex clusters may be best described as a shell of enclosed small-scale vorticity (worms), which exhibits little structure internally, but as an ensemble self-organises in a manner which may resemble more traditional vortex loops. In accordance with Townsend's postulations, del Álamo et al. (2006) found these structures to occur detached from the wall as well as in tall attached clusters, and indeed, the kinematics of these vortex clusters are sufficiently similar in character to use them interchangeably to hairpin packets in the explanation of LSMs and VLSMs. However, a major disparity between the two models is given in terms of the proposed mechanisms of their creation—while hairpin packets are required to form in the inner region of turbulence and subsequently grow to their final size, del Álamo et al. (2006) ascertained that vortex clusters are too short-lived to be created in this way, and hence, must be formed in terms of a sustained large-scale regeneration cycle akin to the near-wall cycle, but far more complex. In fact, Flores and Jiménez (2006) and Flores et al. (2007) demonstrated that the structures of the outer region are essentially independent from the details of the wall, which provides strong evidence of the existence of such an autonomous cycle does not rely on the growth of small near-wall structures.

Up to this point, the coherent eddies which have been discussed were predominantly characterised by means of vorticity. However, a second kind of eddy which has been used in literature to assess coherency in wall-bounded turbulence are connected regions of intense three-dimensional Reynolds stress events (Lozano-Durán et al. 2012a, Jiménez 2013), with the particular focus on ejection and sweep events, which are statistically and dynamically more relevant than their counterparts with a positive product of $u_f' v_f'$. As previously mentioned, sweeps and ejections are important mechanisms in the inner region, though they also occur at larger scales in the outer region. Similar to vortex clusters, large-scale sweeps and ejections may be classified into wall-attached or wall-detached structures, whereby detached structures are observed more frequently, but attached structures occupy a vastly larger region within the flow and contribute considerably more towards the total Reynolds-stress budget (Lozano-Durán et al. 2012a). On average, large-scale sweeps and ejections appear side-by-side in spanwise direction as pairs, and since their streamwise dimensions are typically long, continuity demands that a streamwise-aligned large-scale roller structure connects the two in the conditionally-averaged point-of-view. In the instantaneous view, vortex clusters are typically observed to emerge in the space between sweeps and ejections (Lozano-Durán and Jiménez 2014b), with a statistically significant tendency of being located closer to the former (Jiménez 2013), and hence, vortex clusters are likely to be—at least partially—manifestations of the roller (Lozano-Durán and Jiménez 2014b).

The coherent structures discussed so far are all associated to flows in the presence of walls with no-slip boundary conditions. While this implies that all of these structures are also present in open-channel flows, the additional free-slip boundary condition in this configuration gives rise to some peculiarities caused by the dampening of vertical velocity fluctuations while simultaneously allowing for non-vanishing velocity components in streamwise and spanwise direction. Nagaosa (1999) identified two dominant types of vortical structures in the vicinity of the free surface—near-horizontal quasi-streamwise vortices, and near-vertical surface-attached vortices. The former type of vortices induce regions near the boundary where fluid impinges on ("splats") or is transported away from ("antisplats") the surface, which where demonstrated to be of significance for heat and mass transfer across the boundary (Nagaosa 1999, Nagaosa and Handler 2003, Pinelli et al. 2022). Concerning the interactions of structures originating from the no-slip boundary with the free surface, Nagaosa and Handler (2003) demonstrated via conditional averaging that hairpin-like vortices are morphed ring-like shapes as they approach the free surface. Moreover, there exists evidence that the free surfaces promotes VLSMs (Duan et al. 2020), and indeed, Bauer et al. (2022) found that VLSMs in open channels are roughly twice as long and twice as wide as to their counterpart in closed channels.

The literature on coherent structures in wall-bounded turbulence is vast, and consequently, the preceding brief summary does not raise a claim on comprehensiveness. For more complete recapitulations on the structural aspects of turbulence, the reader is referred to the renowned overview articles of Cantwell (1981), Robinson (1991), Adrian (2007), Smits et al. (2011), Adrian and Marusic (2012), Jiménez (2013), McKeon (2017), Jiménez (2018) and Marusic and Monty (2019).

2.2.4 Chaos and exact coherent states

The investigation of coherent structures in turbulence as discussed in the preceding section was predominantly driven by the availability of experimental data, observations and phenomenological modelling. Nonetheless, the emergence of coherency and structure in turbulence is not an unforeseeable event, as its existence must be contained within in the framework of the Navier–Stokes equations. While the implications of coherent motions on the flow statistics are well-established by now, it can in fact be argued that the understanding of turbulence remains incomplete as long as the emergence of these structures cannot be explained from first principles, i.e. solely based on the fundamental equations without any observation-based assumptions—or to put it in the words of Eberhard Hopf:

> "The ultimate goal, however, must be a rational theory of statistical hydrodynamics where [...] properties of turbulent flow can be mathematically deduced from the fundamental equations of hydromechanics."
>
> — Hopf (1948)

The possibility of eventually obtaining such an understanding is closely connected to the fact that turbulence is deterministic, and therefore follows a strict set of rules without any stochasticity. The "randomness" which is commonly attributed to turbulence is hence by no means random in the stochastic sense, but is rather a result of deterministic chaos—the state of a nonlinear dynamical systems where "the present determines the future, but the approximate present does not approximately determine the future", as Edward Lorenz adequately phrased it—ergo the sensitivity of the evolution of a turbulent flow with regards to the details of its initial condition.

Turbulence, or chaos, only occurs at sufficiently high values of a controlling parameter R, which in case of channel flow is an integral Reynolds number such as Re_b, cf. §2.2.1, but may correspond to another nondimensional number depending on the flow scenario of interest. The route through which turbulence is achieved with increasing values of R varies among the canonical flows, and can roughly be divided into two general types of transition—supercritical and subcritical transition. In order for supercritical transition to occur, the flow must be linearly unstable above a critical value of the control parameter, which is denoted by R^* in the following. Transition may then occurs by a series of bifurcations, i.e. abrupt changes in the flow topology which become increasingly complex the further R^* is exceeded and may introduce time-periodic behaviour to the flow by means of a Hopf bifurcation, and eventually becomes chaotic. While it was originally envisioned by Landau (1944) and Hopf (1948) that the successive bifurcations introduce evermore frequencies to the flow which implies turbulence to be quasi-periodic, Ruelle and Takens (1971) demonstrated that turbulence is in fact aperiodic, and is hence initiated by a sudden transition as well. In the case of subcritical transition, bifurcations occur below the critical value of the control parameter by means of finite-amplitude perturbations, and it is thus the only scenario by which linearly stable flows such as plane Couette flow (Davey 1973, Romanov 1973) or pipe flow (Salwen et al. 1980) may become turbulent. However, subcritical bifurcations are frequently also

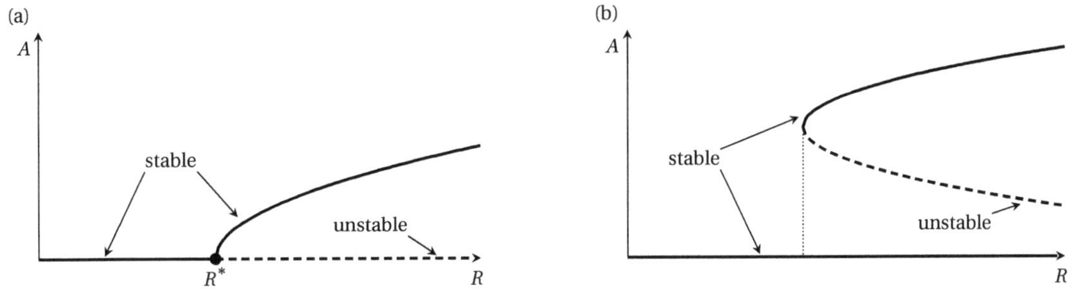

Figure 2.5: Examples of bifurcations as a function of the control parameter R in terms of a one-dimensional observable A. (a) The new state bifurcates supercritical above a critical value R^* for the control parameter. (b) Subcritical bifurcation in form of a saddle-node bifurcation. Solid lines denote stable states, whereas dashed lines denote unstable ones. The dotted vertical line in (b) marks the value of R above which a subcritical bifurcation may occur.

the dominant mechanism for the transition to turbulence in linearly unstable flows such as channel flow (Sano and Tamai 2016, Shimizu and Manneville 2019, Gomé et al. 2020), cf. §2.2.1. While various routes to chaos do exist, one route which is routinely observed for such flows is the intermittent appearance of spatially-confined turbulent regions which are denoted as spots, slugs, puffs or bands, and add additional complexity to the analysis of transition due to their spatiotemporal character (Wygnanski and Champagne 1973, Pomeau and Manneville 1980, Gollub and Benson 1980, Bottin and Chaté 1998, Eckhardt et al. 2007, Sano and Tamai 2016, Gomé et al. 2020).

A more systematic view on flow topologies can be obtained by introducing the concept of a state space, which denotes the space of all states that may be realised by the fluid flow, or consequently, the set of all divergence-free velocity fields which satisfy the appropriate boundary conditions. The trajectories upon which the flow evolves in this space are determined by eq. (2.2), and this temporal evolution of states is called solution flow. The state space itself is vast as only little constraints are posed towards the states which are comprised by it. However, it is straightforward to acknowledge that the subspace within which the solution flow occurs is restricted by the values of the control parameter of system. For instance, whenever the control parameter is sufficiently low, the solution flow will eventually converge towards the laminar state irrespective of the initial state of the system, and hence, the subspace in which the solution flow occurs at long times only comprises the laminar solution in this case. Such a characteristic state towards which all sufficiently close solution flows converge is called an attractor (Lanford 1982).

It is apparent that attractors may emerge or disappear with changing values of the control parameter. In fact, the bifurcation scenarios which have been describe previously in the context of transition to turbulence can be viewed as a transformation of attractors in the system. Moreover, the existence of subcritical bifurcations makes it obvious that multiple attractors may coexist for a given R, and thus, that the solution flow may converge towards different states depending on the initial state or perturbations. Various types of attractors exist and the most relevant for the present discussion are summarised in the following. The time-invariant laminar attractor constitutes a so-called fixed point, which is a singular element in the set of states. In fact, laminar flow is the fixed point of lowest dimensionality, and as will be elaborated at a later point, more complex fixed points may occur in fluid flows. The simplest attractor with time-dependent behaviour are closed trajectories in the state space which are called limit cycles with a well-known example being the Kármán vortex street. Limit cycles may consist of multiple frequencies, and whenever the ratio between two of these frequencies is irrational, the cycle is no longer closed, and the resulting quasi-periodic trajectories are denoted as a limit torus. Fixed points, limit cycles and limit torii are geometrically simple objects in the state space, but more complex attractors also exist. An important instance is the strange attractor, a fractal object in the state space which is neither periodic nor quasi-periodic (Guckenheimer and Holmes 1983, Strogatz 2018), and was first discovered by Edward Lorenz in his seminal work on simplified cellular convection (Lorenz 1963) and given its name by Ruelle and Takens (1971). Strange attractors are often associated to chaotic behaviour, and as such, it was only natural to assume that turbulence itself belongs into this category (Ruelle and Takens 1971, Lanford 1982, Guckenheimer 1986). However, as turbulence is not necessarily persistent and may decay to the laminar state, it cannot be an attractor, but is more adequately described by a strange saddle—an object in state space which attracts solution flows from some directions, but also possesses unstable directions which allow for decay (Eckhardt et al. 2007).

The fluid states which correspond to attractors or saddles of the dynamical system are termed invariant solutions. The counterpart to a fixed point in physical space is called an equilibrium solution and fulfils the requirement

$$\mathbf{u}_f(\mathbf{x}, t) = \mathbf{u}_{EQ}(\mathbf{x}), \tag{2.34}$$

which implies that the equilibrium state $\mathbf{u}_{EQ}(\mathbf{x})$ is time-invariant.[11] In streaming flows, non-trivial time-invariant solutions take the form of a travelling wave, viz.

$$\mathbf{u}_f(\mathbf{x} - \mathbf{c}t, t) = \mathbf{u}_{TW}(\mathbf{x}), \tag{2.35}$$

i.e. they are equilibrium solutions within a frame of reference which is comoving with the structure which propagates with a constant propagation velocity $\mathbf{c}$. The equivalent of limit cycles and torii are periodic orbits, which are cyclic flow states where the initial condition is exactly recovered after a period T, and are hence given by

$$\mathbf{u}_f(\mathbf{x}, t + T) = \mathbf{u}_{PO}(\mathbf{x}, t). \tag{2.36}$$

It is not uncommon for a periodic orbit to not return to its exact original state, but rather to a possibly translated or mirrored version of itself. If this is the case, the resulting invariant solution is termed a relative periodic orbit. Since this subtle distinction is of little importance in the following, the notion of periodic orbits will also include relative ones.

Aside from supercritical transition, invariant solutions are not commonly observed in their purest form in real-life fluid flows. However, there exists compelling evidence that these characteristic states are embedded in the turbulent subspace, and that the dynamics of turbulence can be at least partially understood in terms of trajectories around and between invariant saddles and attractors (Chandler and Kerswell 2013, Suri et al. 2017, Budanur et al. 2017), i.e. in terms of homoclinic and heteroclinic connections in the state space, respectively. In fact, when the structure of some invariant solutions is examined, similarities with the coherent structures discussed in §2.2.3 may be found. In literature, equilibrium solutions resembling quasi-streamwise vortices (Nagata 1990, Waleffe 2001), hairpins (Itano and Generalis 2009, Shekar and Graham 2018), or attached eddies (Park and Graham 2015, Yang et al. 2019) have been reported. Figure 2.6 provides a visual example of the former two in a channel flow. These similarities are not solely restricted to the topology of the flows, but also manifest themselves in flow statistics which may resemble those of turbulent flows (Kawahara et al. 2012).

Despite the similarities to known coherent structures, equilibrium solutions differ from turbulence in that they do not possesses any temporal dynamics by themselves—they are static objects in the state

[11] Although some solutions are indeed stable with respect to perturbations, it is generally more expedient to not demand stability for the definition of invariant solutions. In fact, many of the invariant solutions which will be present in the following are unstable, which implies that they correspond to saddles rather than pure attractors.

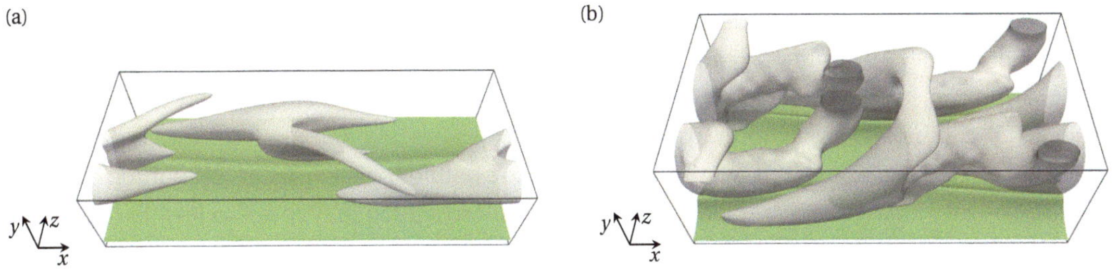

Figure 2.6: Examples of travelling waves in (open-)channel flow. An isosurface of streamwise velocity is shown in green, whereas vortical structures are depicted in grey in terms of the Q-criterion of Hunt et al. (1988). (a) The solution of Waleffe (2001) which comprises a wavy velocity streak and quasi-streamwise vortices. (b) A hairpin-like structure first discovered by Shekar and Graham (2018).

space. While this fact might be compensated to a degree by potential homoclinic and heteroclinic connections (Halcrow et al. 2009), arguments have been made that periodic orbits might be dynamically more relevant and their theoretical understanding might serve as the "rational theory of statistical hydrodynamics" that Eberhard Hopf was aiming at (Auerbach et al. 1987, Christiansen et al. 1997, Chandler and Kerswell 2013, Cvitanović 2013). In fact, a periodic orbit in Couette flow with dynamics reflecting the near-wall cycle portrayed in fig. 2.2 astonishingly well was famously identified by Kawahara and Kida (2001). Other periodic orbits showcasing bursting dynamics have been detected by Toh and Itano (2003) and Viswanath (2007), and it is subject to speculation that turbulence can indeed by understood as being made up by "building blocks" of equilibrium solutions and periodic orbits which are recurrently visited by the solution flow and are likely the origin of the coherent structures observed in turbulence. For this reason, invariant solutions within the turbulent subspace are sometimes referred to as exact coherent structures or exact coherent states.

Although the presence of exact coherent states can be detected in experiments (Hof et al. 2004), their identification is almost exclusively achieved by a numerical procedures which are surprisingly straightforward. To begin with the Navier-Stokes momentum equations are formulated in terms of a generalised dynamical system,

$$\frac{\mathrm{d}\mathbf{q}}{\mathrm{d}t} = \mathbf{f}(\mathbf{q},\mathbf{p}), \tag{2.37}$$

where $\mathbf{f}$ is a vector field stemming from the Navier–Stokes operator, $\mathbf{q}$ is the vector of unknowns, e.g. the velocity field and parameters such as the propagation velocity for travelling waves or the period for orbits, and $\mathbf{p}$ is the vector of parameters which fully describe the problem, e.g. the Reynolds number and geometry of the flow domain. The vector space spanned by $\mathbf{q}$ is infinite-dimensional for the continuous flow problem, and must hence be replaced by a finite dimensional manifold by means of discretisation for numerical treatment, cf. Ruelle and Takens (1971) a discussion on the implications. The problem then becomes are matter of finding solutions of eq. (2.37) for which $\partial\mathbf{q}/\partial t = 0$ to obtain equilibria, or $\mathbf{q}(T) = \mathbf{q}(0)$ to obtain periodic orbits, which can be solved using standard root-finding algorithms such as the Newton–Raphson method.

In order for root-finding algorithms to converge to a non-trivial invariant solution, an appropriate initial guess needs to be provided which turns out to be the most intricate problem in obtaining exact coherent states. An effective approach of generating initial guesses is homotopy—a successive variation of the vector of parameters $\mathbf{p}$ starting from a pre-existing invariant solution. Naturally, the question arises on how the initial condition for this procedure is obtained, and the answer is that homotopy may be initiated from a supercritical bifurcation of a related linearly unstable flow geometry. The most seminal example of this approach is likely the work of Nagata (1990) who continued the Taylor-vortex flow state between a set of co-rotating cylinders, which is readily obtained from a linear instability (Coles 1965, Gollub and Swinney 1975), towards the limit of zero rotation which corresponds to the linearly stable flow between two surfaces in relative motion (plane Couette flow). Based on Nagata's solution, Waleffe (1998, 2003) continued this solution even further to obtain exact coherent states for channel flow by varying the pressure gradient and relative velocities of the boundaries. Evidently, homotopy can also be employed to vary the control parameters for a fixed geometry which may result in further bifurcations, and hence, large networks of related exact coherent states may be generated which are termed "families of solutions".

Initial guesses for the root-finding algorithm may also be obtained by a mindful inspection of turbulent flow fields obtained by direct numerical simulation of the NSEs. As exact coherent states regularly possess intrinsic symmetries, it is often effective to restrain the numerical simulations by imposing symmetries or minimal periods of the simulation domain. A prominent example for this approach is the periodic orbit of Kawahara and Kida (2001). Obtaining initial guess from restrained simulations is a protracted procedure, yet it has the advantage that new families of solutions may be discovered which are inaccessible from homotopy. Recently, more efficient methods of searching turbulence for periodic orbits have emerged based on Koopman theory (Page and Kerswell 2020), which hopefully help pave the path towards the achievement of Hopf's ultimate goal.

2.3 Particles in turbulence

Whereas the preceding discussion on fluid flows was predominantly focused at the interaction of the fluid phase with a stationary solid phase—the outer boundaries of the flow domain—the focus is now shifted towards flows with mobile solid objects (particles) immersed in it, which are consequently denoted as particle-laden flows. Similar to the wall interaction in wall-bounded flows, the forces and stresses on the fluid-solid interphase are a result of the impermeability and no-slip boundary conditions, which are given by eq. (2.18) and eq. (2.19), respectively. In contrast to it, the solid phase may exhibit dynamics on its own as the particles are mobile, and hence, interactions are reciprocal resulting in a fluid-structure interaction problem. In order to describe this coupled system, equations of motion for the solid phase need to be introduced. By balancing the linear and angular momentum attributed to a singular rigid object, the Newton–Euler equations are obtained, which read

$$\frac{\mathrm{d}(\rho_p V_p \mathbf{u}_p)}{\mathrm{d}t} = \oint_{S_p} \left(\overline{\overline{\boldsymbol{\sigma}}} \cdot \mathbf{n}\right) \mathrm{d}S_p + \rho_p V_p \mathbf{g} + \mathbf{F}_{p,col}, \tag{2.38}$$

$$\frac{\mathrm{d}(\mathbf{I}_p \cdot \boldsymbol{\omega}_p)}{\mathrm{d}t} = \oint_{S_p} (\mathbf{x} - \mathbf{x}_p) \times \left(\overline{\overline{\boldsymbol{\sigma}}} \cdot \mathbf{n}\right) \mathrm{d}S_p + \mathbf{T}_{p,col}. \tag{2.39}$$

Here, $\mathbf{x}_p$ is the centroid location of the particle, $\mathbf{u}_p$ its translational velocity, $\boldsymbol{\omega}_p$ its rotational velocity, V_p its volume which is enclosed by the fluid-solid interphase S_p with $\mathbf{n}$ being the surface's outward-pointing normal vector and $\mathrm{d}S_p$ being an infinitesimal surface element, ρ_p is the particle's density, and $\mathbf{I}_p$ its moment of inertia. The exchange of momentum between the two phases is governed by the surface stresses with $\overline{\overline{\boldsymbol{\sigma}}}$ being the stress tensor defined in eq. (2.3) which includes buoyancy. The second term on the right-hand side of eq. (2.38) reflects the force due to gravity, which is the only external force considered in the following. Particles may collide with other solid objects such as the fluid domain's bounding walls or other immersed particles. The resulting collision force and torque are denoted by $\mathbf{F}_{p,col}$ and $\mathbf{T}_{p,col}$, respectively, and are yet to be defined.

From the translational and angular velocities obtained by eq. (2.38) and eq. (2.39), the trajectory of the particle is obtained by integrating the kinematic equations

$$\frac{\mathrm{d}\mathbf{x}_p}{\mathrm{d}t} = \mathbf{u}_p, \tag{2.40}$$

$$\frac{\mathrm{d}\boldsymbol{\Omega}_p}{\mathrm{d}t} = \boldsymbol{\omega}_p, \tag{2.41}$$

given adequate initial conditions, where $\boldsymbol{\Omega}_p$ denotes the angular position. For the important case of spherical particles, which are described by a single length scale chosen to be the diameter d_p in the following, the volume of the particle is given by $V_p = \pi d_p^3/6$ and its moment of inertia reduces to a scalar value which reads $I_p = \pi \rho_p d_p^5/60$ if the mass is distributed homogeneously across its volume. Moreover, since spherical bodies are rotationally symmetric, the angular position is irrelevant for the dynamics, and hence, it is typically not required to advance eq. (2.41) in time.

Disregarding the nature of the collision forces and torques, the Newton–Euler equations comprise three fundamental units—length, time and mass—and six independent dimensional parameters—the fluid and particle densities, the viscosity and a characteristic magnitude of the gradients of the fluid velocity which are both embedded in the stress tensor, the diameter of the spherical particle, and the magnitude of gravity. The Buckingham π theorem, cf. §2.1.2, therefore demands three dimensionless groups to fully parametrise the problem. One ad hoc parametrisation is achieved by the solid-to-fluid density ratio, ρ_p/ρ_f, a ratio between the particle diameter and a characteristic length scale of the fluid flow, and the Galileo number[12]

$$Ga := \frac{d_p v_{p,g}}{\nu_f}, \tag{2.42}$$

where $v_{p,g} := \sqrt{(\rho_p/\rho_f - 1)\,|\mathbf{g}|\,d_p}$ is a buoyancy-based velocity scale termed the gravitational velocity. This parametrisation has the advantage of being solely dependent on parameters which are known a priori, however, in the context of turbulent particle-laden flows, other parametrisation may be more convenient and will be introduced whenever they are meaningful.

2.3.1 Sediment-laden open-channel flow

Let us consider a horizontal open-channel flow containing a sufficient number of particles to cover the bottom wall, and for reasons of simplicity, assume that all particles are spherical and monodisperse such that they can be parametrised by single values for Ga, ρ_p/ρ_f and d_p/δ_ν, respectively. When the gravitational velocity of the particles is sufficiently high compared to the characteristic velocity scale of the fluid flow, the particles will accumulate on the lower boundary of the flow domain and form a sediment bed. The accumulation of particles effectively acts as a modified boundary to the flow domain, which is rough, permeable and deformable in contrast to the actual wall, and this modification is not without consequences for the structure and statistics of turbulence, as will be discussed in the following. Sediment beds are widespread in natural fluvial environments, and thus, sediment-laden open-channel

[12] The Galileo number is named in honour of the Italian polymath Galileo Galilei, who is commonly simply referred to as Galileo. There exists some disagreement whether the nondimensional number should rather be termed "Galilei number" as the naming convention is generally based on surnames. In the context of chemical engineering, the square of the Galileo number is also referred to as the Archimedes number.

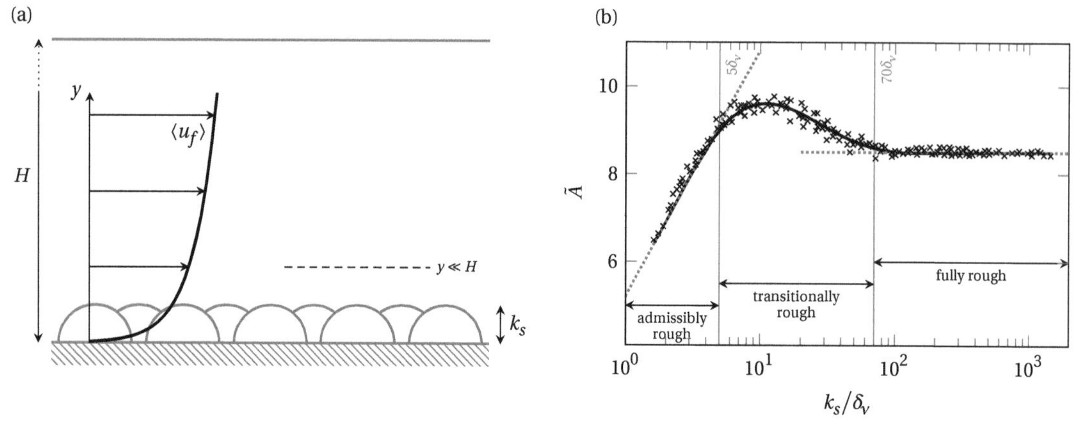

Figure 2.7: (a) Conceptual sketch of the mean flow over regular hemispherical roughness elements in open-channel flow. (b) Classification of roughness regimes depending on the ratio of effective roughness k_s and inner units of turbulence. The layers within which different scaling laws for the mean streamwise velocity can be applied. The dashed lines indicate the laws stated in eq. (2.44) and eq. (2.45), respectively. Experimental data (×) was obtained by Nikuradse (1933) for rough pipe flows.

flow may be regarded as a more sophisticated idealised description of rivers and canals than canonical channel flow.

The effect of roughness depends predominantly on the size, shape and arrangement of the roughness elements, see e.g. Jiménez (2004) for a discussion. A good consolidation of these effects can frequently be achieved by determining an equivalent "sand" roughness k_s, which in the case of spheres is of $\mathcal{O}(d_p)$, cf. Chan-Braun et al. (2011) and references therein. The mean velocity profile is again found to be described by a logarithmic law similar to eq. (2.31), but is now given by the length scale introduced by the roughness rather than viscous units, viz.

$$\frac{\langle u_f \rangle_{xzt}}{u_\tau} = \frac{1}{\kappa} \ln\left(\frac{y}{k_s}\right) + \tilde{A}. \tag{2.43}$$

The constant of integration $\tilde{A}$ is depends on the size of the roughness elements with respect to the viscous length scale, and in the limit where k_s is small compared to δ_ν, eq. (2.43) reduces to the log-law for smooth walls, and hence $\tilde{A}$ reads

$$\tilde{A} = A + \frac{1}{\kappa} \ln\left(\frac{k_s}{\delta_\nu}\right) \qquad \text{for} \quad k_s \ll \delta_\nu. \tag{2.44}$$

On the other hand, when the roughness elements are large in comparison to the inner region, the fully rough regime is approached where $\tilde{A}$ is constant irrespective of the value of k_s, that is,

$$\tilde{A} \approx 8.5 \qquad \text{for} \quad k_s \gg \delta_\nu. \tag{2.45}$$

The values of k_s for which neither eq. (2.44) nor eq. (2.45) holds is termed transitionally rough regime. The exact thresholds for which the flow is transitionally rough depend on the details of the rough surface (Bradshaw 2000, Jiménez 2004, Shockling et al. 2006). A sketch of the roughness regimes observed by Nikuradse (1933) for carefully sieved sand grain roughness in pipe flow is given in fig. 2.7.

Apart from some exceptional configurations, rough surfaces tend to increase the shear stress on the domain boundary, and hence, increase the value of the friction velocity. The modification of the wall shear stress may be attributed to two mechanisms. First, the shape of the roughness elements increases the viscous drag on the wall, and if the elements are sufficiently large, also induce an additional form drag due to flow separation (Jiménez 2004, Chan-Braun et al. 2011). Second, when k_s/δ_ν is large enough to affect the buffer layer, the turbulent near-wall cycle is modified up to until its breakdown in the fully rough regime, which leads to a reduction in the wall shear and annihilates the peak of the streamwise velocity fluctuations associated to the near-wall streaks (Jiménez 2004). In most flows, however, the former phenomenon outweighs the latter resulting in a net increase in u_τ. It has been subject to debate whether roughness also affects the structure of turbulence in the outer layer (Bandyopadhyay and Watson 1988, Jiménez 2004, Flack et al. 2005, Marusic et al. 2010), and while Jiménez (2004) suggested that universality is only achieved for $k_s \lessapprox H/40$, the study of Flack and Schultz (2014) indicates that modifications of lower order statistics are only minor even for significantly protruding roughness elements.

Permeability is also known to increase the wall friction, even in the absence of roughness effects (Jiménez et al. 2001). This increase originates in part directly from the additional dissipation introduced by the non-zero fluid velocity at the wall, but is mainly attributed to the formation of large spanwise-aligned flow structures which develop due to a shear layer instability of Kelvin–Helmholtz type (Jiménez et al. 2001, Suga et al. 2017). These spanwise rolls are found to significantly alter the distribution of Reynolds stresses near the permeable wall, but also affect the outer layer of turbulence by introducing sweep and ejection events with wavelengths of $\mathcal{O}(H)$ (Suga et al. 2018). The effect of spanwise rolls on the transfer of heat and mass has been subject to recent investigations, and it was found that their presence enhances heat transfer in the same way as momentum transfer, unlike roughness were flow separation only enhances the latter (Nishiyama et al. 2020, Motoki et al. 2022).

Regarding the mobility of the sediment, various modes of sediment transport can be distinguished (Bagnold 1956, van Rijn 1984a,b). In close vicinity to the stationary bed, particles predominantly stay in contact with the bed and move along by rolling, sliding or saltation, i.e. the momentary lift-off of a particles with subsequent redeposition upon the bed. These modes of transport are summarised as bedload, and the region in which they are dominant is termed the bedload layer. Whenever the lift forces acting on a particle are sufficiently high, particles may resuspend into the outer region of the flow and be transported alongside other already suspended particle as suspended load. Although the suspended load is sometimes further compartmentalised into a "washload" of particles which stay permanently suspended, this distinction will not be made in the following as its definition comprises some arbitrariness (Khullar et al. 2010) and is not of importance in the following. A sketch of the modes of sediment transport and the definition of transport loads can be found in fig. 2.8(a).

The vertical concentration profile of particles in the quasi-steady state in a turbulent environment is governed by the balance of fluid-induced vertical forces acting on the particles and gravitational forces. The predominant parameter determining this distribution is the relative turbulence intensity, which is defined as

$$I_\tau := \frac{u_\tau}{v_{p,\infty}}, \tag{2.46}$$

where $v_{p,\infty}$ is the settling velocity of the particle in a quiescent environment. In the traditional theory of Rouse (1937), which is derived from the balance of turbulent diffusion and settling effects, I_τ acts as the exponent of a power law describing the particle concentration as a function of vertical position. Although Rouse's theory should be complemented by additional terms in horizontal wall-bounded turbulence (Bragg et al. 2021a), it is plausible that any additional terms describing near-wall turbulent effects should scale in inner units, hence upholding the significance of u_τ as the relevant velocity scale.

River beds are seldomly flat, but rather self-organise in macroscopic patterns such as ripples, dunes or antidunes (Engelund and Fredsoe 1982, van Rijn 1984c). Therefore, knowledge on such features is essential to predict sediment transport and deposition in rivers, cf. Best (2005), Seminara (2010) or Charru et al. (2013) for reviews on patterns in fluvial environments. The aforementioned features tend to orientate themselves perpendicular to the flow direction and typically span many grain diameters in vertical elevation. While they are understood to be caused by instabilities which depend only little on the actual structure of the flow above, there exists a type of bedform which is most notably driven by turbulent effects—the self-organization in long streamwise-aligned streaks of crests and troughs which are known as ridges. Ridges are thought to be closely related to secondary currents of Prandtl's second kind which occur in laterally-bounded flows (Casey 1935, Ikeda 1981, Nezu and Nakagawa 1984, Colombini 1993, Colombini and Parker 1995), and hence, their occurrence was long believed to be induced by a knock-on effect beginning at the wall. However, it was already demonstrated by Colombini (1993) that lateral boundaries are not required for their occurence and only recently it was shown by Scherer et al. (2022) that ridges are indeed the footprints of long-lived large-scale velocity streaks, which as described in §2.2.3 also occur in turbulent channel flows which lack a lateral confinement. While ridges can be regularly observed if large-scale turbulent structures are present in the flow, their amplitude is typically restricted to several particle diameters, which usually implies that ripples or dunes will be the dominant pattern whenever they are able to form. The conditions under which this is the case are summarised in fig. 2.8(b), and depend on the relative turbulence intensity, as well as on the size of the particles in terms of viscous units, i.e.

$$d_p^+ = \frac{d_p}{\delta_\nu} = \frac{d_p u_\tau}{\nu_f}, \tag{2.47}$$

which may also be understood as a particle Reynolds number based on the wall shear. It should be noted that it is also common to parametrise the bedform regimes in terms of the Shields parameter (Shields 1936) given by

$$\theta := \left(\frac{u_\tau}{v_{p,g}}\right)^2 \tag{2.48}$$

whose essential difference to I_τ is that it is based on the gravitational velocity rather than the terminal settling velocity. However, it can be demonstrated (Jenny et al. 2004) that for sufficiently low Galileo numbers, these two quantities are linked by the relation

$$\frac{v_{p,\infty}}{v_{p,g}} = \frac{2}{\sqrt{3C_d}}, \tag{2.49}$$

where C_d is the steady-state drag coefficient of the particle, which has to be determined empirically in the general case. Hence, the advantage of the Shields parameter over the relative turbulence intensity lies in the ability to straightforwardly determine its value a priori.

Concerning the structure of turbulence over mobile beds, it has been demonstrated that at sufficiently high rates of bedload transport, the near-wall regeneration cycle may be suppressed completely (Scherer et al. 2022). In fact, while for stationary roughness elements this breakdown is expected to occur for $k_s \approx 50\text{–}100\delta_\nu$ (Jiménez 2004), Scherer et al. (2022) found that a particle size of $d_p \approx 30\delta_\nu$ suffices if the rate of bedload transport is considerable. Little is known so far about the differences in the behaviour of LSMs and VLSMs over mobile beds compared to the smooth-wall theory discussed in §2.2.3, although Peruzzi et al. (2020) suggests that the distinctions are only minor. Vice versa, these large-scale structures have been identified to be relevant for the erosion of particles from the bed (Cameron et al. 2020, Scherer et al. 2022), as this process is thought to be strongly driven by ejections and sweeps (Cellino and Lemmin 2004). In the presence of spanwise-aligned sediment patterns such as ripples or dunes, which may

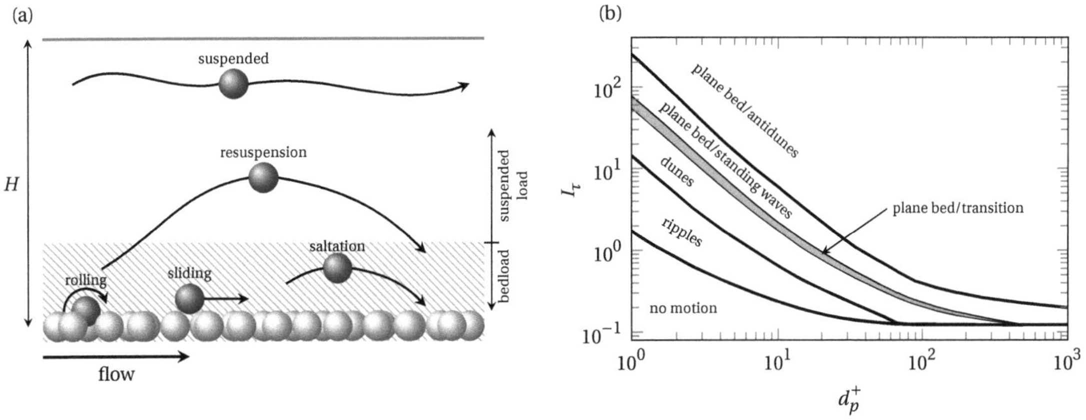

Figure 2.8: (a) Conceptual sketch of modes of sediment transport. The illustration is inspired by Dey (2014). (b) Classification of bedforms as a function of the nondimensional particle size and relative turbulence intensity as proposed by Liu (1957) and extended by Simons and Richardson (1961). The data is taken from Dey and Ali (2020) under the premise that $k_s = d_p$.

extent vertically over a significant fraction of H, a turbulent free shear layer is formed behind the crest which contains spanwise vorticity (Kadota and Nezu 1999, Cellino and Graf 2000, Kidanemariam 2016), i.e. the bedform acts like an obstacle to the mean flow. Further discussions on the role of coherent structures in sediment-laden flows are provided by García (2008) and Adrian and Marusic (2012).

2.3.2 Preferential concentration and clustering

The focus is now shifted towards particles suspended in turbulence, with an emphasis on those particle parameter points for which only negligible gravitational effects are exhibited ($Ga \to 0$), but still inertial effects are important ($\rho_p/\rho_f > 1$). Under the assumption that there are no immediate length scale effects, these inertial particles are fully parametrised by a single nondimensional number,

$$St := t_p / t_f, \tag{2.50}$$

which is known as the Stokes number. In this definition, t_p is a characteristic inertial time scale of the particle, which under the aforementioned assumptions and in the case of vanishing relative velocities is given by $t_p = \rho_p d_p^2 / (18 \rho_f \nu_f)$, and t_f is a characteristic fluid time scale, such as the Kolmogorov time scale η / u_η. The time scale t_p is also known as the particle response time, and hence, St may be interpreted as the ability of a particle to adapt its motion to a varying fluid environment.

Within an apposite range of the Stokes number, the distribution of particles in turbulence is found to be spatially intermittent even in the absence of statistical inhomogeneities within the turbulent flow (Squires and Eaton 1991, Elghobashi and Truesdell 1992, Eaton and Fessler 1994, McLaughlin 1994). This implies that even if particles are uniformly scattered across the flow domain initially, turbulence will exert a de-mixing effect and create regions of preferential concentration or clusters. The earliest explanation for this effect was provided by Maxey (1987), who, based on an earlier work on particle trajectories in artificial cellular flows Maxey and Corrsin (1986), argued that heavy particles are expelled from coherent vortices due to a centrifugal mechanism, and hence, preferentially concentrate in strained regions as they tend to avoid regions of high vorticity. In fact, assuming that the fluid acceleration can be approximated by the pressure gradient, it can be demonstrated that for $St \ll 1$ the rate of accumulation of particles in the Eulerian sense is negatively correlated to the local value of the second invariant of the velocity gradient tensor (Maxey 1987), which is given by

$$Q(\mathbf{x},t) := \frac{1}{2}\left(\overline{\overline{\boldsymbol{\Omega}}}^2 - \overline{\overline{\mathbf{S}}}^2\right) = -\frac{1}{2}\frac{\partial u_{f,i}}{\partial x_j}\frac{\partial u_{f,j}}{\partial x_i} \tag{2.51}$$

where $\overline{\overline{\mathbf{S}}}$ and $\overline{\overline{\boldsymbol{\Omega}}}$ are the symmetric and antisymmetric part of $\nabla \mathbf{u}_f$, respectively, and Einstein summation notation has been employed. This so-called Q-criterion (Hunt et al. 1988) may be interpreted as the relative importance of vorticity versus strain with positive(negative) values indicating that vorticity(strain) prevails, and thus, provides a quantification of the implications of the centrifugal mechanism.

The preceding inferences are only valid in the limit of very small values of the Stokes number, and as preferential concentration can also be observed at significantly higher Stokes numbers, and moreover also in randomly generated flow fields (Bec 2003, Duncan et al. 2005) where the centrifugal mechanism is inactive (Bragg et al. 2015), extensions for the theory are required in order to fully explain the phenomenon. A Lagrangian theory based on the interaction between pairs of particles was proposed and expanded by Zaichik and Alipchenkov (2003, 2007, 2009) to predict radial distribution functions based on the statistics of the turbulent flow. For $St \ll 1$, this theory is in its essence an extension of the centrifugal mechanism (Bragg and Collins 2014), however, for $St = \mathcal{O}(1)$ it predicts that the velocity dynamics lose their locality, and suggest another non-local mechanism which leads to clustering—the inward drift mechanism. Here, by non-locality it is referred to the fact that the inward drift mechanism does not depend on a biased sampling of local regions of strain, but should rather be understood as a result of the trajectory history and an asymmetry in the probability of converging and diverging trajectories of particle pairs. The validity of the Zaichik and Alipchenkov theory was corroborated empirically by Bragg and Collins (2014) using numerical simulations, and it was subsequently recognised by Bragg et al. (2015) that the intensity of the inward drift depends on the fluid velocity gradients previously experienced by the particles, which explains the local accumulation of particles in high strain regions even for Stokes numbers of $\mathcal{O}(1)$ as has been reported by e.g. Squires and Eaton (1991).

An alternative mechanism causing preferential concentration was introduced by Goto and Vassilicos (2008) and Coleman and Vassilicos (2009) originating from similar arguments which lead to Maxey's centrifugal mechanism. The sweep-stick mechanism is based on the idea that particles are driven away from regions of non-vanishing fluid acceleration, and subsequently get stuck in regions where

$$\mathbf{a}_f(\mathbf{x},t) := \frac{\partial \mathbf{u}_f}{\partial t} + \nabla\cdot\left(\mathbf{u}_f \otimes \mathbf{u}_f\right) \tag{2.52}$$

is nil, which themselves are coherent and swept away by the local fluid velocity. While these regions are abundant in two-dimensional turbulence and preferential concentration is well explained by sampling regions where $\mathbf{a}_f = \mathbf{0}$ (Chen et al. 2006, Goto and Vassilicos 2006), they are rarely observed in three-dimensional turbulence (Goto and Vassilicos 2008). However, the driving force of the convergence mechanism has been shown to be proportional to the divergence of the fluid acceleration vector, and thus, still acts even in the absence of regions where $\mathbf{a}_f$ vanishes. The sweep-stick mechanism has been generalised recently by Oka and Goto (2021) using a coarse-graining method in order to account for the fractal nature of the inhomogeneous particle distribution, which yields a self-similar mechanism over multiple scales and extends the theory for $St > 1$.

The aforementioned theories have been formulated under the premise of small and heavy particles, and therefore it is necessary to discuss potential deviations if this precondition is not satisfied. To begin with, it is expedient to consider the asymptotic behaviour in terms of the Stokes number. For $St = 0$, the particles adapt instantaneously to any changes in the local fluid velocity and consequently behave as fluid tracers which cannot preferentially concentrate in incompressible flows. For $St \to \infty$, the particle motion is decoupled from the fluid velocity, and thus, turbulence is unable to induce any accumulation of particles. If the length scale of the particle is comparable to the length scales of the fluid flow, the Stokes number is not the only relevant parameter anymore, as there is empirical evidence that variations in size and density are not interchangeable (Qureshi et al. 2007, Lucci et al. 2011, Fiabane et al. 2012, 2013). However, there are indications that the previously described mechanisms are nonetheless relevant for particles of finite size (Uhlmann and Chouippe 2017). Indeed, in order perpetuate the concept of the Stokes number as the only parameter of interest for clustering, various definitions have been proposed for a particle time scale which accounts for finite size and nonlinear drag effects with one prominent example being

$$t_p = \frac{1+2\rho_p/\rho_f}{36}\frac{d_p^2}{\nu_f}\frac{24/Re_{p,\infty}}{C_d}, \tag{2.53}$$

introduced by Balachandar (2009), where

$$Re_{p,\infty} := \frac{d_p\, v_{p,\infty}}{\nu_f} \tag{2.54}$$

is the particle Reynolds number based on the terminal settling velocity in an ambient container, whose value fully determined by the Galileo number and can be approximated by eq. (2.49).

While the terms clustering and preferential concentration have been used largely interchangeable in the preceding discussion, both phenomena are not exactly the same. Preferential concentration denotes a spatially non-uniform distribution of the probability of individual particles to be located within a certain region of the flow, whereas clustering describes the process of multiple particles approaching each other. As a consequence, while preferential concentration can be observed in a single particle system, clustering inherently involves multiple particles. As a consequence, clustering may be the result of preferential concentration in multiparticle systems, but preferential concentration does not necessarily lead to clustering, and clustering can occur without preferential concentration for instance through non-local interactions.

For more information on the formation of preferential concentration and particle clustering in turbulence, the reader is referred to the reviews of Balachandar and Eaton (2010), Monchaux et al. (2012) and Gustavsson and Mehlig (2016).

2.3.3 Settling under turbulent conditions

In the discussion on turbulence-induced preferential concentration, the focus has been laid on clustering under microgravity conditions. The presence of a significant gravitational force demands the introduction of an additional parameter which compares the characteristic particle settling velocity scale to a fluid velocity scale. The former is typically chosen to be the terminal settling velocity of the particle in an ambient fluid, which is linked to the particle response time by

$$v_{p,\infty} = \left(\frac{\rho_p/\rho_f - 1}{\rho_p/\rho_f + \frac{1}{2}}\right) t_p\, |\mathbf{g}| \tag{2.55}$$

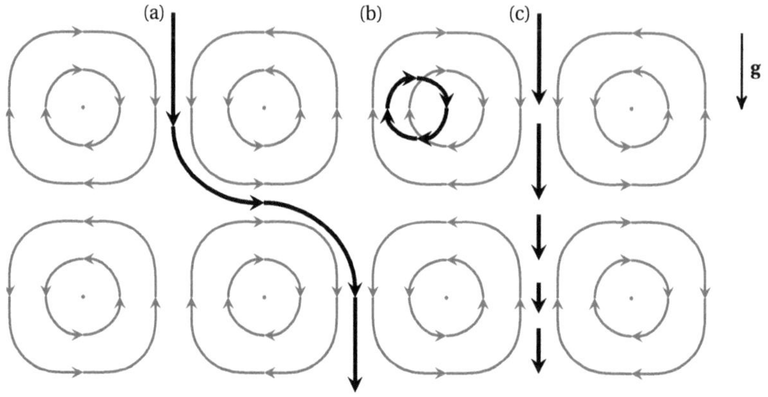

Figure 2.9: Conceptual sketch particle-vortex interactions in a cellular flow similar to Maxey and Corrsin (1986). The fluid flow is indicated by grey arrows, while particle trajectories are drawn as black arrows. (a) The fast tracking mechanism increases the particle settling velocity by preferentially sweeping the downward moving sides of vortices. (b) Vortex trapping may keep a particle in suspension. (c) Loitering reduces the average settling velocity as a particle spends more time facing a counteracting flow.

using eq. (2.53) as definition of t_p.[13] The latter is typically taken to be the velocity scale associated to t_f, and hence, for the common choice of Kolmogorov scales, the resulting nondimensional parameter is the relative turbulence intensity $I_\eta = u_\eta / v_{p,\infty}$, or in the case of wall-bounded flows where viscous units are more practical, I_τ defined in eq. (2.46). It is straightforward to acknowledge that in the case of vanishing relative turbulence intensity, the particle motion is largely decoupled from the fluid motion, similar to the behaviour at large Stokes numbers. On the contrary, for $v_{p,\infty} \ll u_\eta$, particles can be expected to behave similar to the microgravity case, though it has been argued by Bragg et al. (2021b) for wall-bounded flows that as long as I_τ is finite, there always exists a region where settling cannot be neglected.

Regarding the centrifugal mechanism responsible for preferential concentration, gravity exhibits a symmetry breaking effect which has a striking implication on the interaction with vortices which are not coaligned with the direction of gravity—particles tend to preferentially sweep the downward moving side of vortices. This so-called fast tracking phenomenon was first noted by Maxey and Corrsin (1986) and is a result of the particles' tendency to avoid high vorticity regions and to adapt to the local fluid motion, see fig. 2.9(a) for a simplified sketch of this effect. In a follow-up study, Maxey (1987) deduced that a net increase of the average settling velocity compared to $v_{p,\infty}$ can be expected even if the flow is less organised, and indeed, an increase of mean settling velocity due to preferential sweeping has since been reported by various authors (Wang and Maxey 1993, Yang and Lei 1998, Aliseda et al. 2002, Yang and Shy 2003, Cuthbertson and Ervine 2007, Baker et al. 2017, Petersen et al. 2019).

Under certain conditions, particles may become trapped in vortices and orbit on approximately closed paths (Manton 1974, Tooby et al. 1977). This phenomenon is termed vortex trapping, and is rather common in fluvial systems with an example being the swirling of particles due to the leeside vortex of bedforms (Bijker et al. 1976). The mechanism itself depends only little on the specific vortex type (Nielsen 1984, 1992), and thus, it has been hypothesised to also occur for turbulent eddies (Nielsen 1993).

[13] Please note that for a given value of the gravitational acceleration, $v_{p,\infty}$ can only be varied independently of the particle response time by a change in the density ratio. In the context of a particle accelerating from rest to its terminal velocity in an undisturbed environment, this statement may be interpreted as follows: For a given terminal velocity, the density ratio controls the time scale of the particle velocity's asymptotic approach towards the terminal velocity.

When the mobility of the particle is insufficient to follow the variations in fluid velocity rapidly enough, preferential sweeping does not occur and the particle will sample both downward and upward fluid motions. Nielsen (1993) argued that the particle will, on average, spend more time in regions of counteracting flow than assisting flow, which consequently reduces the average settling velocity, see fig. 2.9(c) for a sketch. This phenomenon is known as *loitering*, and its effect becomes most relevant when the particle Reynolds number is in the nonlinear regime, such that *nonlinear drag* effects can contribute to a further reduction of the settling speed (Murray 1970, Mei 1994, Ireland and Collins 2012).

In general, the various mechanisms of particle-turbulence interaction may be active concurrently, and interactions may depend on the eddy scale under consideration. Both net increases and net reductions of the average velocity have been reported (Kawanisi and Shiozaki 2008, Good et al. 2012, Ireland and Collins 2012, Chouippe and Uhlmann 2019), and hence, it is not trivial to predict the net effect of turbulence on settling for a given flow geometry and parameter point. Overview articles on the turbulent effects on particle settling velocity have been provided by Nielsen (1993), and more recently, by Dey et al. (2019).

2.3.4 Particle wakes and collective effects

So far the phenomena discussed were primarily caused by turbulence modulating the dynamics of particles, albeit the interaction is generally mutual, and hence, one also has to take into account the fluid motion induced by particles and mutual interaction in between multiple particles. To put it more systematically, there are two primary mechanisms to inject kinetic energy into a turbulent particle-laden flow—the direct forcing of the fluid flow in form of a pressure gradient or gravitational force acting on a slope for instance, and the gravitational force acting on the solid particles which in turn transfers their energy to the fluid by means of a boundary layer or wake. While in the former case the energy is injected at the large scales and subsequently cascades towards the dissipation range, cf. §2.1.3, the latter injects energy at scales at the order of the particle size which may lead to modifications of the cascade. In general, these two aspects are thus not decoupled, and background turbulence can significantly alter the characteristics of the particles' wakes (Wu and Faeth 1994, 1995, Bagchi and Balachandar 2004, Legendre et al. 2006, Amoura et al. 2010, Rind and Castro 2012), energy may be transferred between the two phases (Dritselis and Vlachos 2011, Zhao et al. 2013), turbulence may be attenuated by additional dissipation (Gore and Crowe 1989, Squires and Eaton 1990, Elghobashi and Truesdell 1993, Kulick et al. 1994), and even the inherent mechanism of wall-bounded turbulence may be modulated by the presence of particles (Wang et al. 2017, 2018a, Wang and Richter 2019, Scherer et al. 2022).

In order to fathom this interaction, it is beneficial to first ascertain the characteristic of the wakes of freely settling mobile particles in the absence of background turbulence. For spherical particles with $\rho_p/\rho_f > 1$, it can be distinguished between four wake regimes which emerge during the supercritical transition to chaos of the flow around the sphere, with the controlling parameter being the Galileo number. At low Galileo numbers the wake is axisymmetric and steady, and correspondingly the direction of falling and gravity are coaligned. This changes at $Ga \approx 155$ where the first bifurcation is observed and the wake becomes oblique which results in a nonvertical falling direction. The oblique state undergoes a Hopf bifurcation at $Ga \approx 195$, and hence, the wake begins to oscillate which induces periodically varying fluctuations in the vertical and lateral hydrodynamic forces on the particle leading to an oscillating trajectory. Chaotic fluid motion occurs starting at $Ga \approx 240$ where statistical symmetry of the wake is recovered, and hence, particles settle vertically on average. The exact values of Ga at which the bifurcations occur depend on the value of the solid-to-fluid density ratio, and a comprehensive study is given by Jenny et al. (2004). The various regimes differ distinctly in their ability to transfer heat and mass, and

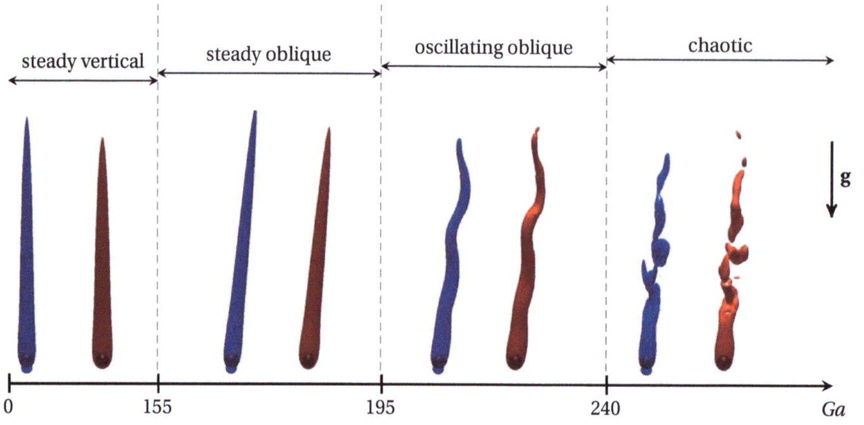

Figure 2.10: Wake regimes of freely settling particles in undisturbed conditions. The illustrations depict the parameter points $Ga = [150, 170, 200, 300]$ at $\rho_p/\rho_f = 10$. Blue isosurfaces represent disturbances in terms of the vertical fluid velocity, whereas red isosurfaces show the induced temperature field for a heated sphere. The critical Galileo numbers for bifurcation depend on the density ratio (Jenny et al. 2004), and the values shown correspond to $\rho_p/\rho_f = 10$. The illustrations have been first published in Chouippe et al. (2019).

a discussion thereof can be found in Chouippe et al. (2019). Figure 2.10 summarises the wake regimes described and provides a visual clarification.

It should be noted that the wake is generally not turbulent until the critical transition occurs at Galileo numbers of $\mathcal{O}(10^5)$, as the wake can still be considered quasi-periodic until this point (Clift et al. 1978). Moreover, the four regimes previously introduced are not necessarily uniform which means that less prominent changes in flow characteristics may still occur within a given regime. For instance, in case of the steady axisymmetric regime, the flow gradually develops a front-rear asymmetry for finite values of Ga which ushers into the formation of a toroidal vortex in the rear of the sphere which induces a zone of recirculating flow where fluid is transported towards the rear stagnation point. The onset of recirculation occurs around $Ga \approx 30$ and the downstream extent of the zone of recirculation increases with increasing Galileo number (Clift et al. 1978). Photographs of the formation and evolution of the toroidal vortex ring can be found in Taneda (1956).

The modification of the fluid flow in the mediate vicinity provides a mechanism for multiple particles to interact with each other. One prominent example for such a mechanism is the "drafting–kissing–tumbling" interaction (Fortes et al. 1987, Wu and Manasseh 1998) in which the trailing particle of a particle pair is attracted towards the position of the leading one, comes into (near) contact, and subsequently positions itself alongside its partner. A consequence of this interaction is the possibility of cluster formation even in the absence of background turbulence (Kajishima and Takiguchi 2002, Kajishima 2004a), which is in the case of spherical particles further promoted by nonvertical settling at sufficiently high Galileo numbers (Uhlmann and Doychev 2014). An interesting consequence of wake-induced particle clusters is that the ensemble of particles may settle faster than the terminal velocity of an individual particle (Uhlmann and Doychev 2014, Huisman et al. 2016). Moreover, since a cluster can be essentially regarded as a coherent object itself, a mechanism is provided to inject energy into the fluid flow at length scales significantly larger than the particles themselves (Kajishima and Takiguchi 2002).

Ensembles of particles may also interact in a way which reduces the average settling velocity. This hindrance effect was first described by Richardson and Zaki (1954) who derived a parametrisation for the reduction in settling velocity for randomly distributed spheres at low Galileo numbers which implies an approximately linear relationship with the inverse of the solid volume fraction. The hindrance is a result of fluid being displaced from below the front of settling particles which induces a counteracting flow and consequently a hydrodynamic force opposed to gravity. For this reason, the hindrance is strongest if the particles are arranged cross-stream, i.e. perpendicular to the falling direction, as the backflow then acts most effectively (Yin and Koch 2007). Such a cross-stream orientation may result from drafting–kissing–tumbling at low volume fractions, but the same mechanism may also promote vertical organization, as has been previously discussed, which minimises hindrance. The considerable dependence of the hindrance effect on the microstructure of particle orientation makes it difficult to judge its importance under turbulent flow conditions, although Fornari et al. (2019) found that it may play a major role in homogeneous isotropic turbulence at volume fractions of $\mathcal{O}(1\text{–}10\%)$.

2.3.5 Suspended particles in horizontal channels

The majority of the aforementioned phenomena have been primarily investigated in the absence of walls and in statistically homogeneous domains. While the inferences derived from these configurations are certainly also relevant for flow configurations where these prerequisites are not given, it is important to acknowledge that the settling of particles in horizontal wall-bounded flows differs from that in domains which are unbounded in the direction of gravity due to three major properties:

Boundedness As the direction of gravity is bounded by walls, particles may only travel vertically over a limited distance. As a result, three different scenarios of settling may be considered in such flows—the quasi-steady state where particles are continuously removed from the bottom of the domain and reseeded at the top resulting in a sustained net downward flux; the quasi-steady state where resuspension and deposition are in equilibrium and the net flux is zero; the initial value problem where particles are distributed out of equilibrium and transiently settle until a quasi-steady state is obtained.

Anisotropy The presence of mean shear in the turbulent background flow leads to a preferred alignment of coherent structures with the streamwise directions. As gravity acts perpendicular to this direction, particles are more likely to approach these structures from a preferred direction as compared to isotropic turbulence. Moreover, characteristic structures such as Reynolds-stress events do not occur in isotropic flows.

Inhomogeneity As the flow statistics depend on the wall-normal position, phenomena such as turbophoresis and further turbulent dispersion effects based on gradients of flow statistics may become relevant for the analysis of the problem (Reeks 1983, 1991, 2005, Bragg et al. 2012, 2021a).

Many of the early works on suspended particles were concerned with the statistics of dispersion by sheared and turbulent flows. Based on his theory of turbulent diffusion (Taylor 1922), G.I. Taylor laid the groundwork for this analysis with his contributions on the dispersion of soluble matter in pipe flows (Taylor 1953, 1954) which were later applied to channel flow by Elder (1959). Using the moment transformation of Aris (1956), it was Sayre (1967) who first extended this continuous Eulerian treatment of mass transport towards settling particles by applying Taylor's approach to a modified transport equation. Indeed, one of the motivations behind Sayre's work was the prediction of dispersion and interaction of silt particles with the sediment of rivers (Sayre and Hubbell 1965, Sayre 1969)—a problem which is akin to the issue of bacteria transport (Jamieson et al. 2005) as discussed in §1.2. Utilizing the same approach and assumptions on near-wall turbulence, Sumer (1974) was able to analytically determine the concentration profile, mean velocity and longitudinal dispersion coefficient for heavy particles in smooth-wall open-channel flow. With the application to settling basins in mind, Sumer (1977) subsequently applied the method to particles transiently settling from the free surface of an open-channel and analytically obtained the equilibrium spatial distribution of particles, the distribution of retention time, as well as the statistics of the longitudinal settling length under various assumptions.

The problem of transiently settling particles in channels has also been investigated using a Lagrangian random walk model assuming normally distributed fluid velocities by Li and Shen (1975) who concluded that the settling velocities of heavy particles in turbulent open-channel flow are likely different from their corresponding value in a quiescent environment, and that the settling length follows a log-normal distribution. Their work was extended by Hoyal et al. (1995) towards a higher range in relative turbulence intensities and to model particle deposition by means of a fully absorbing lower boundary resulting in a net particle flux. Their results indicate that the settling velocity is enhanced for relative turbulence intensities larger than unity and are otherwise approximated well by the quiescent terminal velocity.

Experimental evidence for turbulent enhancement of the settling velocity during the transient was first provided by Jobson and Sayre (1970) who studied the settling of sand particles of varying size by continuously injecting them from a spanwise-aligned line source at the free surface of a flume. For all sizes investigated, they observed an enhancement in settling velocity compared to its still water value and attributed it to clustering effects in case of small sand particles and to turbulent effects in case of larger particles, although the enhancement in the latter case was significantly weaker. Cuthbertson (2001) and Cuthbertson and Ervine (2007) experimentally studied the settling of fine sand particles with terminal velocities of the order of unity over a porous bed, and observed that the particles accumulate on the downward moving side of large vortical structures of the flow. This behaviour was attributed to the preferential sweeping mechanism which also serves as an explanation for the observed enhancement in settling velocities. Moreover, it was found that an enhancement of the settling velocity often coincides with the sampling of fluid sweeps, whereas in contrast reduced settling velocities are observed in ejections. The role of these Reynolds-stress events on the settling velocity has also been recognised by Breugem and Uijttewaal (2007) who found downward moving particles during the transient phase preferentially located in sweeps, whereas in the fully developed state upward moving particles sampled regions where $u_f' < 0$ and downward moving particles regions of $u_f' > 0$. Wang et al. (2018b) conducted settling experiments in a channel with grid-induced turbulence and focused on the role of turbulent length scale in the settling process. They conjectured that large-scale eddies lead to an increase in the particle settling velocity through preferential sweeping, whereas small-scale fluctuations have a counteracting effect.

Numerical simulations on the settling of particles in horizontal channel flow for the scenarios involving a net downward flux are scarce. In fact, only the recent point-particle study of Bragg et al. (2021a) could be identified to belong in this category. In their cases with net downward particle fluxes, the average settling velocity was found to considerably exceed the ambient terminal velocity and this enhancement could be attributed for the most part to the preferential sweeping mechanism, with the exception of the near-wall region where turbophoresis was found to gain significance.

The literature on the quasi-steady regime with a zero net flux of particles is ample, and some of its key characteristics are briefly summarised in the following: In the near wall region of smooth-wall channels, particles which are small compared to the size of the inner region of turbulence are known to accumulate in low-speed streaks as a result of their interaction with coherent structures of the buffer layer (Rashidi et al. 1990, Kaftori et al. 1995a,b, Kiger and Pan 2002, Marchioli and Soldati 2002, Kidanemariam et al. 2013). As a result, their ensemble-averaged streamwise velocity lags behind the corresponding fluid velocity due to preferential sampling (Kaftori et al. 1995b, Kiger and Pan 2002, Kidanemariam et al. 2013). Concerning the resuspension, it has been conjectured that turbulent ejections are a major contributor apart from wall- and shear-induced lift forces (Kiger and Pan 2002, Marchioli and Soldati 2002, Niño et al. 2003, Vinkovic et al. 2011). Similar mechanisms are active over sediment beds and give rise to sand ridges, although the driving mechanisms may be due to large scale turbulent structures if the buffer layer is nonexistent (Scherer et al. 2022), cf. §2.3.1.

The discussion on particles in channel turbulence is concluded by an outline of clustering effects and turbulence interactions in the outer layer. Fessler et al. (1994) have experimentally studied the preferential concentration of heavy particles near the centreline of a closed channel, and found that their distribution is far from random and resembles that of homogeneous isotropic turbulence. They related the self-organization of the solid phase to the Stokes number and observed the strongest deviation from randomness at $St \approx 1$. This inferences was corroborated by Rouson and Eaton (2001) who studied a similar parameter point using numerical simulation of point particles, but also observed long streamwise-aligned patterns. The recent microgravity simulations of Motoori et al. (2022) strikingly demonstrated using a coarse-graining method that preferential concentration in channels is a multi-scale phenomenon. They observed that vortices of all scales and types deplete particle concentrations in their core in accordance with Maxey's arguments, and that this depletion can be described by respective scale-dependent Stokes numbers. Moreover, it was shown that wall-detached tubular vortices lead to the formation of approximately isotropic clusters around them, whereas wall-attached vortices transport particles predominantly into nearby low-speed streaks.

2.4 Knowledge gaps

For the river contamination problem which serves as a motivation for this thesis, the transient settling of suspended particles in horizontal channels is a paramount problem. The hydrodynamic literature on this configuration is scarce, and albeit the very relevant experimental study of Cuthbertson and Ervine (2007) and the more idealised theoretical and numerical treatment of Bragg et al. (2021a) could be identified, various questions remain unanswered. Of particular interest are the interactions between coherent flow structures and the settling particles which give rise to the enhancement in settling velocity reported by both studies, but also collective mechanisms such as clustering which have not received much attention in this configuration yet. Up until this point, the transient settling problem has not been examined using fully-resolved numerical simulations yet, which have the advantage over traditional experiments that the full state of the system is easily accessible at all times, and over unresolved numerical simulations that they do not require extensive modelling which deems the results more reliable and allows for a more extensive variation in parameters. Moreover, little is known on the effect of bedforms on the settling and long-term deposition of particles. While Cuthbertson and Ervine (2007) conducted their experiment in the presence of a sediment bed, a systematic analysis of its influence is lacking and needs to be addressed due to its relevance for fluvial systems (Wilkinson et al. 1995).

3 Problem formulation and methodology

3.1 Mathematical framework

While the fundamental equations used for this numerical study study were already introduced throughout §2, their definite form describing the system to be numerically approximated will be summarised in this section for the sake of convenience and completeness. Open-channel flow is considered within a domain $\Omega \in \mathbb{R}^3$ with periodic boundary conditions in x and z, a no-slip boundary condition (eq. (2.19)) for the lower wall and a free-slip boundary condition (eq. (2.20)) for the upper wall. Both walls are impermeable (eq. (2.18)) and non-deformable, thus corresponding to the $Fr \to 0$ limit of free-surface flows. The flow is particle-laden, and the problem domain can thus be divided into disjoint subdomains,

$$\Omega = \Omega_f \cup \left(\bigcup_{i=1}^{N_p} \Omega_p^{(i)} \right), \tag{3.1}$$

where Ω_f denotes the subdomain occupied by the fluid phase, $\Omega_p^{(i)}$ is the subdomain occupied by the i-th solid particle and N_p is the total number of particles in Ω. The interfaces between the fluid and the solid subdomains are treated by impermeable and no-slip boundary conditions.

Within the fluid domain, the system is described by the continuity and Navier–Stokes momentum equations displayed in eq. (2.1) and eq. (2.2). The fluid density is considered to be constant throughout the flow, and the only external force considered is gravity which yields the definite form

$$\nabla \cdot \mathbf{u}_f = 0, \tag{3.2}$$

$$\frac{\partial \mathbf{u}_f}{\partial t} + \nabla \cdot \left(\mathbf{u}_f \otimes \mathbf{u}_f \right) = -\frac{1}{\rho_f} \nabla p + \nu_f \nabla^2 \mathbf{u}_f. \tag{3.3}$$

Here, the gravitational acceleration has been expressed as the gradient of the gravitational potential, $\mathbf{g} =: -\nabla \Psi$, and a new pressure variable

$$p := P + \rho_f \Psi \tag{3.4}$$

has been defined which signifies the pressure without its hydrostatic part.

The linear and angular momentum balances for the particles are given by the Newton–Euler equations displayed in eq. (2.38) and eq. (2.39). The particles are considered to be spherical and their density to be distributed uniformly throughout $\Omega_p^{(i)}$, which yields

$$\frac{\rho_p^{(i)}\pi d_p^{(i)3}}{6}\frac{\mathrm{d}\mathbf{u}_p^{(i)}}{\mathrm{d}t} = \rho_f \oint_{S_p^{(i)}} \left(2\nu_f \overline{\overline{\mathbf{S}}}\cdot\mathbf{n}\right)\mathrm{d}S_p^{(i)} - \oint_{S_p^{(i)}} \left(p\overline{\overline{\mathbf{I}}}\cdot\mathbf{n}\right)\mathrm{d}S_p^{(i)} + \frac{\left(\rho_p^{(i)}-\rho_f\right)\pi d_p^{(i)3}}{6}\mathbf{g} + \mathbf{F}_{p,col}^{(i)}, \tag{3.5a}$$

$$\frac{\rho_p^{(i)}\pi d_p^{(i)5}}{60}\frac{\mathrm{d}\boldsymbol{\omega}_p^{(i)}}{\mathrm{d}t} = \rho_f \oint_{S_p^{(i)}} (\mathbf{x}-\mathbf{x}_p^{(i)}) \times \left(2\nu_f \overline{\overline{\mathbf{S}}}\cdot\mathbf{n}\right)\mathrm{d}S_p^{(i)} + \mathbf{T}_{p,col}^{(i)}, \tag{3.5b}$$

for the i-th particle with $\overline{\overline{\mathbf{S}}}$ being the rate-of-strain tensor defined as

$$S_{ij} := \frac{1}{2}\left(\frac{\partial u_{f,i}}{\partial x_j} + \frac{\partial u_{f,j}}{\partial x_i}\right). \tag{3.6}$$

Moreover, the surface integral for the hydrostatic pressure has been evaluated to yield Archimedes's principle for buoyancy,

$$\oint_{S_p^{(i)}} \left(\rho_f \Psi \overline{\overline{\mathbf{I}}}\cdot\mathbf{n}\right)\mathrm{d}S_p^{(i)} = -\frac{\rho_f \pi d_p^{(i)3}}{6}\mathbf{g}. \tag{3.7}$$

While the rate-of-strain and pressure distribution on the particles' surface at the fluid-solid interface is completely determined by eq. (3.2), eq. (3.3) and the boundary conditions, the interaction between multiple particles or between a particle and the wall needs to be modelled using a collision model which supplies the collision force $\mathbf{F}_{p,col}^{(i)}$ and torque $\mathbf{T}_{p,col}^{(i)}$. The trajectory of the i-th particle is then obtained by integrating the kinematic equation

$$\frac{\mathrm{d}\mathbf{x}_p^{(i)}}{\mathrm{d}t} = \mathbf{u}_p^{(i)}, \tag{3.8}$$

whereas it is not required to track the angular position of the spheres in time due to their rotational symmetry.

The transport of suspended bacteria only plays a subordinate role in this thesis, however, whenever its distribution in Ω_f is tracked it is modelled by the advection-diffusion equation

$$\frac{\partial C_{SB}}{\partial t} + \mathbf{u}_f \cdot \nabla C_{SB} = \mathcal{D}_{SB}\nabla^2 C_{SB}, \tag{3.9}$$

which corresponds to eq. (2.4) with homogeneous diffusivity and in the absence of any mechanisms for decay and growth for the sake of generality. Here, C_{SB} is the concentration of suspended bacteria and $\mathcal{D}_{SB}$ its diffusion coefficient. Suspended bacteria are introduced to the flow by either the initial condition or by means of Dirichlet boundary conditions on the fluid-solid interface. In case of the former, the particles are also occasionally modelled as inert with respect to C_{SB} by imposing the Neumann-type boundary condition $\partial C_{SB}/\partial \mathbf{n} = 0$ on $S_p^{(i)}$. The concentration of particle-attached bacteria is not modelled explicitly as the primary focus of this thesis lies on bacteria transport and not their fate, and hence, the knowledge of particle dynamics is sufficient for this purpose.

3.2 Computational method

Although the mathematical framework previously described is in itself complete and only demands for constitutive equations for the collision forces and torques, the nature of differential equations and the lack of analytical solutions requires the framework to be solved numerically on a computer. For this purpose, the system of equations needs to be discretised in space and time while keeping in mind the range of scales of the particle-laden flow problem. This thesis utilises the approach of direct numerical simulation (DNS) which means that all relevant spatial and temporal scales of the problem are captured by the discretisation. Therefore, no further assumptions have to be made and the simulations can be considered to be derived “from first principles”, i.e. directly from the most fundamental principles of hydromechanics.

3.2.1 Immersed boundary method

While it is certainly possible to discretise Ω_f using a computational mesh which excludes Ω_p, such a body-conforming approach is generally not feasible if a large number of mobile particles is immersed into the fluid, as this would require the mesh to be continuously regenerated as the particles change their position within Ω in time which comes at an immense computational cost (Prosperetti and Tryggvason 2009). This dilemma can be resolved by the use of *fictitious domain methods*. Fictitious domain methods take advantage of the lower geometric complexity of the complete domain and discretise Ω by a stationary and regular mesh which hence comprises both the fluid domain and the particle domains, i.e. the equations (3.2), (3.3) and (3.9) are solved in Ω rather than Ω_f. However, this comes at the cost that the boundary conditions on the fluid-particle interface cannot be imposed directly as this interface is not captured by the discretisation. Thus, in order to account for the presence of solid particles, an artificial body force is introduced to the transport equations whose sole purpose lies in modifying the fields in a manner which adequately reflects Ω_p.

Various instances of the fictitious domain method exist and they can be categorised by the formulation of the artificial force, the positions at which it is active, and the discretisation schemes within the method is employed. The approach utilised in this thesis belongs to the group of *immersed boundary methods* (IBMs), which are based on spatial discretisations of Ω in physical space and employ a localised force field which is restricted to the boundaries of Ω_p denoted as Γ_p in the following. The origin of IBMs can be traced back to Peskin (1972) who numerically studied the flow through the mitral valve in the human heart and modelled the interaction between the blood flow and the valve by this approach. Ever since, this approach has been adapted for various fluid-structure interaction problems, and have been eventually applied to particle-laden flows, cf. Mittal and Iaccarino (2005) for a review on IBMs and Maxey (2017) a review of fictitious domain methods for particle-laden flows.

The formulation of the IBM employed in this thesis is provided by Uhlmann (2005) and is now summarised briefly for the Navier–Stokes momentum equation. Using a finite-difference approach on a regular Cartesian staggered grid, the complete domain Ω is discretised to yield discrete nodes $\mathbf{x}_{ijk} = (i\Delta x, j\Delta y, k\Delta z)^T$ on which the velocity and pressure fields are evaluated, with i, j, k denoting the integer indices of the nodes and $\Delta x = \Delta y = \Delta z$ being the mesh spacing in x, y, z direction, respectively, which is chosen to be uniform. For each particle, $N_m^{(i)}$ marker points are distributed as evenly as possible at position $\mathbf{x}_{p,\Gamma}^{(i,m)}$ on $\Gamma_p^{(i)}$, and each marker point is attributed a volume of

$$\Delta V^{(i,m)} = \frac{\pi \Delta x}{3N_m}\left(3d_p^{(i)2} + \Delta x^2\right), \tag{3.10}$$

where the superscript (i,m) denotes the m-th marker point of the i-th particle. In the first step, the fluid velocity at $\mathbf{x}_{p,\Gamma}^{(i,m)}$ is gathered from the discretised velocity fields by means of interpolation, viz.

$$\mathbf{u}_f^{(i,m)}(t) = \sum_{ijk} \mathbf{u}_f(\mathbf{x}_{ijk}, t)\, \delta_{ibm}\left(\mathbf{x}_{ijk} - \mathbf{x}_{p,\Gamma}^{(i,m)}\right) \Delta x \Delta y \Delta z. \tag{3.11}$$

Here, the interpolation kernel δ_{ibm} is a regularised Delta function—a discrete version of the Dirac delta function with finite support—introduced by Peskin (1972, 2002) in the variant of Roma et al. (1999) which has a support of three grid points in each direction. The velocity which is to be imposed at the position of the marker points is dictated by the state of motion of the individual particles, and is given by

$$\mathbf{u}_{p,\Gamma}^{(i,m)}(t) = \mathbf{u}_p^{(i)} + \boldsymbol{\omega}_p^{(i)} \times \left(\mathbf{x}_{p,\Gamma}^{(i,m)} - \mathbf{x}_p^{(i)}\right). \tag{3.12}$$

By comparing the gathered fluid velocity $\mathbf{u}_f^{(i,m)}$ to the value to be imposed $\mathbf{u}_{p,\Gamma}^{(i,m)}$, the volume force required at each marker point is determined by

$$\mathbf{f}_{p,\Gamma}^{(i,m)}(t) = \frac{\mathbf{u}_{p,\Gamma}^{(i,m)} - \mathbf{u}_f^{(i,m)}}{\Delta t}, \tag{3.13}$$

where Δt is the time step of the temporal discretisation. In the second step, the gathered forces are applied to the fluid and solid domains. In case of the former, $\mathbf{f}_{p,\Gamma}^{(i,m)}$ is spread to the regular grid by

$$\mathbf{f}_{ibm}(\mathbf{x}_{ijk}, t) = \sum_i^{N_p} \sum_m^{N_m^{(i)}} \mathbf{f}_{p,\Gamma}^{(i,m)}\, \delta_{ibm}\left(\mathbf{x}_{ijk} - \mathbf{x}_{p,\Gamma}^{(i,m)}\right) \Delta V^{(i,m)}, \tag{3.14}$$

where $\mathbf{f}_{ibm}$ is the additional forcing term added to the momentum equation as previously stated. Concerning the solid domains, the surface integrals in equations (3.5a) and (3.5b) which represent the hydrodynamic forces and torques are approximated by

$$\rho_f \oint_{S_p^{(i)}} \left(2\nu_f \bar{\bar{\mathbf{S}}} \cdot \mathbf{n}\right) \mathrm{d}S_p^{(i)} - \oint_{S_p^{(i)}} \left(p\bar{\bar{\mathbf{I}}} \cdot \mathbf{n}\right) \mathrm{d}S_p^{(i)} \approx -\rho_f \sum_m^{N_m^{(i)}} \mathbf{f}_{p,\Gamma}^{(i,m)} \Delta V^{(i,m)}, \tag{3.15a}$$

$$\rho_f \oint_{S_p^{(i)}} \left(\mathbf{x} - \mathbf{x}_p^{(i)}\right) \times \left(2\nu_f \bar{\bar{\mathbf{S}}} \cdot \mathbf{n}\right) \mathrm{d}S_p^{(i)} \approx -\rho_f \sum_m^{N_m^{(i)}} \left(\mathbf{x}_{p,\Gamma}^{(i,m)} - \mathbf{x}_p^{(i)}\right) \times \mathbf{f}_{p,\Gamma}^{(i,m)} \Delta V^{(i,m)} + \frac{\rho_f \pi d_p^{(i)5}}{60} \frac{\mathrm{d}\boldsymbol{\omega}_p^{(i)}}{\mathrm{d}t}, \tag{3.15b}$$

thus satisfying Newton's third law of motion.

The DNS code employed in this thesis which implements the IBM of Uhlmann (2005) uses a standard central finite-difference scheme for the spatial derivatives and a semi-implicit temporal discretisation based on the Crank-Nicolson scheme for the viscous term and a three-step low-storage Runge-Kutta scheme for the nonlinear term, similar to Verzicco and Orlandi (1996). Implicit coupling between the continuity constraint and the pressure in the momentum equation is achieved by means of the fractional step method (Kim and Moin 1985). A comprehensive description on how the IBM is implemented in this framework can be found in Uhlmann (2005)

3.2.2 Particle collision model

In order to supply collision forces and torques to equations (3.5a) and (3.5b), the framework needs to be complemented by constitutive equations describing the collision process between pairs of particles and particles with the walls. In this thesis, a *soft-sphere* approach is employed which resolves the collision process by allowing for small overlaps and finite durations of contact, which will be outlined in the following for the case of interparticle collisions.

The normal vector for the collision of particle (i) and particle (j) is defined by the line connecting the centre point locations of the two spheres, viz.

$$\mathbf{n}^{(i,j)}(t) := \frac{\mathbf{x}_p^{(j)} - \mathbf{x}_p^{(i)}}{|\mathbf{x}_p^{(j)} - \mathbf{x}_p^{(i)}|}. \tag{3.16}$$

The distance of overlap along this line is given by

$$\delta_{col}^{(i,j)}(t) := \tfrac{1}{2}\left(d_p^{(i)} + d_p^{(j)}\right) - \left|\mathbf{x}_p^{(i)} - \mathbf{x}_p^{(j)}\right|, \tag{3.17}$$

which attains positive values when the two particles are currently in contact, and negative values otherwise. The relative velocity at the point of contact is determined from the surface velocity of the solid spheres on the line connecting the centre points and reads

$$\mathbf{u}_{col}^{(i,j)}(t) := \left(\mathbf{u}_p^{(i)} + \tfrac{1}{2} d_p^{(i)} \boldsymbol{\omega}_p^{(i)} \times \mathbf{n}^{(i,j)}\right) - \left(\mathbf{u}_p^{(j)} - \tfrac{1}{2} d_p^{(j)} \boldsymbol{\omega}_p^{(j)} \times \mathbf{n}^{(i,j)}\right). \tag{3.18}$$

From the relative velocity and the normal direction, a tangential vector can be constructed by

$$\mathbf{t}^{(i,j)}(t) := \frac{\mathbf{u}_{col}^{(i,j)} - \left(\mathbf{u}_{col}^{(i,j)} \cdot \mathbf{n}^{(i,j)}\right)\mathbf{n}^{(i,j)}}{\left|\mathbf{u}_{col}^{(i,j)} - \left(\mathbf{u}_{col}^{(i,j)} \cdot \mathbf{n}^{(i,j)}\right)\mathbf{n}^{(i,j)}\right|}, \tag{3.19}$$

which remains undefined if $\mathbf{u}_{col}^{(i,j)}$ solely possesses a normal component.

The collision forces are modelled by a spring-damper system. The normal component of the interparticle collision force is given by

$$\mathbf{F}_{p,norm}^{(i,j)}(t) := \begin{cases} -k_{elas}\left(\delta_{col}^{(i,j)} + \Delta_{col}\right)\mathbf{n}^{(i,j)} - c_{d,norm}\left(\mathbf{u}_{col}^{(i,j)} \cdot \mathbf{n}^{(i,j)}\right)\mathbf{n}^{(i,j)} & \text{if } \delta_{col}^{(i,j)} \geq -\Delta_{col} \\ 0 & \text{otherwise} \end{cases}, \tag{3.20}$$

where the k_{elas} is the stiffness of the spring component, and $c_{d,norm}$ is the dissipation coefficient of the damper component of the system. Moreover, the point of contact has been shifted by an offset such that collision forces are active already at a small finite distance Δ_{col} which reflects the smeared out nature of the immersed boundary interface. This admitted gap is chosen to be $\mathcal{O}(\Delta x)$ similar to the support of the regularised delta function. The tangential component of the collision force is given by

$$\mathbf{F}_{p,tang}^{(i,j)}(t) := \begin{cases} -\min\left(\mu_{fric}\left|\mathbf{F}_{p,norm}^{(i,j)}\right|, c_{d,tang}\left(\mathbf{u}_{col}^{(i,j)} \cdot \mathbf{t}^{(i,j)}\right)\right)\mathbf{t}^{(i,j)} & \text{if } \delta_{col}^{(i,j)} \geq -\Delta_{col} \wedge \mathbf{t}^{(i,j)} \text{ is defined} \\ 0 & \text{otherwise} \end{cases}, \tag{3.21}$$

and thus comprises only a dampening component with a dissipation coefficient $c_{d,tang}$, which is limited by the the condition of traction with the Coulomb friction coefficient μ_{fric}. The overall interparticle collision force and torque acting on the i-th particle is then given by the sum of forces and torques of all collision pairs, viz.

$$\mathbf{F}_{p,col}^{(i)}(t) = \sum_{j=1,j\neq i}^{N_p} \left(\mathbf{F}_{p,norm}^{(i,j)} + \mathbf{F}_{p,tang}^{(i,j)}\right) \tag{3.22a}$$

$$\mathbf{T}_{p,col}^{(i)}(t) = \frac{1}{2} d_p^{(i)} \sum_{j=1,j\neq i}^{N_p} \left(\mathbf{n}^{(i,j)} \times \mathbf{F}_{p,tang}^{(i,j)}\right) \tag{3.22b}$$

The collision between a particle and the wall is treated analogously to interparticle collisions, and, as the region of contact for a sphere with a plane is also a single point, is formally very similar to the aforementioned description. A more comprehensive description of the collision model in the same context can be found in Kidanemariam and Uhlmann (2014b) and Kidanemariam (2016).

The collision model is parameterised by the four physical parameters $\{k_{elas}, c_{d,norm}, c_{d,tang}, \mu_{fric}\}$ which need to be supplied. In accordance with previous work on sediment transport using the same simulation code (Kidanemariam 2016), the normal and tangential dissipation coefficients are set to be equal, and their value can be related to the stiffness coefficient (Crowe et al. 1998) by

$$c_{d,norm} = c_{d,tang} = -2\sqrt{m^{(i,j)} k_{elas}} \frac{\ln(\varepsilon_{rest})}{\sqrt{\pi^2 + \ln(\varepsilon_{rest})^2}} \tag{3.23}$$

where the dry restitution coefficient ε_{rest} has been introduced, which describes the ratio between the total linear momentum of the particle pair post- and pre-collision in a vacuum, and is therefore a measure for the dissipation due to collision. Moreover, the reduced mass $m^{(i,j)} := m^{(i)} m^{(j)} / \left(m^{(i)} + m^{(j)}\right)$ of the pair has been introduced, with $m^{(i)}, m^{(j)}$ being the masses of the collision partners. The alternative parameterisation involving ε_{rest} is beneficial in the sense that it is easily accesible from experiments. For the two main series which will be introduced at a later point (LR,HR), the collision parameters are given by $\mu_{fric} = \{0.4, 0.5\}$, $\varepsilon_{rest} = \{0.3, 0.9\}$ and $k_{elas} = 20/\left(d_p^2 \frac{\pi}{6}\left(\frac{\rho_p}{\rho_f} - 1\right)\right)$.

From a numerical perspective, the time scale of collision is typically shorter than the required time step for the advancement of the flow problem. In order retain a reasonable overall value of Δt, the collisions are resolved using a sub-stepping procedure, which, broadly speaking, implies that the hydrodynamic forces are kept constant over N_{sub} sub-iterations of the Newton-Euler equations while the collision force are reiterated. In this thesis, the number of sub-steps is chosen to be in the range $N_{sub} \in [150, 240]$ in accordance to previous studies, cf. Kidanemariam (2016) for more details.

3.2.3 Transport of suspended bacteria

Similar to the Navier–Stokes momentum equation, eq. (3.9) is solved on Ω rather than than solely on Ω_f, and the boundary conditions are imposed using the immersed boundary method. Two boundary conditions for the bacteria concentration on the particles are investigated—a Dirichlet-type boundary condition imposing a constant value effectively acting as a source of suspended bacteria, and a Neumann-type boundary condition imposing zero bacteria flux homogeneously on Γ_p. In case of the former, a constant, particle-dependent value is imposed at the location of the marker points, viz.

$$C_{SB,\Gamma}^{(i,m)} = C_{SB,\Gamma}^{(i)}. \tag{3.24}$$

For the latter, the value of the scalar field is evaluated at a normal distance of Δ_{ibm} from the marker points, and the values to be imposed on Γ_p are determined by a one-sided finite-difference approximation of the scalar flux, which yields

$$C_{SB,\Gamma}^{(i,m)}(t) = \sum_{ijk} C_{SB}\left(\mathbf{x}_{ijk}, t\right) \delta_{ibm}\left(\mathbf{x}_{ijk} - \mathbf{x}_p^{(i)} - \left(\tfrac{1}{2} d_p^{(i)} + \Delta_{ibm}\right) \mathbf{n}^{(i,m)}\right) \Delta x \Delta y \Delta z \tag{3.25}$$

where

$$\mathbf{n}^{(i,m)} := \frac{\mathbf{x}_{p,\Gamma}^{(i,m)} - \mathbf{x}_p^{(i)}}{\left|\mathbf{x}_{p,\Gamma}^{(i,m)} - \mathbf{x}_p^{(i)}\right|}. \tag{3.26}$$

is the normal vector for marker point (m) of the i-th particle. An artificial flux is added to eq. (3.9) analogous to the artificial forcing term in the momentum equations, and its value is determined by

$$\mathbf{q}_{ibm}\left(\mathbf{x}_{ijk}, t\right) = \sum_{i}^{N_p} \sum_{m}^{N_m^{(i)}} \mathbf{q}_{p,\Gamma}^{(i,m)} \delta_{ibm}\left(\mathbf{x}_{ijk} - \mathbf{x}_{p,\Gamma}^{(i,m)}\right) \Delta V^{(i,m)}, \tag{3.27}$$

$$\mathbf{q}_{p,\Gamma}^{(i,m)}(t) = \frac{C_{SB,\Gamma}^{(i,m)} - C_{SB}^{(i,m)}}{\Delta t}, \tag{3.28}$$

$$C_{SB}^{(i,m)}(t) = \sum_{ijk} C_{SB}\left(\mathbf{x}_{ijk}, t\right) \delta_{ibm}\left(\mathbf{x}_{ijk} - \mathbf{x}_{p,\Gamma}^{(i,m)}\right) \Delta x \Delta y \Delta z. \tag{3.29}$$

No balance equation is solved for the particles which implies that the flux of bacteria is small compared to the number of bacteria attributed to the particle.

The nodes of the concentration field are collocated with the pressure field. A central finite-difference scheme is employed to approximate the spatial derivatives, whereas the temporal derivative is treated fully explicitly, as stability analysis suggests that the diffusive stability limit is typically abided by the timesteps commonly chosen for the flow solver. It has been observed that the use of a central scheme may lead to oscillations in the concentration field in the presence of steep gradients which can be avoided by reducing the grid spacing, see Ferziger et al. (2002) for a simple one-dimensional analysis of this problem. For this reason, a dual-mesh method similar to that of Kubrak et al. (2013) has been implemented in which the computational mesh of the concentration field may be refined such that its spacing corresponds to $\Delta x / N_{ref}$ with N_{ref} being the refinement factor which is set to $N_{ref} = 2$ for the concentration fields presented in §9.4. Regarding the advective term, trilinear interpolation is employed for the fluid velocities. The implementation has been validated using various test cases, and a comprehensive description of the method and validation can be found in Uhlmann (2019).

3.3 Simulation setup

The underlying flow configuration of the primary series of simulations conducted within this thesis is open-channel flow with a Cartesian coordinate system chosen such that x, y, z denote the streamwise, wall-normal and spanwise direction, respectively. The alignment of the channel is horizontal implying that $\mathbf{g} = (0, -g, 0)^T$, and hence, the wall-normal direction is commonly also referred to as the vertical direction. The bottom of the domain, i.e. the location an object which is solely subject to gravity moves towards, corresponds to the no-slip wall, whereas the top of the domain corresponds to the free-slip wall. The domain periodically repeats in the streamwise and spanwise directions with the respective periods being denoted by L_x and L_z. In the context of river flows, this is to be understood in the sense

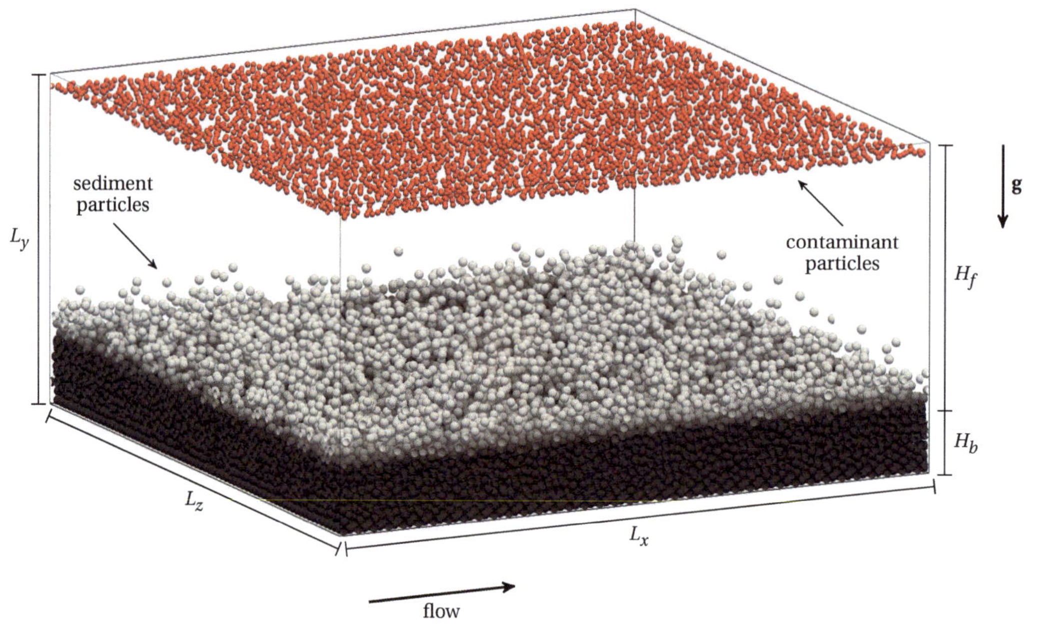

Figure 3.1: Sketch of the main numerical experiments which discussed and referenced throughout this thesis. The simulations feature a turbulent open-channel flow over a fully-resolved mobile sediment bed. Additionally, contaminant particles are added to the fully-developed flow at $t = 0$ and then transiently settle downwards in wall-normal direction.

that there is no influence exerted by any walls normal to these directions, and thus, as an approximation to the core of a river where the region of interest is within a reasonable distance to the river shore such that its effect can be neglected. The flow is driven by a constant mass flux in streamwise direction, which is imposed by means of variable body force corresponding to the mean pressure gradient in eq. (3.3).

A sketch of the simulation setup is provided in fig. 3.1. Two sets of particles are considered in the simulations—*sediment particles* which accumulate at the bottom of the domain to form a sediment bed, and *contaminant particles* which represent the entering contaminating solids. All of these particles are fully resolved with respect to the microscopic flow scales in their vicinity using the immersed boundary method described in §3.2.1, and are subject to the collision model introduced in §3.2.2.

The sediment particles constitute a sediment bed with an average height of H_b which is already present when the contamination is introduced to the flow. In fact, as a precursor for the main simulations, the configuration is simulated in absence of contaminants until a quasi-steady state for the bedform and turbulence is achieved, such that both are completely established before the contamination is introduced to the flow. All sediment particles are fully mobile as described by eq. (3.8) with the exception of a regularly distributed layer at the bottom of the domain which is kept immobile in order to prevent collective sliding of the bed. The particles are monodisperse in the sense that they all share the same diameter and density. The parameter point is chosen such that the bed is sufficiently mobile to exhibit a significant streamwise particle flux and to freely develop natural patterns such as ripples or ridges, but still heavy enough such that bedload transport stays confined to a layer close to the interface and sediment particles are only rarely ejected into the bulk flow. A thorough description of the parameter range is provided in §3.5.

A contamination is introduced to the flow by means of an initial value problem: contaminant particles which mimic the presence of dispersed solids are seeded close to the top boundary at the beginning of the simulations ($t = 0$), and transiently settle downwards until a statistically steady state is reached, cf. §2.3.5. Again, the set of particles is monodisperse, however, their properties may differ from those of the sediment particles. A summary of the contaminant parameter range is provided in §3.6.

3.4 Definition of fluid-sediment interface

Due to the presence of a sediment bed at the bottom of the domain, the near-wall behaviour of the fluid is not primarily driven by the no-slip wall, but rather by the interactions with sediment particles. In fact, the sediment bed constitutes an irregular virtual wall which is porous, rough and deformable, cf. §2.3.1, and its interface position needs to be defined and characterised.

A straightforward approach of defining the fluid-sediment interface is based on the thresholding of the solid indicator function

$$\chi_p(\mathbf{x}, t) = \begin{cases} 1 & \text{if inside solid phase} \\ 0 & \text{else} \end{cases} \tag{3.30}$$

averaged over statistically homogeneous directions. As an example, Kidanemariam and Uhlmann (2014a) used $\langle \chi_p \rangle_z = 0.1$ as a threshold to determine the x- and t-dependent interface of ripple bedforms, where $\langle \bullet \rangle_z$ denotes the spanwise averaging operator defined analogously to eq. (2.21). While this method certainly provides satisfactory results for the investigation of bedforms, its primary disadvantage for the current work is that any instantaneous inhomogeneity is lost in the direction of averaging. With regards to the solid-solid interactions between the contaminants and the sediment bed, it is desirable to have a two-dimensional definition of the interface in order to retain more details of the interface. For this reason, a novel definition based on geometrical and physical considerations is employed in this thesis, which has been developed together with Scherer (2022). Two-dimensional interfaces generated by this method were already successfully utilised in the study of ripples (Scherer et al. 2020) as well as ridge-type patterns (Scherer et al. 2022). In the remainder of this section, the algorithm to obtain such a interfaces is outlined alongside the visual illustrations provided by fig. 3.2

In a first step, from the complete set of particles (fig. 3.2(a)), those particles which are either suspended of belong to bedload, cf. §2.3.1 for definitions of these terms, are removed using the inclusive disjunction of two criteria (fig. 3.2(b)). The first criterion is based on the vertical component of the contact force of the particles, which is easily accessible from the collision model used the numerical simulations. In order for a particle to belong to the bed, the vertical component of this force has to be non-negligible, which implies that all particles for which the collision force is negligible as compared to their submerged weight likely belong to the suspended load. For this reason, particles for which

$$\frac{\left|\mathbf{F}_{p,col} \cdot \mathbf{g}\right|}{\left(\rho_p - \rho_f\right) \frac{\pi}{6} d_p^3 g^2} < 10^{-5}, \tag{3.31}$$

holds are disregarded for the determination of the interface. The second criterion aims to eliminate particles which belong to the bedload, i.e. the region of vigorous sediment transport where particles may still be in contact, and thus, not be captured by the force criterion. In this case, if a particle exhibits an instantaneous kinetic energy larger than a given threshold, that particle is disregarded. In particular, this criterion is formulated as

$$\frac{|\mathbf{u}_p|^2}{v_{p,g}^2} > 0.17, \tag{3.32}$$

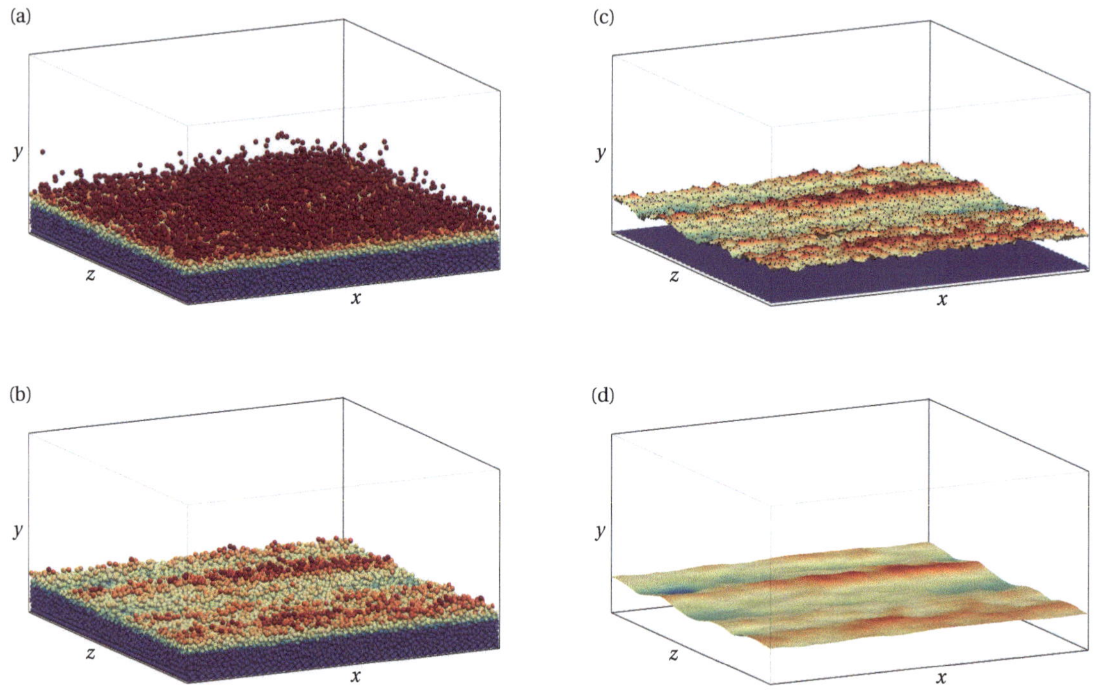

Figure 3.2: Illustration of the individual steps of the algorithm used to extract the two-dimensional interface between the fluid phase and the sediment bed. (a) The complete set of sediment particles. (b) The set of sediment particles after those particles which have been classified as bedload or suspended load are removed. (c) The raw interface obtained by concave shape extraction using the α-shape method by Edelsbrunner and Mücke (1994). The irregular locations of the particle centres which form the surface nodes are indicated by black dots. (d) The final interface $h_b(x, z, t)$ after interpolation to a regular grid and smoothing. In all subfigures, the colouring indicates the height of the particles or interface.

where the threshold may be interpreted as a critical value of the local and instantaneous Shields parameter, and its chosen value is motivated by the value of the global Shields parameter used in the simulations. The thresholds have been varied within a reasonable range and only little sensitivity has been detected for the quantities of interest in this thesis.

The second step requires a definition of the shape of the set of remaining sediment particles. Since this set is generally non-convex—the bed will typically consist of various hills and troughs—and its concaveness is to be taken into account, the intended level of detail has to be prescribed in order to make the solution unique. Here, the α-shape algorithm proposed by Edelsbrunner and Mücke (1994) is applied, which introduces a single free parameter α that defines the minimum length scale for which concaveness is permitted. This algorithm is deployed on the centre point locations of the filtered set of sediment particles, and $\alpha = 1.1d_s$ is chosen to be slightly larger than the diameter of the sediment particles.[1] The result is a triangulation of the topmost and bottommost particles with irregularly distributed vertices corresponding to the centre points of the particles which constitute the interface. Figure 3.2(c) provides a visual example of such a triangulation.

In the last step, the result of the α-shape algorithm is postprocessed. The upper triangulation is interpolated onto a regular grid for reasons of convenience, and additionally, a two-dimensional box filter with a width equivalent to $\sigma_{box} = 5d_s$ in each direction is employed in order to smoothen small fluctuations

[1] The subscript "s" is used here and henceforth as a particularization of the subscript "p" and denotes a quantity which is specific to the set of sediment particles, rather than particles in general.

and reduce noise. The result of this procedure is the instantaneous, two-dimensional fluid-sediment interface $h_b(x, z, t)$ (fig. 3.2(d)).

For the major part of this thesis, it is sufficient to characterise the bed in terms of its statistics. Therefore, the spatio-temporal average of the bed height,

$$H_b := \langle h_b \rangle_{xzt}, \tag{3.33}$$

and its standard deviation

$$\langle h_b'^2 \rangle_{xzt}^{1/2} := \langle h_b^2 \rangle_{xzt} - H_b^2 \tag{3.34}$$

are introduced. Correspondingly, the mean clear fluid height, which denotes the distance between the sediment-interface and the top boundary of the domain, is given by

$$H_f := L_y - H_b \tag{3.35}$$

and constitutes an adequate definition of the vertical extent of the flow domain. As for the transient settling this distance is highly relevant, the coordinate system is chosen such that $y \in [-H_b, H_f]$, and hence, the mean position of the interface is located at $y = 0$.

3.5 Flow over the sediment bed in the absence of contaminants

Now that the fluid-sediment interface has been defined, the base configurations to which contaminants are added can be discussed. Due to the role of the sediment bed as a virtual wall, the statistics of the turbulent flow in the open-channel are strongly tied to the characteristics of the sediment bed, and hence, need to be discussed jointly. Henceforth, the combination of the turbulent flow and the associated sediment bed will be denoted as *background flow*, and its description is crucial for the determination of the contaminant parameter points. In this thesis, three distinct background flows are considered which differ in their value of the bulk Reynolds number, which is now defined as[2]

$$Re_b := \frac{u_b H_f}{\nu_f}, \tag{3.36}$$

and the patterns exhibited by the sediment bed. For the lower Reynolds number of $Re_b = 3009$, two bedforms are considered—one with a macroscopically flat bed which solely exhibits turbulence-induced ridges (henceforth denoted by "LR"), and one with ripple features (henceforth denoted by "RP"). For both background flows, the exact same physical parameters are imposed which enables a direct assessment of the influence of bedforms on the spreading of contaminants. This was achieved by artificially suppressing the formation of ripples in the LR background flow by reducing the streamwise period of the simulation domain below the minimal value required to trigger the instability, see Kidanemariam and Uhlmann (2017) for a detailed discussion on this procedure. Indeed, LR and RP have been initialised from cases H3 and H6 presented in Kidanemariam and Uhlmann (2017) with the distinction that the

[2] The current definition of the bulk Reynolds number is the same as eq. (2.23) with the newly defined clear fluid height used as the length scale. It should be noted that the value of the clear fluid height is not known a priori as it depends on the interaction between the flow and the sediment bed. It can, however, be estimated with reasonable accuracy.

Table 3.1: Summary of the physical parameters associated with the turbulence and sediment bed which constitute the background flow of the transient settling simulations.

case	Re_b	Re_τ	Ga	ρ_s/ρ_f	H_f/d_s	H_b/d_s	$\langle h_b'^2 \rangle_{xzt}^{1/2}/d_s$	d_s^+	θ	bedform
LR	3009	241	28.4	2.5	27.4	11.0	0.34	8.8	0.10	flat
RP	3009	340	28.4	2.5	27.7	10.7	2.08	12.3	0.19	ripple
HR	9483	838	56.7	2.5	28.6	7.0	0.52	29.3	0.27	flat

Table 3.2: Summary of the numerical parameters concerning the computational domain and the resolution of the fluid discretisation and sediment particles. The observation time corresponds to the time interval over which statistics on the background flow have been gathered. N_x, N_y, N_z denote the number of grid nodes in x, y, z direction, respectively, and N_s is the total number of sediment particles.

case	$[L_x \times L_y \times L_z]/H_f$	$[N_x \times N_y \times N_z]$	Δx^+	N_s	$d_s/\Delta x$	t_{obs}/t_b
LR	[2.80×1.40×2.80]	[1152×576×1152]	0.58	65359	15	217
RP	[5.60×1.40×2.80]	[2304×576×1152]	0.82	127070	15	214
HR	[2.49×1.24×2.49]	[2560×1280×2560]	0.81	45612	36	28.3

mesh has been refined by a factor of 1.5 in all directions in order to accommodate a wider range of contaminant sizes. In contrast, the higher Reynolds number flow (HR) has been developed from scratch in a high-performance computing project associated to this thesis (GCS 2021, Gedenk 2022). Its value of the bulk Reynolds number is significantly higher at $Re_b = 9483$, and consequently, only few experiments could be performed for this background flow due to its high demand of computational resources. In particular, the contaminants are only seeded over a featureless sediment bed such that a direct comparison of the Reynolds number effect is only possible for macroscopically flat beds.

For the purpose of defining the wall shear stress on the virtual wall, the method of Chan-Braun et al. (2011) is applied in which the fluid shear τ_f defined in eq. (2.24) is extrapolated from the bulk of the channel onto the mean position of the sediment-fluid interface at $y = 0$. The resulting wall shear stress is then used to define the friction velocity from eq. (2.26) and the remaining inner scales. The friction Reynolds number is then given by

$$Re_\tau := \frac{u_\tau H_f}{\nu_f} \tag{3.37}$$

in accordance with eq. (2.29) using H_f as the outer length scale, and its value reads $Re_\tau = \{241, 340, 838\}$ for the LR, RP and HR background flows, respectively. The corresponding values of the Shields parameter introduced in eq. (2.48) are $\theta = \{0.10, 0.19, 0.27\}$ and bedload transport can be observed for all background flows considered. The particle diameters scaled in inner units read $d_p^+ = \{8.8, 12.3, 29.3\}$ for LR, RP and HR such that the sediment particles are significantly larger than the smallest turbulent scales in all cases. Summaries of relevant physical and numerical parameters regarding the turbulent flow and the sediment bed are provided in table 3.1 and table 3.2, respectively.

In order to characterise the temporal evolution of the sediment bed topology, the instantaneous fluid-sediment interface is presented in streamwise-averaged and spanwise-averaged form in the following. The streamwise-averaged perspective enables an evaluation of turbulence-induced ridges, i.e. the streamwise-aligned streaks of elevations and troughs with relatively low-amplitude, whose length is typically sufficiently long to span the entire streamwise length of the domains considered in this thesis. Figure 3.3 displays the spatio-temporal evolution of $\langle h_b(x, z, t) \rangle_x$ for the two macroscopically flat cases LR

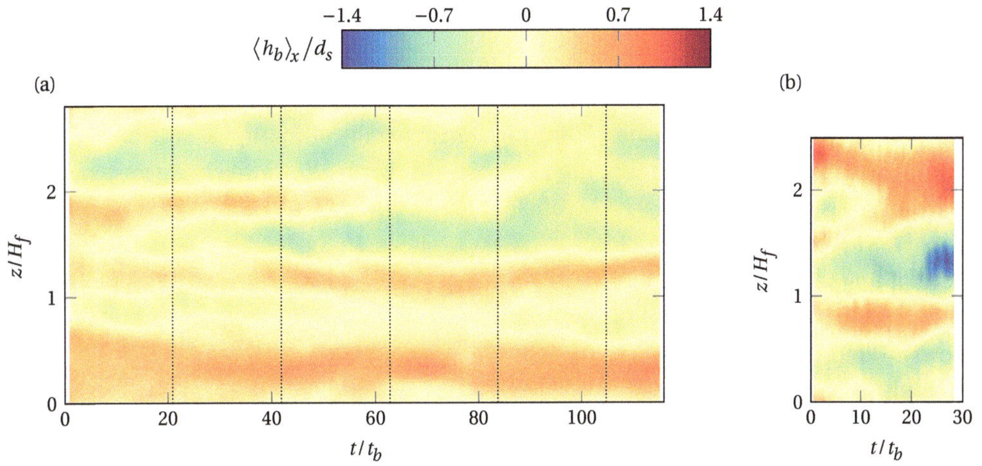

Figure 3.3: Temporal evolution of the streamwise-averaged fluid-sediment interface $\langle h_b(x,z,t) \rangle_x$ for the featureless background flows. (a) LR, $Re_\tau = 241$. (b) HR, $Re_\tau = 838$. The dotted vertical lines mark the initial conditions for the transient settling problem aside from $t = 0$, cf. §3.6. The evolution of the bed is shown in absence of contaminant particles.

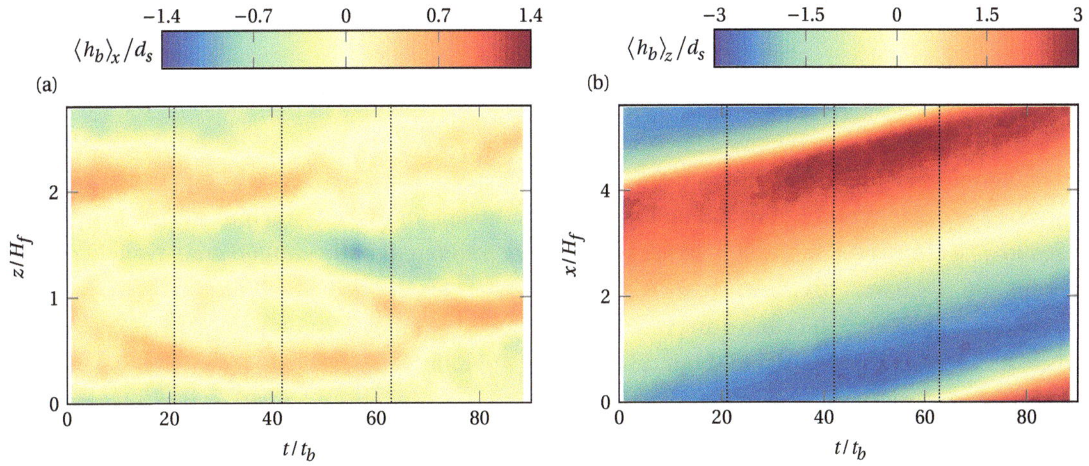

Figure 3.4: Temporal evolution of the averaged fluid-sediment interface for the background flow featuring ripples (RP). (a) streamwise-average $\langle h_b(x,z,t) \rangle_x$ similar to fig. 3.3 indicating the presence of superimposed ridges. (b) spanwise-average $\langle h_b(x,z,t) \rangle_z$ showing the propagation of the ripple. Contaminants are added to the flow at $t = 0$ as well at the times marked by the dotted vertical lines, cf. §3.6. The evolution of the bed is shown in absence of contaminant particles.

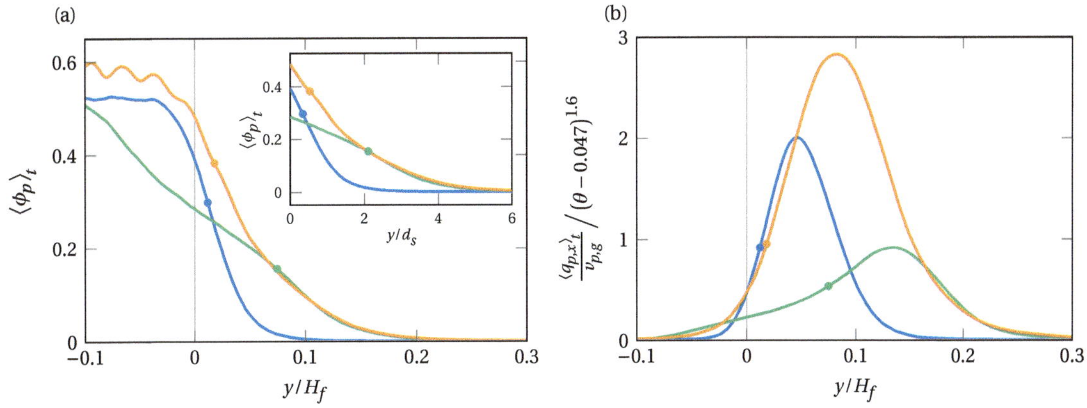

Figure 3.5: Characterization of the sediment bedload layer. (a) Local solid volume fraction as a function of vertical position. (b) Streamwise flux density of particles in inertial scaling and normalised by the expected scaling of $\int q_{p,x}\,\mathrm{d}y / (v_{p,g} d_s)$ according to the Wong and Parker (2006) version of the Meyer-Peter and Müller (1948) formula. The vertical grey lines indicate the mean position of the sediment-fluid interface, whereas the dots on the curves indicate its standard deviation. Legend: LR (——), RP (——), and HR (——).

and HR. Albeit the sediment ridges exhibit some dynamics, their spanwise position is relatively stable in time—an attribute which is likely inherited from the large-scale fluid motion they are generated by. In general, their time scale of meandering and their lifetime are substantially longer than the time scales of the transient to be investigated in this work which are generally of $\mathcal{O}(10t_b)$ as will be unveiled in §3.6. Sediment ridges can also be observed in the presence of other bedforms as is suggested by fig. 3.4(a) which displays the streamwise-averaged bed height for the case RP which features a single ripple. To characterise the spanwise-aligned ripple, $\langle h_b(x,z,t)\rangle_z$ is shown in fig. 3.4(b). The amplitude of this feature is several times larger than that observed for ridges, and hence, ripples dominate the appearance of the fluid-sediment interface. As indicated by the spatio-temporal evolution, the feature propagates in streamwise direction with a nearly constant speed in the time interval under observation. This speed can be quantified by defining the two-point, two-time correlation of the interface fluctuation (Kidanemariam and Uhlmann 2014a), viz.

$$R_{h_b}(\delta x, \delta t) := \left\langle \langle h_b' \rangle_z(x,t)\ \langle h_b' \rangle_z(x+\delta x, t+\delta t) \right\rangle_{xt}, \tag{3.38}$$

and a subsequent linear regression of the streamwise shift δx for which R_{h_b} reaches its maximum at a given temporal shift δt. The propagation velocity obtained for RP in the time interval shown in fig. 3.4(b) reads

$$c_{RP} = 0.023 u_b, \tag{3.39}$$

which is somewhat lower than the value reported by Kidanemariam and Uhlmann (2017) for this case ($c_{RP} = 0.032u_b$). One source of deviation lies in the method of interface extraction, as the method proposed in §3.4 is observed to result in larger values of H_f compared to the method employed by Kidanemariam and Uhlmann (2017), which implies smaller values for u_b at constant flow rate. However, this scaling effect tends to reduce the observed deviation, and thus, the discrepancy still persists when the same extraction method is used for both cases. Ruling out differences in scaling, the reduction of c_{RP} is presumably a result of the grid refinement, as all physical parameters and all other relevant numerical parameters are matched between the simulations. In particular, the distance over which the particle collision force acts before two particles overlap is set to $\Delta_{col} = \Delta x$ in both cases, and thus, the effective diameter $d_p + 2\Delta_{col}$ of the particles in the refined simulation is approximately 5.5% smaller than that in Kidanemariam and Uhlmann (2017). This results in a more compact bed[3], for which it appears reasonable that the mobility of particles, and hence c_{RP}, is reduced.

As stated earlier, the shape of the virtual wall is only one way with which the sediment bed interacts with the carrier flow. In the following, the bedload will be characterised in order to fathom how much the direct influence of the sediment protrudes into the flow. To begin with, a local solid volume fraction is defined as

$$\phi_p(y,t) := \left\langle \chi_p(\mathbf{x},t) \right\rangle_{xz}, \tag{3.40}$$

which is a function of vertical position. Figure 3.5(a) provides the spatial evolution of ϕ_p and demonstrates that for all background flows considered a fair amount of sediment particles can be still be found at distances of $\mathcal{O}(0.1H_f)$ from the interface. In accordance with their Shields number difference, HR extends further into the flow than LR and sediment particles can regularly be found outside the inner layer of turbulence. For RP, the increased variance in bedform height leads to a similar effect and causes

[3] The refined simulations exhibit a mean bed height which is about 7.5% reduced compared to the unrefined simulations when the extraction method of Kidanemariam and Uhlmann (2017) is used.

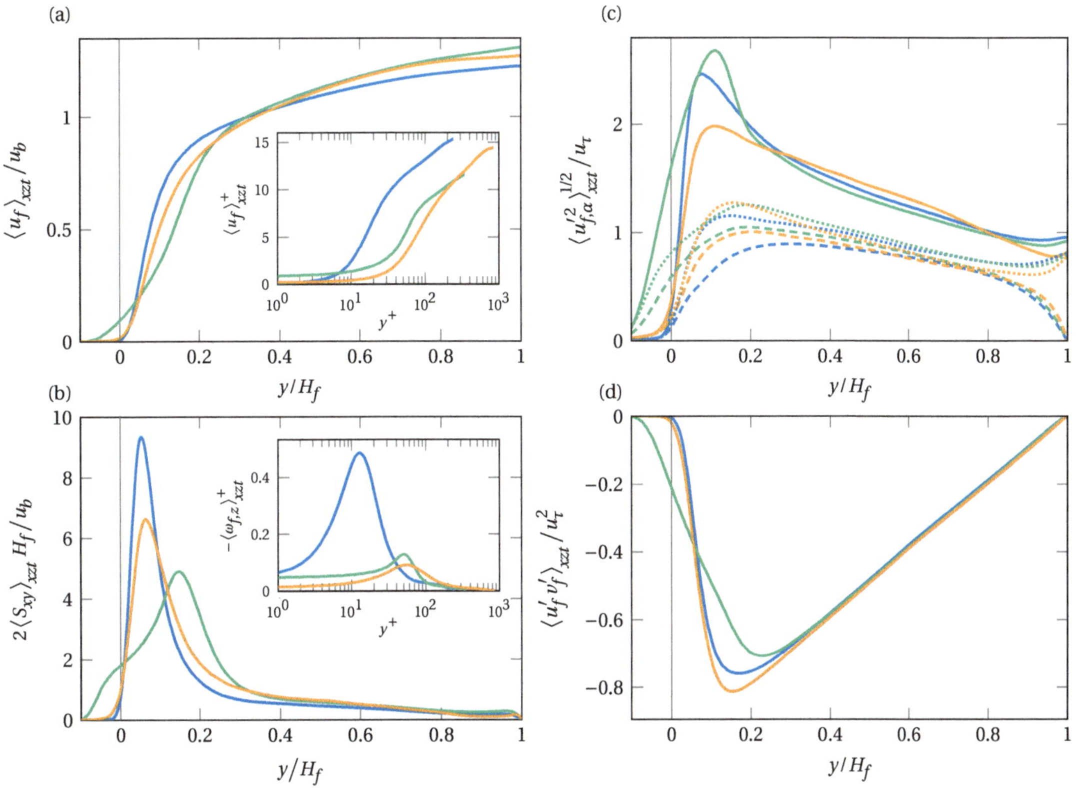

Figure 3.6: Turbulence statistics of the sediment-laden background flows. (a) Mean streamwise velocity. (b) Mean shear induced by the vertical gradient of the streamwise velocity. (c) Root-mean-square of the fluid velocity fluctuations u'_f (——), v'_f (- - -) and w'_f (········). (d) The Reynolds stress $u'_f v'_f$. Legend: LR (——), RP (——), and HR (——).

the solid volume fraction to decrease more gradually in wall-normal position. Aside from the spatial distribution of particles, it is also common to examine the momentum carried by the solid phase. For this purpose, the sediment flux density is defined as

$$q_{p,x}(y,t) := \langle u_p(\mathbf{x},t)\chi_p(\mathbf{x},t)\rangle_{xz}, \tag{3.41}$$

where $u_p(\mathbf{x},t)$ is the streamwise component of the particle velocity evaluated in the Eulerian frame. As suggested by its name, integration of the sediment flux density in the vertical direction yields streamwise sediment flux, and thus, it serves as an adequate proxy to assess bedload transport for the background flows considered. Figure 3.5(b) shows $q_{p,x}$ normalised by the gravitational velocity and scaled across the various background flows in terms of the scaling law of the total sediment flux according to the Wong and Parker (2006) based on the formula of Meyer-Peter and Müller (1948). For all background flows investigated, the bedload layer extends well into the buffer layer and beyond—a characteristic which is ought to be kept in mind when analysing the settling of contaminants.

In §2.3.1, the effect of mobile sediments on the turbulence statistics was discussed in a general context. To conclude the characterization of the background flow, the actual statistics of the turbulent flow are presented for the three background flows. Figure 3.6(a) shows the mean streamwise fluid velocity as a function of the wall-normal coordinate. Due to the porosity of the interface and the vagueness of its location, $\langle u_f \rangle_{xzt}$ does not strictly vanish at the virtual wall. This characteristic is especially pronounced for RP where the considerable variability in bed height causes the velocity to be significantly greater than zero. Another implication of the diffuseness of the interface is that the strongest gradient in $\langle u_f \rangle_{xzt}$ is not observed directly at the (virtual) wall, but rather at a significant distance from it, see fig. 3.6(b).

This effect is most pronounced for RP where the decay of $\langle u_f \rangle_{xzt}$ with decreasing distance from the wall is already initiated at $y \approx 0.3H_f$, whereas the flows over macroscopically flat beds (LR,HR) decay more abruptly starting around $y \approx 0.2H_f$.

Regarding the second moments of the velocity statistics shown in fig. 3.6(c), it can first and foremost be observed that the amplitude of the near-wall peak in $\langle u_f'^2 \rangle_{xzt}$ decreases with increasing Reynolds number, which indicates the absence of turbulent structures associated to the near-wall cycle, cf. §2.2.3 and §2.3.1, and in particular, the velocity streaks of the buffer layer. In presence of ripples, the peak amplitude is increased compared to the featureless case which is presumably due to the bedform itself. Just like the streamwise fluctuations, the wall-normal and spanwise fluctuations differ only little across the background flows above $y \gtrapprox 0.3H_f$. Significant vertical fluctuations protrude closest to the interface in the presence of ripples, and the tendency to do so increases with increasing Reynolds number. Concerning the spanwise fluctuations, the influence of bedforms appears to be more significant than that of the Reynolds number.

The turbulent Reynolds stress is the primary contribution to the fluid shear stress above the height where fluid shear becomes significant, see fig. 3.6(d). In the presence of particles, the balance of mean streamwise momentum formulated in eq. (2.24) is complemented by an additional stress term due to the body forcing term related to the presence of particles (Uhlmann 2008, Kidanemariam et al. 2013). The premature decay of $\langle u_f' v_f' \rangle_{xzt}$ in the presence of ripples can be in part attributed to the presence of this term.

3.6 Description of the contaminants

The principal objective of this thesis is to study the spreading of contaminations which enter a fluvial system from the free surface. Two types of contaminations are distinguished in this thesis—solid contaminants which are subject to the equations of motion of particles, and dissolved contaminants whose transport is governed by the advection-diffusion equation. However, due to the paramount role of particle-attached bacteria in the spreading of microbial contaminations (McDonald et al. 1982, Sherer et al. 1992, Wilkinson et al. 1995, Jamieson et al. 2005) and the complexity of the dynamics of particles in turbulent flows, the primary focus of this thesis lies in the investigation of the former type, and the discussion of dissolved contaminations is restricted to §9.

As described in §2.3.3, the seeding of heavy particles near the top boundary of the flow domain results in a transient settling problem. While there are endless possibilities for the initial distribution of contaminants, the only configuration considered in this work is the distribution in which all particles are initially located at the same vertical position, see fig. 3.7 for a visualization. This choice was made due to the homogeneous properties of this distribution in streamwise and spanwise direction which gives it the advantage of generality over more localised distributions. Moreover, this initial condition may be interpreted as an equivalent distribution to the line source used in the related experimental work of Cuthbertson and Ervine (2007) for a periodic setup, and thus, enables a comparison to existing data.

The initial centre point location of all contaminant particles is $y_{c,0} = 0.953H_f$.[4] For this location, all contaminant particles considered in the following are fully immersed into the flow, which implies that $y_{c,0} < H_f - d_c/2$. Within this y-normal plane, the particles are distributed randomly while ensuring that they do not overlap. For this purpose, a random Poisson process is implemented using the subsequently

[4] Henceforth the subscript "c" is used to denote properties attributed to contaminant particles.

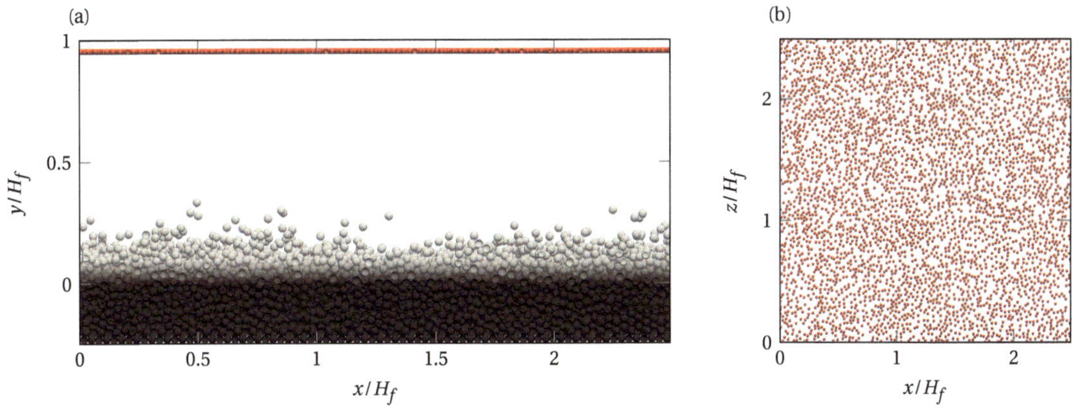

Figure 3.7: Example of the initial distribution of contaminant particles (red spheres). (a) Lateral view of the simulation domain. The contaminant particles are initially located at the exact same vertical position $y_{c,0} = 0.953H_f$. (b) Top view of the simulation domain. Within the xz-plane, the particles are distributed randomly using a random Poisson process while avoiding overlaps. While the global solid volume fraction can be considered as dilute, the distribution in the initial plane is rather dense. This exemplary parameter point corresponds to G30-D15-HR.

Table 3.3: Summary of all contaminant parameter points investigated in this chapter including all relevant physical and numerical parameters. The table includes the value of the characteristic settling time $t_{p,s} = H_f / v_{p,\infty}$ which is introduced at a later point.

case	Ga	ρ_c/ρ_f	d_c^+	H_f/d_c	$Re_{p,\infty}$	I_τ	St^+	N_c	N_{run}	$d_c/\Delta x$	$t_{p,s}/t_b$	t_{obs}/t_b
G30-D10-LR	28.4	2.5	8.8	27.4	18.7	0.47	5.4	1000	{5,1}	15	5.9	{8.3,436}
G15-D10-LR	16.4	1.5	8.8	27.4	8.1	1.08	4.7	1000	6	15	13.5	12.5
G15-D6-LR	15.4	2.5	5.8	41.2	7.3	0.80	3.1	3375	6	10	10.0	12.5
G10-D6-LR	10.9	1.75	5.8	41.2	4.2	1.40	2.7	3375	{5,1}	10	17.5	{12.5,435}
G30-D10-RP	28.4	2.5	12.3	27.7	18.7	0.66	10.5	2000	{3,1}	15	5.8	{12.2,428}
G10-D6-RP	10.9	1.75	8.2	41.5	4.2	1.96	5.3	6750	{3,1}	10	17.3	{12.2,427}
G30-D15-HR	27.1	2.5	17.9	46.8	17.9	1.00	23.4	3913	1	22	11.4	14.1
G10-D10-HR	10.8	1.75	12.2	68.6	4.2	2.88	12.2	12346	1	15	32.6	19.3

described algorithm which is repeated until the desired number of contaminants has been distributed. First, a candidate for the horizontal centre point position is generated by drawing two uniformly distributed pseudorandom numbers from the intervals $[0, L_x)$ and $[0, L_z)$. Second, the distance from this candidate to all previously distributed particles is computed, and if any of those distances is smaller than $d_c + \varepsilon$, a new candidate is generated until this criterion is fulfilled. Here, ε denotes a small distance which is introduced to avoid high initial collision forces and numerical complications, which is set to be of $\mathcal{O}(\Delta x)$. When the distance criterion is fulfilled, the contaminant particle is assigned the position of the candidate and the algorithm is continued until all particles have been distributed. The contaminants' initial velocity is set to the value of the fluid velocity at the position where the particle is placed. Therefore, the mean velocity of the fluid and solid phase is approximately equal at initialization as the contaminants sample the xz-plane randomly. In particular, this implies that the contaminants initially have, on average, zero momentum in the vertical direction, and hence, start settling from rest.

Table 3.3 provides a summary of the contaminant particle parameter points examined in this thesis. The naming convention of the cases consists of three compartments—an indication of the Galileo number, the diameter in inner units, and the background flow—which are separated by hyphens. The input parameters are varied with respect to the Galileo number, the particle diameter, and the density ratio, and the most extensive investigation has been performed in the background flow over a macroscopically flat bed at a low turbulent Reynolds number (LR). These cases are complemented by further simulations involving ripple bedforms (RP) or at a higher Reynolds number (HR), which have been conducted exclusively for the highest and lowest Galileo number under investigation, and are approximately matched in input parameters to the LR simulations.

The total number of contaminant particles per simulation, denoted by N_c, is not constant across the various cases. Instead, N_c is chosen such that the global solid volume fraction of contaminants defined as

$$\Phi_c = \frac{1}{H_f}\int_{-H_b}^{H_f} \phi_c \,\mathrm{d}y = \frac{\pi d_c^3 N_c}{6L_x H_f L_z} \tag{3.42}$$

attains a value of $\Phi_c = 3.2 \cdot 10^{-3}$. In this definition, ϕ_c denotes the local solid volume fraction introduced in eq. (3.40), but only applied to the set of contaminant particles. While the selected value of Φ_c is generally low enough to classify the system as dilute, one has to keep in mind that the local solid volume fraction in the plane of initialization is not small as $\phi_c(y_{c,0}) \in [0.13, 0.33]$, cf. fig. 3.7 for a visualization of the case with an initial area coverage of $\phi_c = 0.23$.

In order to ensure a sufficient number of samples to obtain converged statistics and to minimise the influence of the specific flow conditions at initialization, multiple realizations of the numerical simulations have been performed per parameter point whenever possible. In particular, all simulations at LR have been realised six times and the simulations of the RP series four times. Figures 3.3 and 3.4 indicate the state of the sediment bed at the seeding times of the various realizations. While most realizations only cover a relatively short observation time t_{obs} which constitutes the transient, the LR and RP runs for the highest and lowest Galileo numbers have also been simulated over substantially longer time intervals in order to assess the long-term deposition behaviour. Letting N_{run} denote the number of realizations of the numerical experiment, the total number of samples available for a single parameter point is given by

$$N_{c\Sigma} := N_c N_{run}, \tag{3.43}$$

and its value varies in the range $N_{c\Sigma} \in [3913, 27000]$.

All particles investigated in this thesis are situated in the axisymmetric wake regime, cf. §2.3.4, and for all contaminant particles, the value of the Galileo is sufficiently low such that no noteworthy recirculation can be observed in the rear. The choice of this parameter point is driven by the aim to represent high relative turbulence intensities as the solid matter which is discharged into rivers in CSOs is often pretreated in a manner that only particles with small settling velocities remain, cf. §1.1. In fact, the relative turbulence intensities under investigation vary within $I_\tau \in [0.47, 2.88]$ such that various regimes are covered.

Regarding the particle size compromises had to be made as the limitation in Reynolds number range for DNS restricts the separation between the smallest and largest scales of the flow. The diameter of the particle was chosen such that it is larger than the viscous length scale in order to capture small scale interactions and to avoid computational meshes whose resolution is dictated by the particle scales, i.e. $d_c^+ \gg 1$. At the same time the contaminant particles are ought to be small compared to the clear fluid height in order to provide a reasonable vertical distance for the settling process, i.e. $d_c/H_f \ll 1$. The ability to fulfil both of these demands simultaneously is directly linked to the value of the friction Reynolds number, cf. §2.2.2, and thus, the longest available settling distance can be found for the HR

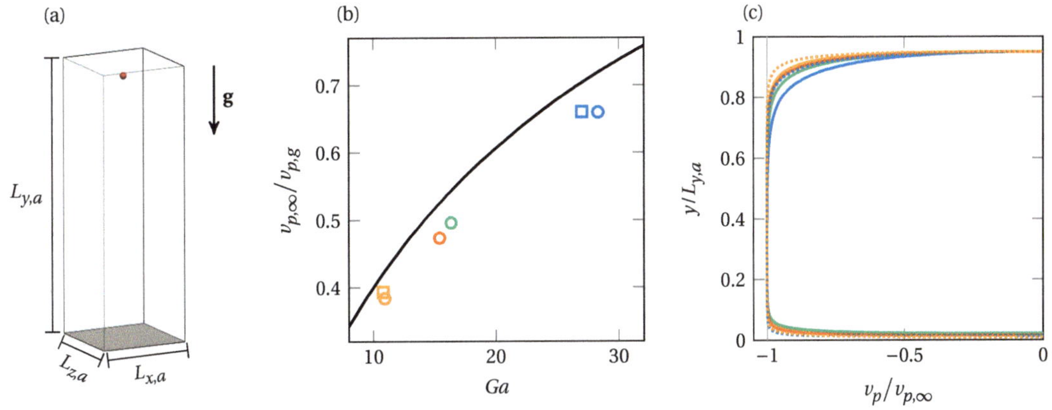

Figure 3.8: Setup and results from the reference simulations to determine the terminal velocity. (a) Visualization of the computational domain showing G10-D6-LR as an example. (b) The resulting terminal velocity as a function of the Galileo number for all parameter points (markers). Circles indicate parameter points with settling distance and numerical resolution of the LR simulations, while squares correspond to the HR runs. The black line shows a reference curve determined using eq. (2.49) and the empirical drag law of Schiller and Naumann (1933). (c) The settling velocity of the isolated particles as a function of wall-normal position showing that the terminal velocity is reached well within the finite settling distance. Legend: G30-D15-HR (······), G10-D10-HR (······), G30-D10-LR (——), G10-D6-LR (——), G15-D10-LR (——) and G15-D6-LR (——).

Table 3.4: Numerical parameters of the auxiliary simulations conducted to determine a reference value for the ambient terminal velocity.

case	$[L_{x,a} \times L_{z,a}]/L_{y,a}$	$[L_{x,a} \times L_{y,a} \times L_{z,a}]/d_c$	$[N_{x,a} \times N_{y,a} \times N_{z,a}]$	$d_c/\Delta x_a$
G30-D10-LR	[0.333×0.333]	[8.5×25.6×8.5]	[128×384×128]	15
G15-D10-LR	[0.333×0.333]	[8.5×25.6×8.5]	[128×384×128]	15
G15-D6-LR	[0.333×0.333]	[12.8×38.4×12.8]	[128×384×128]	10
G10-D6-LR	[0.333×0.333]	[12.8×38.4×12.8]	[128×384×128]	10
G30-D15-HR	[0.267×0.267]	[11.6×64.0×11.6]	[256×960×256]	22
G10-D10-HR	[0.267×0.267]	[17.1×43.6×17.1]	[256×960×256]	15

series of simulations, albeit one would expect considerably larger values of H_f/d_c and Re_τ in actual rivers.

3.7 Determination of the ambient terminal velocity

The definition of some of the relevant particle parameters such as the relative turbulence intensity and the particle Reynolds number require the knowledge of the terminal velocity of the particle in an undisturbed ambient container. The value of $v_{p,\infty}$ depends on the details of the flow developing around the particle, and is thus not known a priori. In fact, it has already been stated in eq. (2.49) that knowledge of the drag coefficient in the quasi-steady settling state is sufficient to derive its value from the gravitational velocity which is known at the outset. Various correlations exist to determine C_d for a given Galileo number, however, in the following the terminal velocity will be determined directly from auxiliary simulations, which has the advantage of providing a more exact reference due to the avoidance of inevitable inaccuracies in the correlations and gauging potential systematic errors related to the numerical method employed here.

The reference simulations contain a single isolated particle in a domain with the same boundary conditions as the open-channel, but without any background flow, see table 3.4 for a summary of the numerical parameters. Similar to the main simulations, the particle is initially located at $y_{c,0} = 0.95L_{y,a}$. The lateral dimensions are kept short to reduce computational cost, but are generally larger than those of Uhlmann and Dušek (2014) such that the results can be expected to be independent of the lateral boundary conditions. The numerical resolution of the particle is matched with that of the turbulent simulation for each case, which entails that the parameter points in the RP background flow do not need to be investigated separately as their relevant numerical parameters exactly match those of LR. This is done in order to reduce systematic errors when normalizing the particle velocities later on, although the data obtained for G10-D6-LR and G10-D10-HR, which approximately match in Galileo number and should therefore attain the same terminal velocity, suggests that this is not a major concern, see fig. 3.8(b). However, the figure also suggests that the results differ from those obtained by an empirical drag law by a few percent, highlighting the benefit of conducting auxiliary simulations.

The reference simulations are also beneficial to understand how quickly the terminal velocity is attained. Generally, it appears favourable for the results to not be significantly affected by the initial acceleration from rest. In order to estimate the significance of the initial transient, the settling distance in terms of particle diameters, i.e. $L_{y,a}/d_c$ is matched approximately to that of the turbulent cases. Figure 3.8(c) shows the particle velocity as a function of wall-normal position under quiescent conditions. For all cases, terminal velocity is essentially attained in the upper quarter of the channel.

3.8 Computational resources

It should not go unnoticed that the direct numerical simulation of the Navier–Stokes equations requires considerable resources in terms of computing and data storage for the problem sizes considered in this work. As may be deduced from table 3.2, the number of grid points required for each field is of $\mathcal{O}(10^9–10^{10})$, and even a factor of eight larger for each transported scalar field which is considered due to the refined dual mesh approach. To put this into perspective, a snapshot of the velocity and pressure fields for a single instance in time for the HR simulation series requires 282 GB of storage capacity, and in order to analyse the flow conditions during the transient in postprocessing, several finely spaced snapshots have to be stored. The total data size accumulated within the computational projects related to this work is of $\mathcal{O}(100\ \text{TB})$, and consequently, most of the analyses which have been performed would not have been realisable without the availability of appropriate data storage infrastructure. The storage facilities utilised are namely the "Large Scale Data Facility" (LSDF)—a hot storage system with a capacity of several petabytes—and the "bwDataArchiv"—a hybrid disk-and-tape-based system for cold storage purposes. Both systems are maintained by the Steinbuch Centre for Computing (SCC) at the Karlsruhe Institute of Technology.

Naturally, the generation of such immense numerical datasets demands the use of massively parallelised supercomputers. The LR and RP simulation series have been for the most part conducted within the computing project "DNS of transient particle settling in horizontal open-channel flow: investigating the influence of initial conditions" on the machine ForHLR II at SCC within a budget of 8.3 million core-h utilizing 324 cores (LR) and 648 cores (RP), respectively. For the HR series, a large-scale project funding from the Gauss Centre for Supercomputing was acquired, and the corresponding project named "Finite-size particle dynamics and scalar transport in turbulent open channel flow over a mobile sediment bed" was conducted on Hazel Hen and Hawk at the High-Performance Computing Center Stuttgart (HLRS) within a budget of 50 million core-h utilizing 4000 cores for the HR series and 8000 cores for the development of ripple features at high Reynolds number. The latter dataset is not immediately discussed within this thesis, although it contributed to the recognitions made in Scherer et al. (2022) which are referred to throughout this work. Aside from these computational projects, several other HPC systems have been used primarily for postprocessing purposes, which includes bwUniCluster and bwUniCluster 2.0 systems at SCC, and the internal clusters of the Institute for Hydromechanics.

The major computational costs are typically associated to solving the Poisson equation for the pressure related to the problem stated in eq. (2.17), as it requires knowledge on the flow state in the entire domain, and thus, gives rise to a large communication overhead on parallel computers. While the direct cost of the treatment of particle motion itself is rather insignificant in most cases, the numerical methods which can be employed for the particle-laden flow problem are generally less efficient than their singlephase counterparts. As a consequence, the advancement in the field of DNS of particle-laden flows is strongly tied to the advances in high-performance computing, and hence, the approach employed in this thesis is not meant to provide a practical framework to model bacteria transport. Rather it should be regarded as a campaign to provide a database for existing transport models and complement them by identifying potentially unknown small-scale effects which might be of relevance in the bigger picture.

4 Particle velocity during the transient

The velocity with which solid particles propagate in the receiving flow plays a paramount role for the prediction of the spreading of particle-bound bacteria. The vertical velocity component is a parameter which is frequently used in bacteria transport models, and in the majority of studies, its value is assumed to equal the terminal velocity in stagnant water disregarding any interactions with the fluid flow. The purpose of the current chapter is to examine the effect of turbulence on the particle velocity in a phenomenological manner to provide a basis for subsequent more detailed discussions on the interaction between the carrier flow and the dispersed phase.

4.1 Temporal evolution of the transient

In the initial state of the numerical experiment, the system is out of equilibrium. After release, the contaminants collectively settle towards the sediment bed under the action of gravity causing a net downward contaminant flux. As time progresses and contaminants approach the sediment bed, an equilibrium between deposition and resuspension is achieved eventually. At this stage, the net contaminant flux of particles is zero if a sufficient number of contaminant particle is considered, and their vertical distribution is constant in time.

Without providing a formal definition of the global contaminant flux, this behaviour becomes observable in terms of the average vertical position of the contaminants,

$$\langle y_c \rangle_{\mathbb{T}}(t) := \frac{1}{N_{c\Sigma}} \sum_{i=1}^{N_{c\Sigma}} y_c^{(i)}(t), \tag{4.1}$$

where here and in the following $\langle \bullet \rangle_{\mathbb{T}}$ denotes the operation of ensemble-averaging over all contaminants and realizations at a given instance in time. The statements made on the global flux then translate into the following: initially, the average position is by definition located at the height of the plane of initialization, i.e. at $y_{c,0}$, whereas at equilibrium, $\langle y_c \rangle_{\mathbb{T}}$ is a constant whose value depends on the characteristics of the contaminants and the details of turbulence. The rate at which the equilibrium is approached is given by the ensemble-averaged settling velocity of the contaminants,

$$\langle v_c \rangle_{\mathbb{T}}(t) = \frac{\mathrm{d}\langle y_c \rangle_{\mathbb{T}}}{\mathrm{d}t}. \tag{4.2}$$

At initialization, the average settling velocity is approximately zero, since the contaminants randomly sample a horizontal plane for which $\langle v_f \rangle_{xz} = 0$ holds. At equilibrium, $\langle v_c \rangle_{\mathbb{T}}$ is required to be zero for sufficiently large samples. Under quiescent conditions and after an initial acceleration phase, this rate

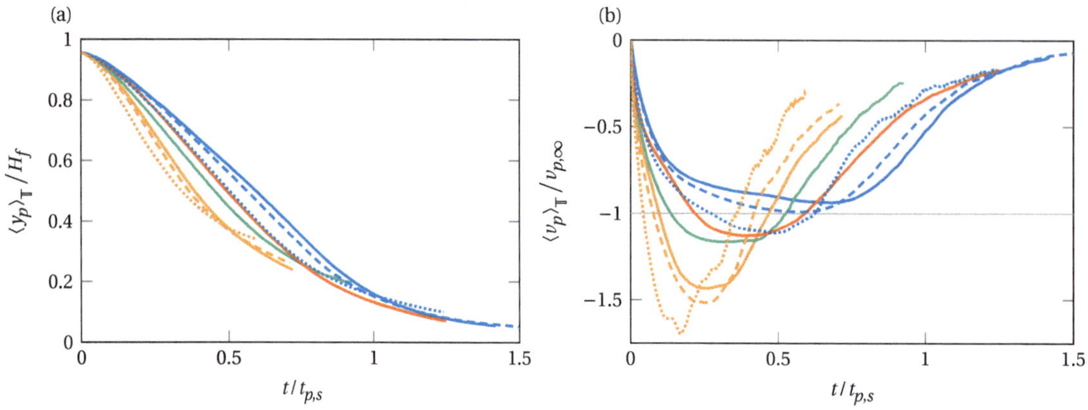

Figure 4.1: Ensemble-averaged (a) vertical position and (b) vertical velocity of the contaminant particles as a function of time normalised by the characteristic settling time. Legend: G30-D10-LR (—), G10-D6-LR (—), G15-D10-LR (—), G15-D6-LR (—), G30-D10-RP (- - -), G10-D6-RP (- - -), G30-D15-HR (······), and G10-D10-HR (······).

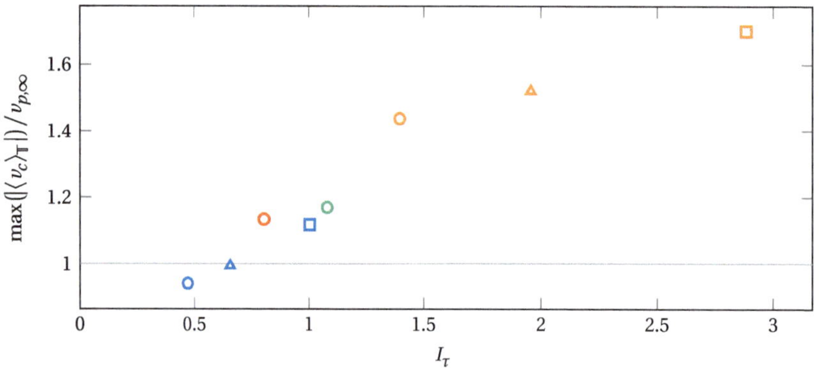

Figure 4.2: The ratio of ensemble-averaged settling velocity of the contaminant particles to their terminal velocity as a function of relative turbulence intensity, $I_\tau = u_\tau / v_{p,\infty}$. Legend: G30-D10-LR (○), G10-D6-LR (○), G15-D10-LR (○), G15-D6-LR (○), G30-D10-RP (△), G10-D6-RP (△), G30-D15-HR (□), and G10-D10-HR (□).

is equal to the terminal velocity. A characteristic time scale at which contaminants reach equilibrium may therefore be defined in terms of the settling distance and the particles' intrinsic velocity, viz.

$$t_{p,s} := H_f / v_{p,\infty}. \tag{4.3}$$

Figure 4.1 shows the ensemble-averaged vertical position and settling velocity for the time interval covering the longest common observation time of all realisations, cf. table 3.3. For all cases, the quasi-steady equilibrium was not reached during the observation time, although the data suggests asymptotic behaviour. More interestingly, the rate at which equilibrium is approached varies significantly between the cases. While G30-D10-LR is the only case for which this rate is consistently slower than its quiescent counterpart, the remaining cases all exhibit a phase where the instantaneous average settling velocity is faster than terminal velocity.

The amplitude of the peak in settling velocity is shown as a function of relative turbulence intensity in fig. 4.2. The degree of enhancement is quite significant: for G10-D10-HR—the parameter point with strongest enhancement—an increase of approximately 70% is observed. This behaviour is notably different from unbounded turbulence (Chouippe and Uhlmann 2015, Fornari et al. 2016, Chouippe and Uhlmann 2019), as well as vertical channel flow (Uhlmann 2008, García-Villalba et al. 2012), where the settling velocity is normally found to be reduced. This discrepancy may be attributed to the substantially higher relative turbulence intensity, anisotropy causing a preferential direction of flow structures, or the

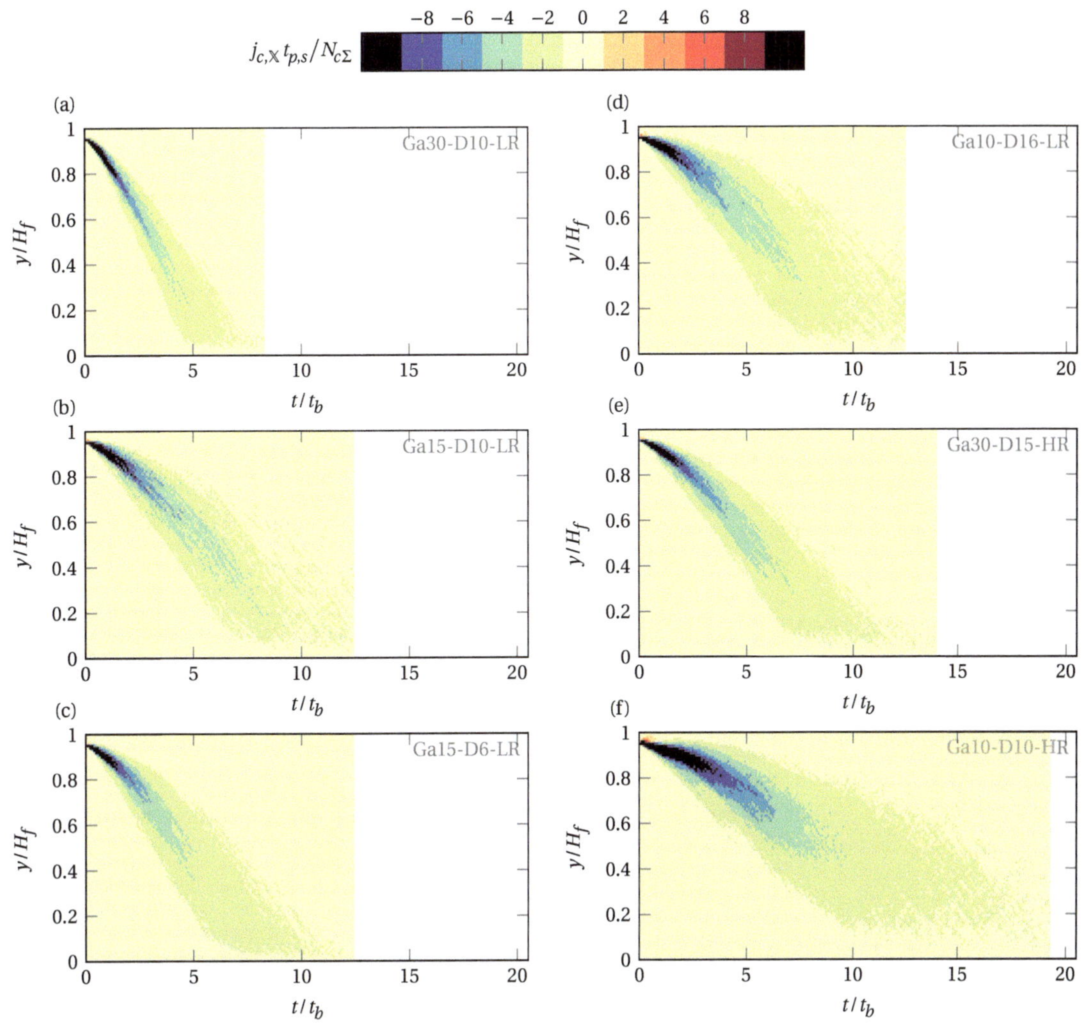

Figure 4.3: The local flux of contaminants (net number of crossings per unit time) as a function of vertical position and time. The various cases are arranged in column-major order: (a) G30-D10-LR, (b) G15-D10-LR, (c) G15-D6-LR, (d) G10-D6-LR, (e) G30-D15-HR and (f) G30-D10-HR.

boundedness of the flow. The finding of a strong enhancement is, however, consistent with the observations made by Cuthbertson and Ervine (2007) and Bragg et al. (2021a) in horizontal channel flows, as will be elaborated in detail at a later point. For all cases under investigation, the settling velocity peak shows the tendency to increase with increasing relative turbulence intensity, and the data points roughly collapse on a single curve for all parameter points investigated. In particular, the effect of ripple bedforms on the settling velocity appears to be captured well by merely considering the modification in u_τ.

While the ensemble-averaged settling velocity is a suitable global indicator for the transient, a more localised quantification may be desirable in order to track the progress. Therefore, the rate at which contaminant particles cross a given wall-parallel plane is defined as

$$j_{c,\mathbb{X}}(y,t) := \frac{1}{\delta t}\int_{t-\delta t/2}^{t+\delta t/2} \left(n_{c,\mathbb{X}\uparrow}(y,t^*) - n_{c,\mathbb{X}\downarrow}(y,t^*)\right)\mathrm{d}t^*, \tag{4.4}$$

where $n_{c,\mathbb{X}\uparrow}(y,t)$ is the number of upwards crossing through such a plane per unit time gathered from all realisations, $n_{c,\mathbb{X}\downarrow}(y,t)$ the corresponding number of downward crossings, and $\delta t \approx 0.088 t_b$ the length of the time interval under consideration. At equilibrium, upward and downward crossings are balanced

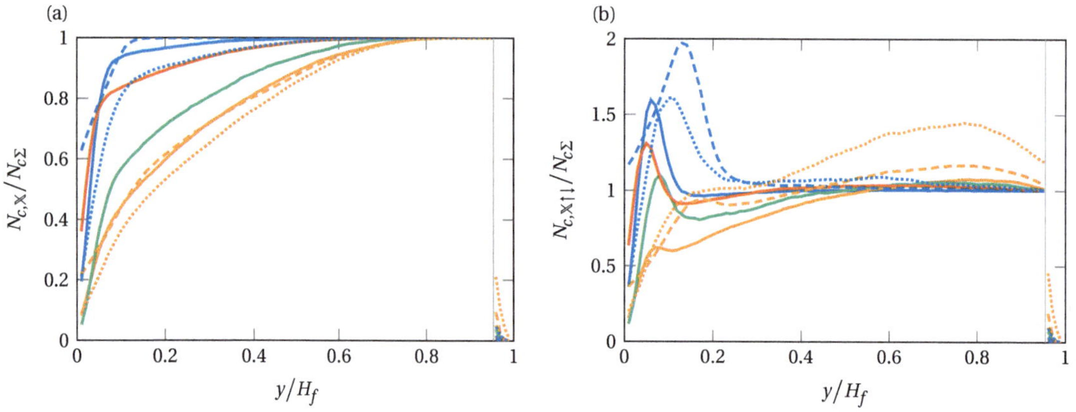

Figure 4.4: The number of contaminant crossings of a given wall-parallel plane as a function of vertical position. The data is normalised by the total number of contaminants, i.e. the number of contaminants per realization times the number of realizations. The vertical grey line denoted the plane of initialization. (a) Only the first crossing of each particle is considered. (b) All upward and downward crossings are considered. Legend: G30-D10-LR (—), G10-D6-LR (—), G15-D10-LR (—), G15-D6-LR (—), G30-D10-RP (- - -), G10-D6-RP (- - -), G30-D15-HR (······), and G10-D10-HR (······).

aside from fluctuations, and thus, $j_{c,\mathbb{X}}$ tends to vanish everywhere. Figure 4.3 displays the temporal and spatial evolution of $j_{c,\mathbb{X}}$, and the inferences on the progress of the transient are in accordance to those made based on the ensemble-averaged settling velocity, with the additional information that the remaining fluctuations are restricted to the lower parts of the channel. Moreover, the data suggests that no significant upward motion across the plane of initialization occurs, with the notable exception of G10-D10-HR where a small upward flux can be perceived at early times due to the low submerged weight.

4.2 Spatial-averaging operator and trajectory crossings

When investigating the time evolution of ensemble-averaged quantities, one has to keep in mind that the contaminants are generally located at various wall-normal positions for a given instance in time, as can be deduced from fig. 4.3, and hence, some information is obfuscated by the averaging procedure. For instance, the rapid decay of $\langle v_c \rangle_{\mathbb{T}}$ after the occurrence of the peak can be attributed to the arrival of fast-paced particles in the near-wall region where they are decelerated. Moreover, spatial information such as the variation in turbulence structure with wall-normal position becomes fuzzy in such a perspective. Therefore, a spatial ensemble-average is introduced in order to investigate position-dependent behaviour with the definition

$$\langle \varphi \rangle_{\mathbb{X}} := \frac{1}{N_{c,\mathbb{X}}} \sum_{i=1}^{N_{c,\mathbb{X}}} \varphi^{(i)}\left(t_{c,\mathbb{X}}^{(i)}\right), \tag{4.5}$$

where $\varphi^{(i)}(t)$ is a placeholder for any quantity attributed to an individual particle (such as its velocity) and $N_{c,\mathbb{X}}(y)$ is the sample size of crossings through a wall-normal plane located at y in all realization at any time. The time at which the i-th crossing occurs is denoted by $t_{c,\mathbb{X}}^{(i)}$, and hence, $\langle \bullet \rangle_{\mathbb{X}}$ ensemble-averages a quantity irrespective of the time at which the particle is located at y.

In order to evaluate eq. (4.5), the cross-through time $t_{c,\mathbb{X}}(y)$ has to be determined from the vertical trajectories of the contaminant particles. In general, $y_c(t)$ is not a bijective function, and thus, the number of crossings does not necessarily correspond to the number of contaminants, i.e. particles may cross a given plane more than once or not at all during the observation time. Therefore, it can be readily acknowledged that averages obtained using this procedure are sensitive to the observation time. The longer the period of observation endures, the more crossings may be attributed to events in the quasi-steady equilibrium rather than the transient phase. This is especially true in the lower parts of the channel where particle concentration is the highest then. In order to avoid this peculiarity, $\langle\bullet\rangle_{\mathbb{X}}$ only considers the first crossing of a given particle through the control plane. This restriction has further advantages which are discussed in §4.4. However, it is occasionally informative (or practical) to loosen this requirement, and whenever a different sample is used it will be denoted by either $\langle\bullet\rangle_{\mathbb{X}\uparrow}$ (all upward crossings), $\langle\bullet\rangle_{\mathbb{X}\downarrow}$ (all downward crossings) or $\langle\bullet\rangle_{\mathbb{X}\uparrow\downarrow}$ (all up- and downward crossings), with $N_{c,\mathbb{X}\uparrow}$, $N_{c,\mathbb{X}\downarrow}$ and $N_{c,\mathbb{X}\uparrow\downarrow}$ denoting the respective sample sizes.[1] In fact, for the inferences made in the present work, $\langle\bullet\rangle_{\mathbb{X}\downarrow}$ only differs little from $\langle\bullet\rangle_{\mathbb{X}}$ as will be demonstrated in §5.2.1.

4.2.1 Number of crossings

The number of crossings provides valuable information on its own. When only the first crossing is considered, $N_{c,\mathbb{X}}/N_{c\Sigma}$ becomes a measure for the sufficiency of the observation time under the assumption that every particle crosses a given plane eventually, which is corroborated by the findings in the subsequent section §4.4. If for a given vertical position its value equals one, then all contaminants have settled through this layer at least once and the transient can be considered as completely captured within the observation time. For $N_{c,\mathbb{X}}/N_{c\Sigma} < 1$, its value provides a quantitative measure for the sufficiency of the observation time. Hence, fig. 4.4(a) implies that the cases G15-D10-LR, G10-D6-LR and G10-D10-HR are somewhat truncated by the finite observation time at vertical positions roughly below $y = \{0.4, 0.5, 0.6\}\, H_f$, respectively, judging by a 90% percentile requirement. Therefore, possible implications of such a truncation need to be discussed.

It seems natural that an incomplete sample is biased towards faster moving particles, as these likely have a higher probability to first cross a given plane within a shorter amount of time.[2] Consequently, the mean of the settling velocity is presumably overestimated and its variance underestimated for positions where $N_{c,\mathbb{X}}/N_{c\Sigma} \ll 1$, and similar ramifications are to be expected for any quantity which is correlated to the settling velocity. Nevertheless, it should be noted that $N_{c,\mathbb{X}}$ close to the sediment bed is also subject to geometric constraints, as $y = 0$ corresponds to the spatio-temporal mean position of the interface and irregularities in the bed will cause particles to be deposited before this position is reached, thus, limiting the validity of this analysis for the near-bed region.

When the total number of crossings is considered, information on the extent of the region of quasi-steady equilibrium can be inferred, see fig. 4.4(b). For the cases with relatively long observation times, a peak in $N_{c,\mathbb{X}\uparrow\downarrow}$ is visible in the vicinity of the sediment bed, whereas $N_{c,\mathbb{X}\uparrow\downarrow} \approx N_{c,\mathbb{X}}$ above. The peak resembles the streamwise flux of sediment particles in shape, cf. fig. 3.5(b), and is presumably correlated to the streamwise flux of contaminants at equilibrium. Furthermore, its location coincides with the sudden drop in $N_{c,\mathbb{X}}$ close to the lower boundary, indicating that resuspension acts as an apparent deposition barrier at this point. The three cases with $Ga \approx 10$ slightly deviate from the behaviour of the remaining

1 The sample size of crossings is linked to the previously defined number density of crossings by means of integration over the entire observation time, e.g. $N_{c,\mathbb{X}\downarrow} = \int_0^{t_{obs}} n_{c,\mathbb{X}\downarrow}\, \mathrm{d}t$.

2 While this statement might appear self-evident at a glance, it really depends on the definition of a "fast" particle and its acceleration history. The correlation between the initial crossing time of a given wall-normal and the corresponding instantaneous settling velocity is examined in §4.3.5.

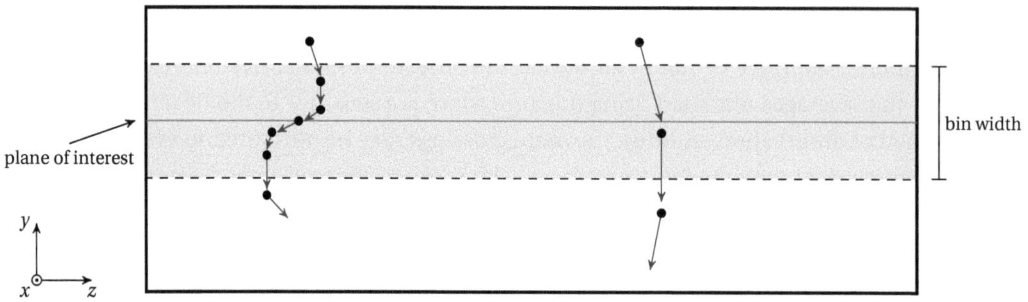

Figure 4.5: Conceptual sketch of the bias in ensemble averaging when plain binning is employed. Slowly settling particles contribute more samples (●) to the average than fast ones, thus causing every ensemble-averaged quantity to be biased towards slow particles. The spatial averaging procedure defined in eq. (4.5) circumvents this problem by only considering the crossings of a two-dimensional plane as samples ("plane of interest").

cases in the sense that a perceivable number of upward crossing can already be observed well above the near bed region. This is likely due to the high relative turbulence intensities in these cases, and is partly concealed by the significant decrease in the total number of crossings in the lower half of the channel and near absence of a near-bed peak. A comparison with fig. 4.4(a) suggests that this is a manifestation of the shorter relative observation time, i.e. the footprint of the transient phase still dominates that of the equilibrium phase for these parameter points.

Both the data for $N_{c,\mathbb{X}}$ and $N_{c,\mathbb{X}\uparrow\downarrow}$ indicate that contaminants rarely move upwards directly after their initial release, and thus, potential non-physical interactions with the upper boundary can safely be neglected. Furthermore, for G10-D10-HR where these events are the most frequent, $N_{c,\mathbb{X}\uparrow\downarrow} \approx 2N_{c,\mathbb{X}}$ which suggests a parabola-like motion with little complexity.

4.2.2 Comparison to binned averaging

The spatial ensemble-average defined in eq. (4.5) is distinct from a plain binned averaging procedure, which is often employed for other flow configurations or, more relevantly, in the relevant experimental study by Cuthbertson and Ervine (2007). This is because binned averaging exhibits a bias with the settling velocity: slowly moving particles spend more time in a given bin than fast moving ones, and thus, for a given sampling frequencies, contribute more samples towards the average, see fig. 4.5 for a conceptual sketch of this phenomenon. In fact, it is demonstrated in §5.2.1 that $\langle \bullet \rangle_{\mathbb{X}\uparrow\downarrow}$ can be approximated by a weighted binned-averaging procedure where the weights correspond to the absolute value of the instantaneous settling velocity, and other variations may be approximated as well in this manner using straightforward adjustments.

In the following, averages will often be presented as a function of wall-normal position. The reader should be aware that these can be interpreted in two different ways. First, since the flow is inhomogeneous in y, variations may be attributed to differences in the local structure of turbulence, e.g. due to differences in the fluid-particle interaction within the buffer layer, logarithmic or wake region. Second, $y_{c,0} - y$ may also be treated as a time-like coordinate for the progress of the transient, and thus, phenomena like acceleration from rest near the plane of initialization or inertia/response time effects are superimposed. While the aim is to separate these effects eventually, a sharp distinction might not always be possible.

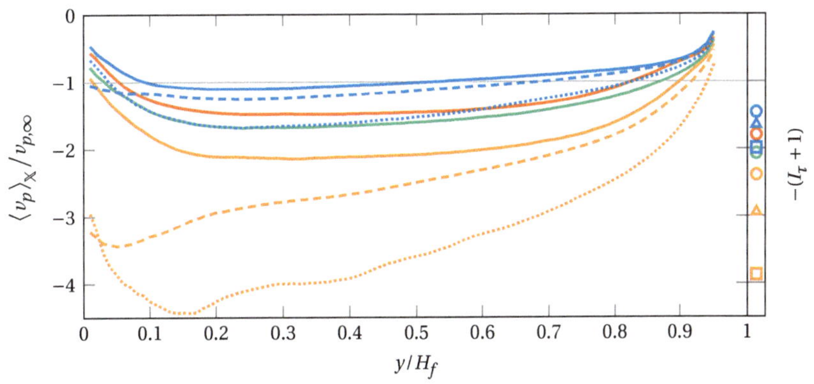

Figure 4.6: Ensemble-averaged vertical component of particle velocity, normalised by the corresponding ambient terminal velocity of an isolated particle, as a function of vertical position. On the right-hand sight, the relative turbulence velocity $I_\tau = u_\tau / v_{p,\infty}$ is indicated for the various cases shown. Legend: G30-D10-LR (—, ○), G10-D6-LR (—, ○), G15-D10-LR (—, ○), G15-D6-LR (—, ○), G30-D10-RP (- - -, △), G10-D6-RP (- - -, △), G30-D15-HR (⋯⋯, □), and G10-D10-HR (⋯⋯, □).

4.3 Instantaneous particle velocity

Now that a spatial averaging procedure has been defined, the question "what is the contaminants' instantaneous velocity at a given vertical position?" is posed. The average particle velocity at position y is given in accordance with eq. (4.5) by

$$\langle \mathbf{u}_c \rangle_{\mathbb{X}} := \frac{1}{N_{c,\mathbb{X}}} \sum_{i=1}^{N_{c,\mathbb{X}}} \mathbf{u}_c^{(i)} \left(t_{c,\mathbb{X}}^{(i)} \right). \tag{4.6}$$

It should be noted that the definition of the sampling procedure gives cause to two peculiarities for sampling the wall-normal component of the particle velocity. First, the velocity cannot be zero because for the crossing of a wall-parallel plane to occur movement is required in its normal direction, and second, the samples are strictly negative for positions below the plane of initialization. While the former is true for all kinds of samples used in this work, the latter only applies to the original definition where solely the first crossing is considered. Furthermore, one has to keep in mind that no precise time can be attributed to $\langle \mathbf{u}_c \rangle_{\mathbb{X}}$ as the arrival times generally differ, which gains relevance with increasing distance of the plane of interest from the plane of initialization.

4.3.1 Mean velocity

When presented in terms of the ambient terminal velocity, it can be observed that the contaminants reach, on average, instantaneous downward velocities which span from approximately unity to values vastly larger than one when normalised by the reference settling velocity, see fig. 4.6. This observation is in accordance with Cuthbertson and Ervine (2007) who experimentally studied a very similar parameter range and likewise reported significant enhancements of the settling velocity for similar configurations. As indicated on the right-hand side of the figure, the excess in velocity[3] is correlated to the relative turbulence intensity, i.e. the higher the value of I_τ, the stronger is the enhancement in settling velocity is, and thus, it can be hypothesised that turbulent fluctuations are the dominant cause of enhancement.

[3] The excess in settling velocity is given by $v_c + v_{p,\infty}$ since the direction of gravity is pointing in negative y-direction, but $v_{p,\infty}$ is positive because it has been defined as the absolute value of the converged settling velocity in the ambient case.

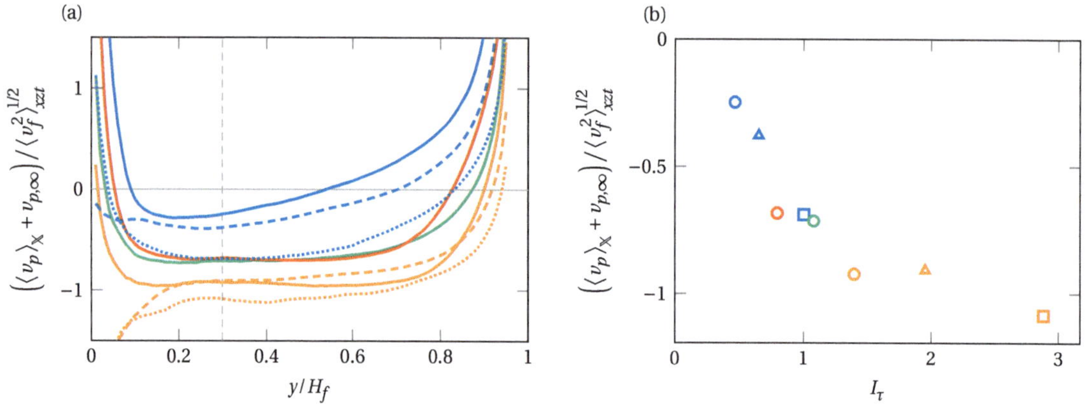

Figure 4.7: Vertical component of the ensemble-averaged particle velocity, normalised by the root-mean-square of vertical fluid velocity fluctuations (a) as a function of vertical position and (b) as a function of relative turbulence intensity at $y = 0.3H_f$. The difference from the terminal velocity under quiescent conditions is shown, where negative(positive) values indicate settling at speeds faster(slower) than $v_{p,\infty}$. Legend: G30-D10-LR (—, ○), G10-D6-LR (—, ○), G15-D10-LR (—, ○), G15-D6-LR (—, ○), G30-D10-RP (- - -, △), G10-D6-RP (- - -, △), G30-D15-HR (······, □), and G10-D10-HR (······, □).

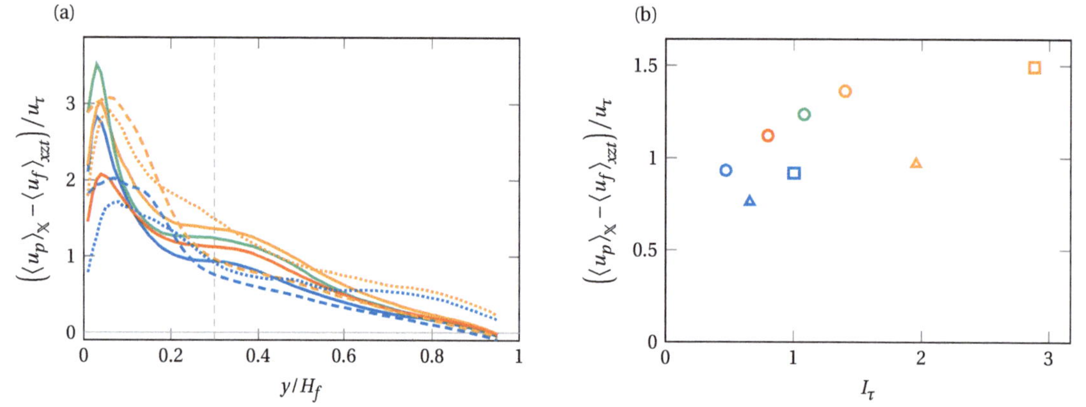

Figure 4.8: Streamwise component of the ensemble-averaged particle velocity normalised by the friction velocity (a) as a function of vertical position and (b) as a function of relative turbulence intensity at $y = 0.3H_f$. Legend: G30-D10-LR (—, ○), G10-D6-LR (—, ○), G15-D10-LR (—, ○), G15-D6-LR (—, ○), G30-D10-RP (- - -, △), G10-D6-RP (- - -, △), G30-D15-HR (······, □), and G10-D10-HR (······, □).

Figure 4.7(a) depicts the same data, albeit now scaled in terms of the root-mean-square of the vertical component of the fluid velocity, which itself is a function of wall-normal position. For all cases apart from G30-D10-LR, a plateau region exists where $\langle v_c \rangle_{\mathbb{X}} / \langle v_f^2 \rangle_{xzt}^{1/2}$ is approximately constant. This indicates that—after an initial phase of adaption—the contaminants are in balance with their turbulent environment, and—through a mechanism to be determined—utilise fluctuations in vertical fluid velocity in order to enhance their mean settling velocity. Furthermore, this mechanism itself becomes more effective as the relative turbulence intensity increases, cf. fig. 4.7(b) where the excess settling velocity in the plateau region is shown with respect to I_τ. In the limit of negligible (relative) turbulence intensity, the enhancement mechanism appears to be inactive, whereas for high values, it can be hypothesised that the effectiveness must be bounded, since the enhancement ought to be constraint by the intensity of the turbulent fluctuations, i.e. it can at most be of the order of the extreme values of v_f'. Evidence for this hypothesis is provided by the apparent convergence seen in fig. 4.7(b) for high values of I_τ. Close to the lower boundary $\langle v_f^2 \rangle_{xzt}^{1/2}$ decreases rapidly, cf. fig. 3.6, and the relationship between vertical fluid fluctuations and settling velocity cannot be sustained, causing a divergence to either large positive or negative values.

The streamwise component of the mean particle velocity exhibits a behaviour which is less universal with respect to the background flow. In particular, the wall-normal evolution does not perceptibly correlate with $\langle u_f'^2 \rangle_{xzt}^{1/2}$ (figure omitted), and the curves suggest a dependence on the flow Reynolds number, as well as on the bedforms even at greater distances from the bed (fig. 4.8). However, for a given background flow, the curves only differ quantitatively by a factor which is approximately constant for a wide range of wall-normal positions. This factor is decidedly a function of relative turbulence intensity, albeit fig. 4.8(b) suggests that other flow dependent parameters need to be considered.

In general, the contaminants travel faster than the mean fluid velocity at the corresponding height. One plausible explanation is a persistent lag in adapting to the local conditions, as the contaminants move from top to bottom, i.e. from regions of high to low average streamwise fluid momentum. Therefore, the particles may keep their momentum attained at higher positions temporarily on their way downwards similar to the fluid momentum transfer in Prandtl's mixing length model. It is apparent that such a lag ought to be proportional to the vertical gradient of $\langle u_f \rangle_{xzt}$, the average settling velocity and the particle response time. Evaluating the product of these parameters yields a curve with a significant peak close to the lower boundary—similar to what can be observed in fig. 4.8(a) at $y \approx 0.05H_f$—but does not explain neither the shape of the curve nor the specific order of the amplitudes for different cases in higher regions (figure omitted). The shift in said order in the near-wall region might suggest that response time effects gain significance very close to the wall where $\langle S_{xy} \rangle_{xzt}$ attains its highest values, as will be further elaborated in §5.2.3.

Another potential mechanism which does not directly depend on the inhomogeneity of the mean flow in vertical direction is a possible bias of the particles to preferentially visit regions of high streamwise momentum while settling. Like for the vertical velocity component, where particles must preferentially visit downward sweeping regions in order to enhance their mean settling velocity from turbulent fluctuations around a zero mean, a similar phenomenon may occur for the streamwise component. In fact, the scaling of both $\langle u_f \rangle_{\mathbb{X}}$ and $\langle v_f \rangle_{\mathbb{X}}$ with relative turbulence intensity possibly suggests that these phenomena might be related, as will be investigated next.

4.3.2 Quadrant analysis

It is a well-established fact in wall-bounded turbulence that fluctuations of the streamwise and vertical fluid velocity are negatively correlated, see e.g. Wallace et al. (1972) or Willmarth and Lu (1972), which results in a negative sign of the tangential Reynolds stress $\langle u_f' v_f' \rangle_{xzt}$. In fact, fluctuations with a negative value of the product $u_f' v_f'$ are about twice as likely to occur than those with a positive value (Pope 2000), and these events are designated by the terms *sweep* ($u_f' > 0$, $v_f' < 0$) and *ejection* ($u_f' < 0$, $v_f' > 0$), cf. §2.2.3. In regard to settling particles, sweep events are particularly interesting, as they combine downwashing events and high streamwise momentum, and thus, are a promising candidate for providing the missing link from the preceding paragraph.

In the following, an analogous analysis is conducted in terms of the particle velocity, however, the fluctuations are in this context defined in terms of deviations from $(\langle u_f \rangle_{xzt}, -v_{p,\infty})$. This enables a direct comparison to the fluid fluctuations, as this formulation likewise corresponds to deviations from the fluid mean while additionally accounting for the inherent gravity-driven velocity difference between the solid and the fluid phase. Figure 4.9 shows the empirical joint probability density function of the velocity of all downward(upward) crossing particles in blue(red), as well as that of fluid fluctuations (black lines). Additionally, a black dashed line indicates the value $v_c = 0$, which bisects the diagram, since no particle crossings can occur at $v_c = 0$, and hence, the joint PDF is required to be zero on this line. This implies

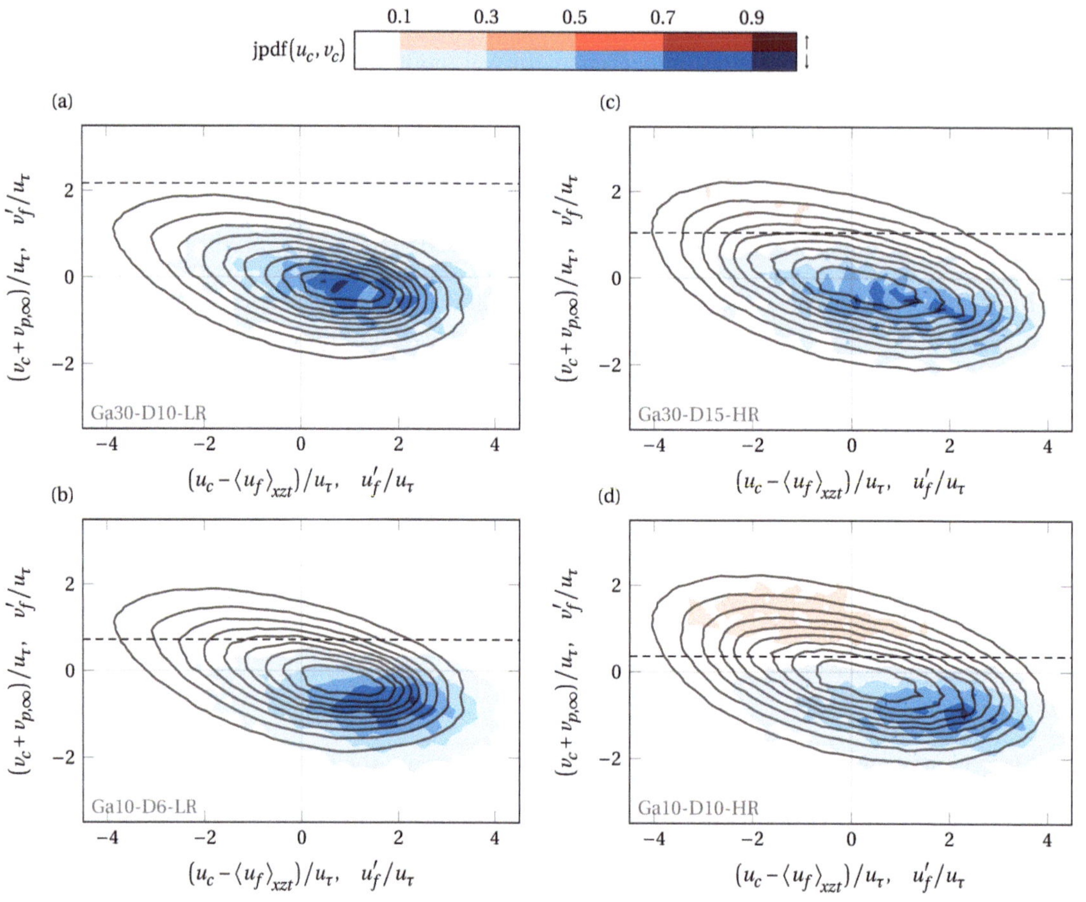

Figure 4.9: Empirical joint probability density function of the streamwise and vertical fluctuations of the particle velocity (shaded regions) and fluid velocity (contour lines) for $y = 0.3H_f$ and all up-/downward crossings. For cases where upward crossings occur, the joint PDF of the particle velocity is bimodal with two regions strictly separated by $v_c = 0$ (- - ·). Upward crossings occur above this bisecting line and are coloured in red, whereas downward crossings occur beneath and are coloured in blue. The colouring corresponds to {0.1, 0.3, 0.5, 0.7, 0.9} times the respective maximum value of the distribution. The joint PDF of fluid fluctuations, viz. jpdf(u_f', v_f'), is superimposed with black contour lines which correspond to [0.1 : 0.1 : 0.9] times the maximum value. The following cases are arranged in column-major order: (a) G30-D10-LR, (b) G10-D6-LR, (c) G30-D15-HR and (d) G30-D10-HR.

that jpdf(u_c, v_c) is bimodal if upward crossings occur. Therefore, the bisecting line defines a strict upper bound for downward crossings to occur, as they require $v_c < 0$ by definition. The opposite is true for upward crossings.

The particle velocity fluctuations for downward crossings predominantly occur in the 4th quadrant ($u_c > \langle u_f \rangle_{xzt}$, $v_c < -v_{p,\infty}$), which comes at no surprise when recalling the earlier findings regarding the mean velocity. What is interesting, however, is that the range of particle velocity fluctuations is approximately restricted by the range of fluid fluctuations, or in other words, the set of likely particle velocity fluctuations seems to be a subset of all possible fluid fluctuations. Although a direct relationship between the local fluid velocity and the particle velocity is yet to be demonstrated, this observation corroborates the hypothesis that the increase in $\langle u_f \rangle_{\mathbb{X}}$ and enhancement of $\langle v_f \rangle_{\mathbb{X}}$ are two manifestations of the same phenomenon and driven by turbulent events rather than a lag in response.

The position of the $v_c = 0$ line provides an appealing heuristic on the subset of fluid fluctuations which act on the downward and upward moving contaminants.[4] As this position is given by $v_{p,\infty}/u_\tau = I_\tau^{-1}$, one can consider the limit behaviours of vanishing and prevailing relative turbulence intensity. For $v_{p,\infty} \gg u_\tau$ the bisector is located well above the space of probable fluid fluctuations, which causes all occurring fluid fluctuations to act on downward moving particles and decreases the likelihood that fluid fluctuations which are intense enough to move a particle upwards occur. Vice versa, for $v_{p,\infty} \ll u_\tau$ the range of fluctuations which can be visited by a downward moving particle is severely restricted, and the probability to encounter a downward fluid fluctuation severely faster than its terminal velocity is rather high. However, this comes at the expense that at the same time the number of upward crossings will increase accordingly. This difference in limit behaviour is well demonstrated by G30-D10-LR and G10-D10-HR shown in fig. 4.9(a,d).

While the effect of relative turbulence intensity on the instantaneous particle velocity conditioned to the crossing direction may be obvious from this perspective, a further inference can be made which is less trivial. Due to the fact that for $v_{p,\infty} \ll u_\tau$ downward moving particles are essentially restricted to quadrants 3 and 4, the probability of encountering $u_f' > 0$ is twice as high as that of encountering $u_f' < 0$ due to the dominance of sweep events. This implies that for a sufficiently high relative turbulence intensity, a statistical bias towards positive streamwise velocity fluctuations exists simply because a particle is moving downwards. Naturally, the opposite is true for upward moving particles. Nevertheless, it should be appreciated that this bias is insufficient in fully explaining the excess in streamwise particle velocity observed, as G30-D10-LR, where effectively no such quadrant restriction occurs, still exhibits a tendency towards positive fluctuations. The boundedness of $u_c - \langle u_f \rangle_{xzt}$ by the range of u_f' still indicates that this phenomenon is driven by turbulence, though its understanding likely requires an examination of the dynamics.

Albeit this heuristic certainly neglects important details—primarily explicit particle-fluid interactions—it provides a convenient way to determine solely from fluid data and apriori known particle attributes whether significant upward particle motion is to be expected and what the approximate magnitude of the average instantaneous velocity might be. Knowledge of both these quantities is essential whenever the overall average settling velocity $\langle v_c \rangle_{\mathbb{X}\uparrow\downarrow}$ is to be estimated for a net flux of particles as it occurs in a transient settling.

4.3.3 Velocity fluctuations

For some applications knowledge of the magnitude of variations in settling velocity around its mean might be valuable. Using the standard deviation of the particle velocity as a measure, viz.

$$\langle u_{c,\alpha}'^2 \rangle_{\mathbb{X}}^{1/2} := \sqrt{\langle u_{c,\alpha}^2 \rangle_{\mathbb{X}} - \langle u_{c,\alpha} \rangle_{\mathbb{X}}^2}, \tag{4.7}$$

with $u_{c,\alpha}$ being a component of the contaminant particle velocity, this variability is presented in fig. 4.10 as a function of vertical position.

For the wall-normal component of the velocity, the spatial trend can again be described well by the standard deviation of the fluid fluctuations, i.e. a normalisation by $\langle v_f'^2 \rangle_{xzt}^{1/2}$ once more yields a pronounced plateau region. In contrast to the mean velocity, the amplitude is now negatively correlated with the

[4] This formulation is not to be taken in the literal sense, since, so far, the only link between the particle velocity and the fluid velocity is that they occupy similar regions in the (u', v')-space.

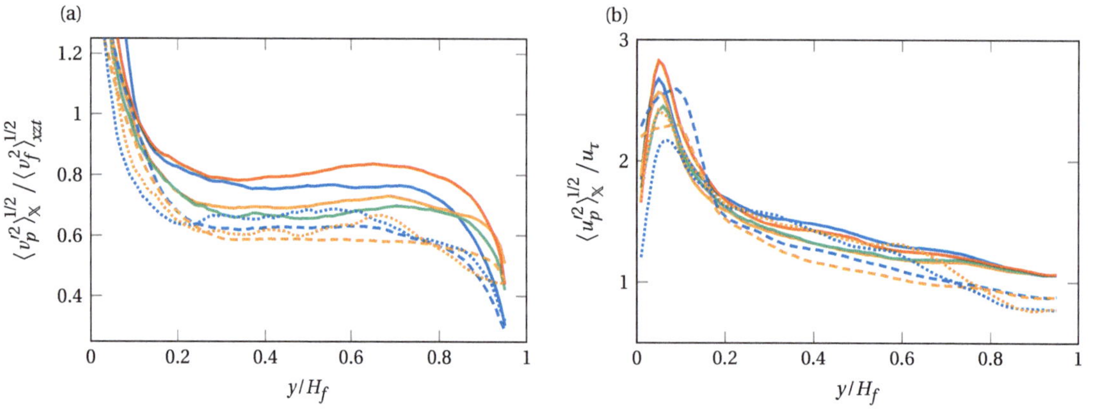

Figure 4.10: Standard deviation of the particle velocity as a function of vertical position. (a) Vertical component normalised by standard deviation of local fluid fluctuations. (b) Streamwise component normalised by friction velocity. Legend: G30-D10-LR (—), G10-D6-LR (—), G15-D10-LR (—), G15-D6-LR (—), G30-D10-RP (- - ·), G10-D6-RP (- - ·), G30-D15-HR (······), and G10-D10-HR (······).

Table 4.1: Approximate parameter ranges of the experiments of Cuthbertson and Ervine (2007) and the numerical experiments conducted within the current work.

	Re_τ	Ga	ρ_c/ρ_f	d_c^+	H_f/d_c	I_τ
Cuthbertson and Ervine (2007)	14000–30000	10–45	2.5	8–20	160–640	0.75–2.25
Current work	241–838	10–30	1.5–2.5	6–18	25–70	0.5–3.0

relative turbulence intensity. While this behaviour perhaps seems counterintuitive at first, it can easily be explained using the heuristic discussed in the context of the quadrant analysis. As previously stated, $I_\tau \gg 1$ restricts the range of fluid fluctuations which can be experienced during downward motion, and thus, the variability is reduced as long as only downward motion is considered or downward motion is dominant due to a net flux of particles.

A negative correlation with turbulence intensity can be observed for the variability of the streamwise component as well. The same argument as for the vertical component can be applied to arrive at the conclusion that increased turbulence intensity decreases the subset of streamwise fluid fluctuations which may be experienced because of the fact that u_f and v_f are correlated. Another observation to be made is that the evolution of the standard deviation with y is akin to that of the mean, cf. fig. 4.8. Thus, if a master curve can be identified for the mean, it likely also serves as a good description of the standard deviation.

4.3.4 Comparison to experiments at high Reynolds numbers

In this paragraph, the data obtained from the present simulations are compared to the experiments of Cuthbertson and Ervine (2007). Despite a considerable difference in the flow Reynolds number—their experiments are conducted around $Re_\tau \approx 2.5 \cdot 10^4$ which is about 30 times as high as that of the HR cases—many similarities exist between their experimental setup and the computational setup employed here. In particular, the experiments are conducted in an open channel over a rough porous bed, settling particles are added at the free surface and the particle parameters are varied within a similar range. A juxtaposition of the fundamental parameters is provided in table 4.1.

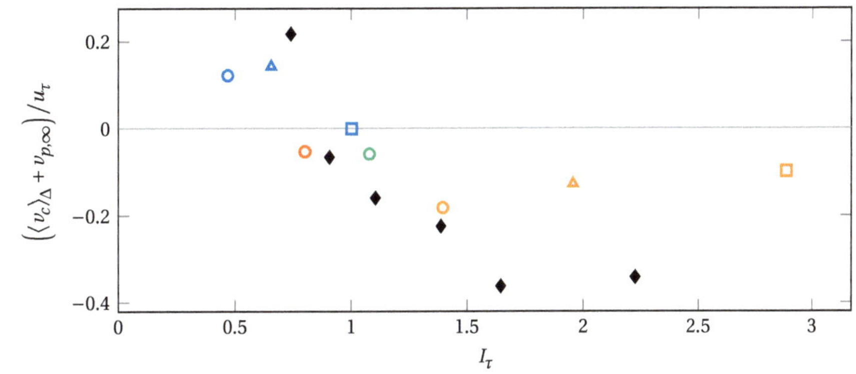

Figure 4.11: Comparison of the average settling velocity obtained from the numerical experiments at $Re_\tau \approx 241$ (○, LR), $Re_\tau \approx 340$ (▲, RP) and $Re_\tau \approx 838$ (□, HR), and the experiments conducted by Cuthbertson and Ervine (2007) at $Re_\tau \approx 25000$ (♦). Contrary to the convention usually employed in this thesis, the average is taken over an ensemble within the finite-width bin $0.2 < y/H_f \leq 0.5$ as is done in Cuthbertson and Ervine (2007). Legend: G30-D10-LR (○), G10-D6-LR (○), G15-D10-LR (○), G15-D6-LR (○), G30-D10-RP (△), G10-D6-RP (△), G30-D15-HR (□), G10-D10-HR (□), and Cuthbertson and Ervine (2007) (♦).

In order to compare their findings to the current data, their figure 5(b), which displays the particle settling velocity averaged over all of their experiments, has been rescaled using the information available in the publication to be presented in terms of the relative turbulence intensity. Although they present averages in three different regions, the comparison is solely made for the region $0.2 < y/H_f \leq 0.5$ as the region above may be subject to a bias caused by differences in settling distance (initial acceleration phase) and the region below is likely sensitive to the specifics of the sediment bed. Moreover, it has to be taken into account that their velocity is obtained by binned averaging, and thus, subject to the bias elaborated in §4.2 and dependent on the observation time. For the purpose of a fair comparison, the numerical data is presented using their definition of the average.

As can be seen in fig. 4.11, the numerical and experimental data are generally in decent agreement, the more so when consideration for the various potential sources of inaccuracies is shown. One conspicuous feature of the data is that the excess in settling velocity appears to peak around $I_\tau \approx 2$. Since the only simulation data point available in this parameter range is one exhibiting ripple features, it is unclear whether or not the observed discrepancy in numerical and experimental data is caused by the presence of bedforms or rather hints at a more substantial inconsistency. There is an argument to be made that a peak in the settling velocity obtained by binned averaging is to be expected: as the turbulence intensity increases, the region where upward and downward crossings are quasi-steady extends further away from the wall to the point that a considerable number of samples in the bin are contributed from the equilibrium phase rather than the transient. Evidence that this effect is indeed active is given by fig. 4.7(b) where only the first crossing, i.e. only the contribution from the transient, is considered and no decline in enhancement can be observed at high values of I_τ. Moreover, fig. 4.4 suggests that the quasi-steady region extends significantly further into the outer flow in the presence of ripples, which reinforces the presumption previously made about the outlier.

Apart from the assurance that the findings from the numerical simulations are in good agreement with the physical world, the comparison discloses that the mechanism which enhances the settling particles' instantaneous velocity is insensitive to variations in the flow Reynolds number over a wide range. It further corroborates the hypothesis that the relative turbulence intensity is the major influencing factor. This realization is highly relevant for the upscaling of the numerical results towards large fluvial systems.

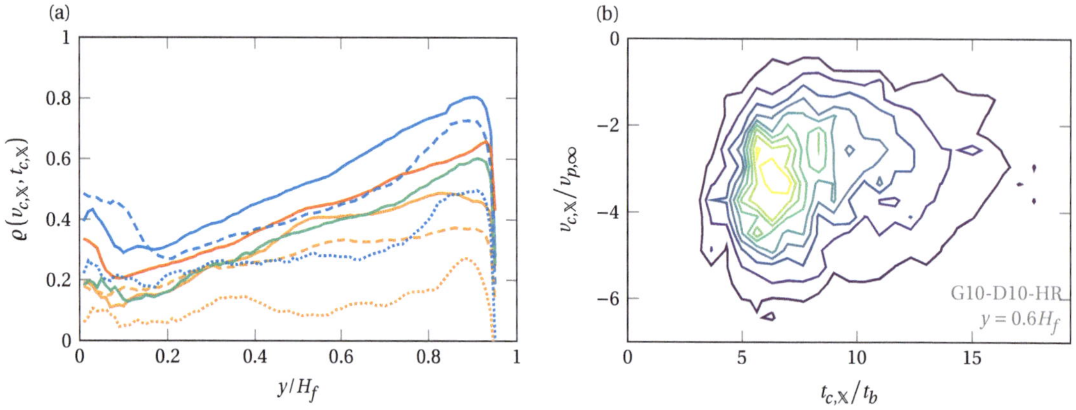

Figure 4.12: Correlation between instantaneous settling velocity and cross-through time. (a) Pearson correlation coefficient as a function of vertical position. (b) Empirical joint probability density function for G10-D10-HR at $y = 0.6H_f$ with contour lines spaced at 0.1 times the maximum value. Legend: G30-D10-LR (—), G10-D6-LR (—), G15-D10-LR (—), G15-D6-LR (—), G30-D10-RP (- - -), G10-D6-RP (- - -), G30-D15-HR (······), and G10-D10-HR (······).

4.3.5 Consequences of truncated observation time

To conclude the discussion on the instantaneous particle velocity, the remark from §4.2 that a truncation of the observation time could possibly lead to a bias in the average towards fast moving particles is revisited. In order for such a bias to exist, the instantaneous particle velocities have to be correlated with the crossing time, or in other words, particles which cross through the control plane at earlier times must also do so generally at higher speeds. A simple measure for (linear) correlation is Pearson's correlation coefficient, which is defined as

$$\varrho\left(v_{c,\mathbb{X}}, t_{c,\mathbb{X}}\right) := \frac{\operatorname{cov}\left(v_{c,\mathbb{X}}, t_{c,\mathbb{X}}\right)}{\operatorname{std}\left(v_{c,\mathbb{X}}\right)\operatorname{std}\left(t_{c,\mathbb{X}}\right)} \tag{4.8}$$

where $\operatorname{cov}\left(v_{c,\mathbb{X}}, t_{c,\mathbb{X}}\right)$ denotes the covariance of the particle velocity and time of crossing. When inspecting this coefficient as a function of vertical position, it can be observed that both quantities are visibly correlated near the plane of initialization, but the correlation is progressively lost with increasing distance to it, see fig. 4.12(a). The intuition behind this is that a greater travelling distance allows for more variations in the acceleration history, i.e. it is more probable that a particle which is instantaneously fast has experienced a phase of slow movement which delayed its cross-through time.

It should be noted that the higher the relative turbulence intensity is, the weaker the correlation is in general. At the same time, the most truncated cases in this work are the ones with high values of I_τ, cf. fig. 4.4. In particular, the most truncated case G10-D10-HR is at the same time the case which has the lowest correlation coefficient. Figure 4.12(b) depicts the joint probability density of $t_{c,\mathbb{X}}$ and $v_{c,\mathbb{X}}$ at $y = 0.6H_f$ for this case. At this position the observation time of G10-D10-HR begins to be significantly truncated. At the same time it can be deduced from the joint PDF that the value of the mean velocity conditioned to late cross-through times is not substantially different from that for early cross-through times. Consequently, no considerable systematic errors are to be expected in the estimation of the mean settling velocity.

In conclusion, while it is indeed true that a truncation in observation time leads to an overestimation of the average settling velocity, this effect is rather unimportant for the current analysis since the factors which benefit truncation are the same which benefit a decorrelation of the cross-through time and instantaneous velocity.

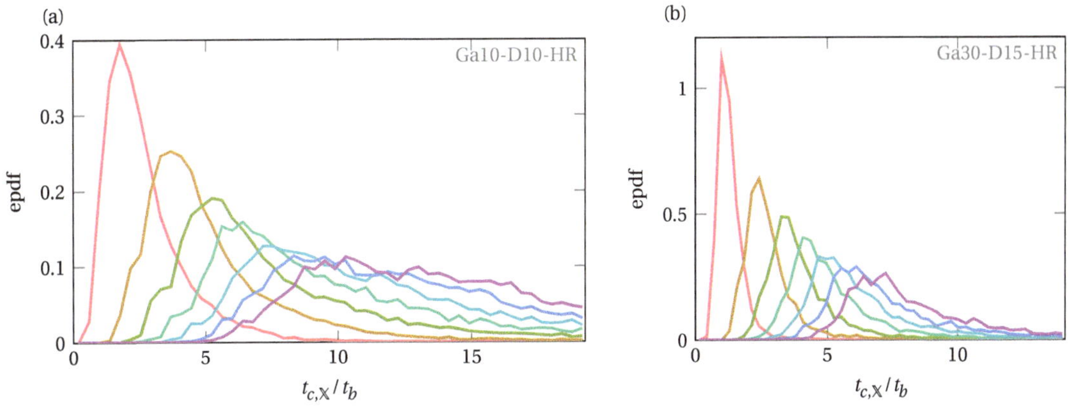

Figure 4.13: Empirical probability density functions of cross-through times at various heights (peaks from left to right: $y/H_f = \{0.9, 0.8, 0.7, 0.6, 0.5, 0.4, 0.3\}$). (a) For G10-D10-HR. (b) For G30-D15-HR.

4.4 Cross-through time

As was emphasised earlier, contaminants cross a given control plane at different times. Since the cross-through time $t_{c,\mathbb{X}}(y)$ can be interpreted as the (incomplete) inverse of the vertical particle trajectory $y_c(t)$, it provides a link between the temporal and spatial perspective on the settling data. Therefore, a detailed analysis of its characteristic seems rewarding.

By means of a histogram, an empirical probability density function, epdf($t_{c,\mathbb{X}}$), is constructed, and fig. 4.13 displays the empirical PDFs for the cross-through times obtained at various wall-normal positions for the two HR cases. As might have been anticipated, the peak of the PDF shifts towards later times and the width of the distribution increases as contaminants advance towards the lower sections of the channel. Furthermore, the difference in characteristic settling time and susceptibility to dispersion is also reflected when comparing the two cases shown. What is striking, however, is that the general shape of the distributions shown varies only little, and hence, it can be hypothesised that they are self-similar.

4.4.1 Self-similarity

The shape of the empirical PDFs is approximated best by the three-parameter generalised Gamma distribution (Stacy 1962), which is defined as

$$\mathrm{pdf}\left(t_{c,\mathbb{X}}\right) := \frac{c b^{a/c}}{\Gamma\left(\frac{a}{c}\right)} t_{c,\mathbb{X}}^{a-1} \exp\left(-b t_{c,\mathbb{X}}^{c}\right) \tag{4.9}$$

where $a, b, c > 0$ are the free parameters of the regression and $\Gamma(s) = \int_0^\infty t^{s-1} e^{-t}\,\mathrm{d}t$ is the gamma function. Eq. (4.9) fulfils the normalisation properties of a probability density function for any valid set of parameters.

The set of free parameters are determined from the empirical PDF for each case and vertical position using the method of least squares. As can be seen in fig. 4.13, some empirical PDFs are visibly truncated due to the finite observation time. Therefore, in general,

$$\int_0^{t_{obs}} \mathrm{epdf}\left(t_{c,\mathbb{X}}\right) \mathrm{d}t_{c,\mathbb{X}} \neq \int_0^{t_{obs}} \mathrm{pdf}\left(t_{c,\mathbb{X}}\right) \mathrm{d}t_{c,\mathbb{X}} = \mathrm{cdf}\left(t_{obs}\right), \tag{4.10}$$

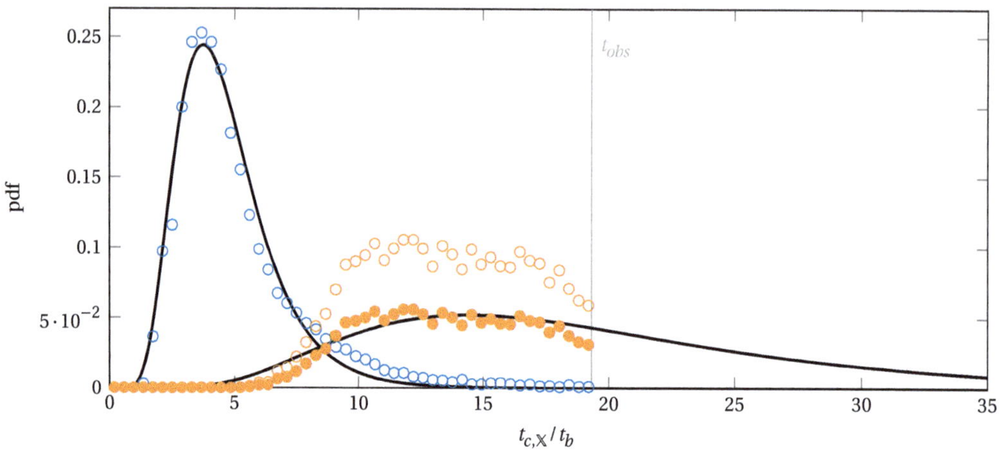

Figure 4.14: Exemplary results for two generalised Gamma distribution fits (—) to the empirical probability density functions (O). The shape fits well for vertical positions close to the plane of initialization (O: $y = 0.8H_f$) as well as for positions closer to the sediment bed (O: $y = 0.2H_f$). The case shown is G10-D10-HR with a strong dispersion in cross-through times where the truncation of the empirical PDF at the observation time (grey line) is clearly visible. The gamma distribution fit enables a reasonable extrapolation of the values. The rescaled empirical PDF as explained in the algorithm is depicted by solid markers (●). Further examples for various cases and at various positions can be found in fig. B.2 in the appendix.

and for this reason the optimization problem has to be formulated in terms of a rescaled empirical PDF. Fortunately, since the data suggests that all contaminants cross a given plane eventually[5] the rescaling factor is known and corresponds to $N_{c,\mathbb{X}} / N_{c\Sigma}$ presented in fig. 4.4. Applying this factor ensures that both the resulting generalised Gamma distribution and the rescaled empirical distribution have an approximately equal value of their cumulative distribution function at the observation time. The curious reader is referred to section §B in the appendix for details on the optimization problem and a summary of the algorithm employed.

Figure 4.14 shows two resulting Gamma distributions fitted at different vertical positions for Ga10-D10-HR in an exemplary manner. The generalised Gamma distribution describes the empirical PDF close to the plane of initialization well without any need of renormalisation (blue, $y = 0.8H_f$). For crossings close to the sediment bed, the observational data is visibly truncated (orange, $y = 0.2H_f$), but may still be described well by the distribution if the empirical data is renormalised (solid markers). Further examples can be found in the appendix in fig. B.2 in conjunction with the residuals for all cases and vertical positions (fig. B.1).

4.4.2 Estimation of required observation time

As has already been noted, the ratio $N_{c,\mathbb{X}} / N_{c\Sigma}$, which is approximately equal to cdf(t_{obs}), provides a measure for the sufficiency of the observation time. If for a given vertical position cdf(t_{obs}) = 1, then all contaminants have settled through this layer at least once and the transient can be considered as completely captured within the observation time. For cdf(t_{obs}) < 1, its value provides a measure for the sufficiency of the observation time. Conversely, given a fit of the Gamma distribution on truncated data, the cumulative density function provides a method to estimate the observation time required to capture the transient within a given accuracy. The reliability of such an estimation, however, is limited by the quality of the empirical data and the uniqueness of the underlying optimization problem.

[5] The integral of the generalised gamma distribution converges to unity, so as long as the assumption that the crossing time follows this distribution holds, all contaminants must ultimately cross a given plane.

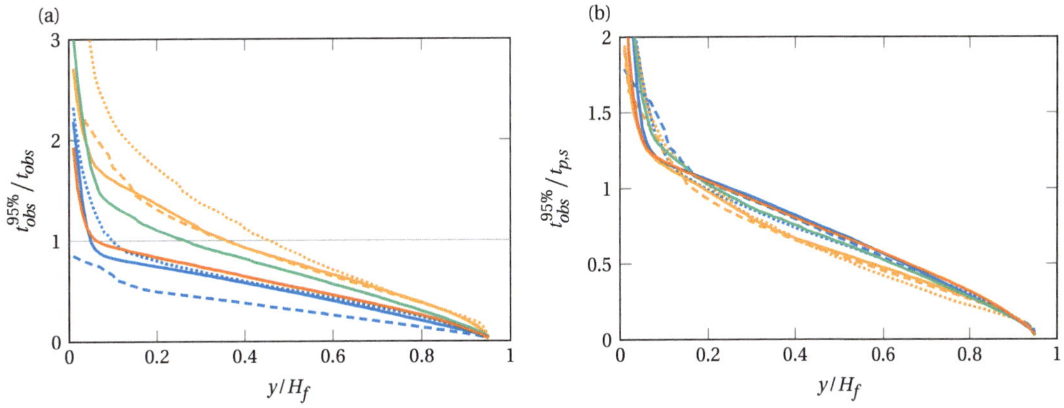

Figure 4.15: Estimation of the required observation time in order to capture 95% of the contaminants' principal crossings. (a) normalised by the actual observation time. (b) normalised by the characteristic settling time. Legend: G30-D10-LR (—), G10-D6-LR (—), G15-D10-LR (—), G15-D6-LR (—), G30-D10-RP (- - -), G10-D6-RP (- - -), G30-D15-HR (······), and G10-D10-HR (······).

A quantitative estimate on the observation time required to capture the transient to a satisfactory degree, here arbitrarily chosen as the 95% percentile, is provided in fig. 4.15. When the data is presented in terms of the actual observation time, it becomes apparent that the required observation time for the truncated cases needs to be 20%–80% longer in order to capture the transient up until the bedload layer. More interestingly, when the required observation time is normalised by the characteristic settling time, the curves for all cases collapse to a reasonable degree, and it is suggested that $t_{p,s}$ is an adequate measure to gauge the required length of the experiment.

4.4.3 Moments of cross-through time distribution

By evaluating the moments of the generalised Gamma functions, the mean and variance of the crossing time distribution data may be estimated even for crossing planes which are truncated by insufficient observation time. This reconstruction requires no assumptions other than that the shape of the data is well described by the generalised Gamma function, for which persuasive evidence from the completed cases exists. Naturally, the quality of the reconstruction is confined by the uncertainty in fit parameters which increases with decreasing amount of empirical data points. The n-th raw moment of the distribution can be calculated explicitly from its parameters by

$$\langle t_{c,\mathbb{X}}^n \rangle_\Gamma = b^{-n/c} \frac{\Gamma\left(\frac{a+n}{c}\right)}{\Gamma\left(\frac{a}{c}\right)}. \tag{4.11}$$

Figure 4.16 shows the (reconstructed) mean of the crossing time and its standard deviation as a function of vertical position. The curves are similar in shape to the 95% percentile (fig. 4.15(b)), which is unsurprising since all these quantities are a direct result of the increasing spread of the Gamma distribution with decreasing height.

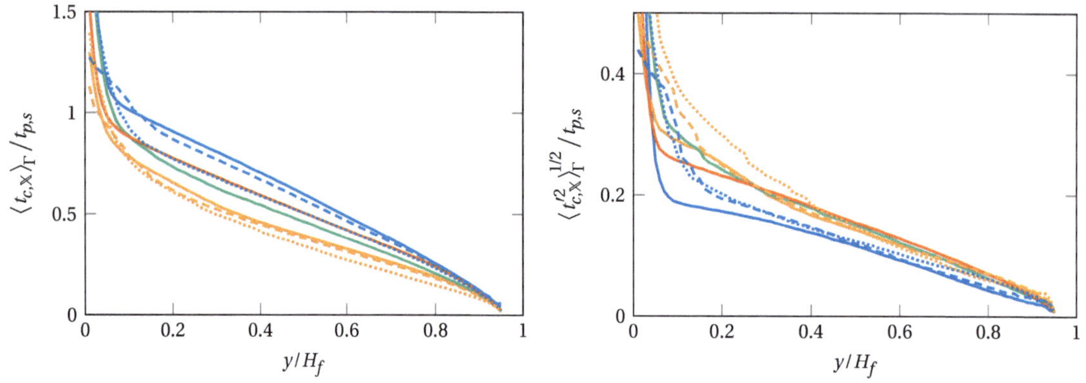

Figure 4.16: The (a) mean and (b) standard deviation of the fitted generalised Gamma distributions of cross-through times as function of vertical position. Legend: G30-D10-LR (—), G10-D6-LR (—), G15-D10-LR (—), G15-D6-LR (—), G30-D10-RP (- - -), G10-D6-RP (- - -), G30-D15-HR (······), and G10-D10-HR (······).

4.5 Group velocity

Based on the preceding analysis of particle cross-through times as a function of vertical position, yet another velocity scale can be defined as

$$\bar{v}_{c,\mathbb{X}}(y) := \frac{y - y_{c,0}}{\langle t_{c,\mathbb{X}} \rangle_\Gamma} \tag{4.12}$$

where $\langle t_{c,\mathbb{X}} \rangle_\Gamma$ is the first raw moment of the generalised Gamma distribution as given by eq. (4.11).[6] This newly defined quantity will be denoted as *group velocity* of the contaminant particles in the following, since it describes the (average) velocity with which the ensemble of particles settles. Although the denomination is motivated by wave phenomena, it should not be understood literally in this sense, but rather in the spirit of a propagation velocity of an envelope, i.e. here in the sense of propagation of the mean of the generalised Gamma distribution. It might be debatable whether the mean or a different measure like the median is of more practical relevance for such a definition, nonetheless, the inferences remain the same. In fact, for all cases and positions, $\mathrm{median}(t_{c,\mathbb{X}}) \approx 0.8 \langle t_{c,\mathbb{X}} \rangle_\Gamma$ is an apposite approximation.

The group velocity is defined as an integral quantity rather than a differential one, i.e. it should be understood as the average velocity of the ensemble from its initial position $y_{c,0}$ to its current position y. In fact, for a fully captured transient, one can show that

$$\bar{v}_{c,\mathbb{X}} = \frac{1}{N_{c\Sigma}} \sum_i^{N_{c\Sigma}} \frac{1}{t_{c,\mathbb{X}}^{(i)}} \int_0^{t_{c,\mathbb{X}}^{(i)}} v_c^{(i)}(t)\, \mathrm{d}t, \tag{4.13}$$

while, in contrast, the spatially-averaged particle settling velocity as introduced in §4.3 reads

$$\langle v_c \rangle_{\mathbb{X}} = \frac{1}{N_{c\Sigma}} \sum_i^{N_{c\Sigma}} v_c^{(i)}(t_{c,\mathbb{X}}). \tag{4.14}$$

[6] Note that $\langle t_{c,\mathbb{X}} \rangle_\Gamma$ is similar to $\langle t_{c,\mathbb{X}} \rangle_{\mathbb{X}}$ if the transient has been fully captured, but differs from it if the observation time was too short. The value of $\langle t_{c,\mathbb{X}} \rangle_\Gamma$ is likely a more accurate estimation of the true value, cf. §4.4.

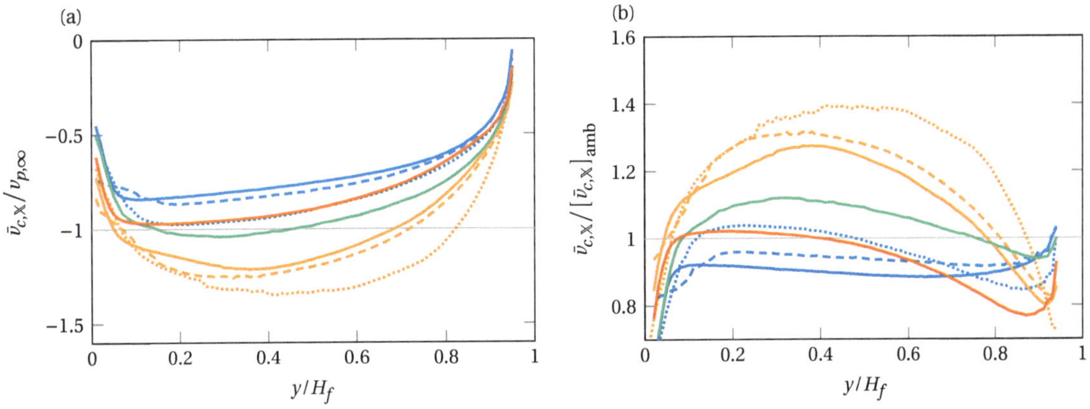

Figure 4.17: Group velocity of contaminant particles as a function of vertical position (a) scaled by the terminal velocity and (b) scaled by the group velocity evaluated for a single particle settling in a quiescent environment. Legend: G30-D10-LR (—), G10-D6-LR (—), G15-D10-LR (—), G15-D6-LR (—), G30-D10-RP (- - ·), G10-D6-RP (- - ·), G30-D15-HR (······), and G10-D10-HR (······).

under the same assumption. Thus, $\bar{v}_{c,\mathbb{X}}$ corresponds to the ensemble-average of the settling velocity time-averaged until the first crossing of a given y-normal plane, whereas $\langle v_c \rangle_{\mathbb{X}}$ is merely the ensemble-average of the settling velocity at the time of crossing. Therefore, the group velocity accounts for the entire history of the particle trajectory—including temporary upward movements or other loitering mechanisms—whereas the spatially-averaged settling velocity is an instantaneous measure.

Figure 4.17(a) shows the group velocity scaled by the corresponding terminal velocity as a function of vertical position. Clearly, the initial acceleration of the ensemble from rest is observable in the upper part of the channel. It then steadily increases with decreasing wall-normal position, but does not reach a plateau within the distance available for settling. Since the value of the group velocity depends on the entire acceleration history since release, comparing it quantitatively to the terminal velocity may be elusive. More appropriately, it should be compared to $\bar{v}_{c,\mathbb{X}}$ evaluated for a single particle settling in a quiescent environment, i.e. the respective values obtained for the accompanying experiments described in §3.7, since this allows for a point-wise, quantitative comparison of the propagation speed under ambient and turbulent conditions. This comparison is shown in fig. 4.17(b). At positions far from the plane of initialization, the values differ only slightly from the simpler normalisation with the terminal velocity. In fact, for long settling distances, $\bar{v}_{c,\mathbb{X}}/v_{p,\infty} \rightarrow \bar{v}_{c,\mathbb{X}}/\left[\bar{v}_{c,\mathbb{X}}\right]_{\text{amb}}$ due to the loss of significance of the acceleration phase in ambient settling. Closer to the initial position, however, a small delay of all cases compared to their ambient counterpart may be observed. It is only for Ga30-D10-LR that this delay persists, and hence, for this case the ensemble of contaminant settles slower than one would expect judging from ambient conditions.

Given that $\bar{v}_{c,\mathbb{X}}/v_{p,\infty}$, and thus also $\bar{v}_{c,\mathbb{X}}/[\bar{v}_{c,\mathbb{X}}]_{\text{amb}}$ at a reasonable distance, is universal to some extent, it is worthwhile to investigate the variation of its scaling factor. In fig. 4.18, the value of $\bar{v}_{c,\mathbb{X}}/[\bar{v}_{c,\mathbb{X}}]_{\text{amb}}$ at $y = 0.3H_f$ is presented as a function of the relative turbulence intensity, which has been identified as the predominant scaling parameter.[7] As turbulent fluctuations become more intense with respect to the particles' terminal velocity, the ensemble tends to settle faster. Interestingly, the transition from delayed to enhanced settling appears to take place around $I_\tau \approx 1$. However, due to the limitation in data, the functional dependency of the group velocity on I_τ is not clear. If only $Re_b \approx 3000$ (LR, RP) is considered, a linear dependency seems to be appropriate. Yet, if data from both flow Reynolds numbers is considered, the enhancement appears to saturate for high relative turbulence intensities. It can be reasoned

7 It may be noteworthy that no trend can be observed when the data is presented in terms of the Stokes number.

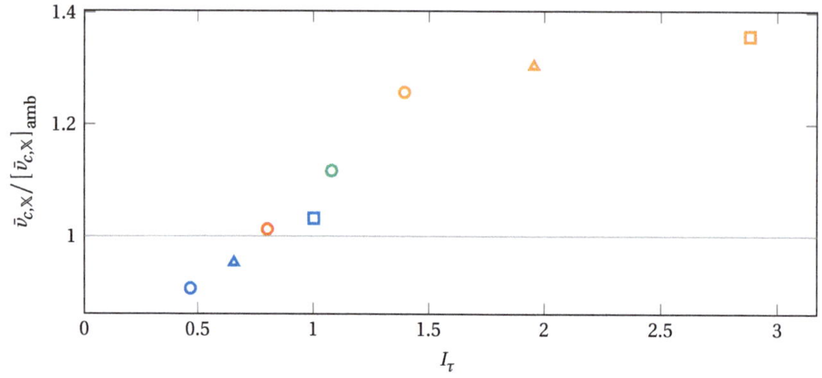

Figure 4.18: Group velocity of contaminant particles as a function of relative turbulence intensity at $y = 0.3H_f$ normalised by its respective value in a quiescent environment. Legend: G30-D10-LR (○), G10-D6-LR (○), G15-D10-LR (○), G15-D6-LR (○), G30-D10-RP (△), G10-D6-RP (△), G30-D15-HR (□), and G10-D10-HR (□).

that for a vanishing turbulence intensity, $\bar{v}_{c,\mathbb{X}} / [\bar{v}_{c,\mathbb{X}}]_{amb}$ should tend to unity under the assumption of negligible collective effects, which reinforces the speculation on a nonlinear behaviour. So far it remains unclear whether the relative turbulence intensity is sufficient to accurately parametrise the group velocity, or other major parameters need to be accounted for, especially in regard to the dependency on flow Reynolds number.

4.6 Particle transport after the initial settling phase

The focus is now shifted towards the phase of quasi-steady deposition and resuspension which is assumed to occur after the initial settling phase. Figure 4.19(a) presents the long-term temporal evolution of the mean vertical particle position for the four cases for which a realisation with longer observation time is available, cf. table 3.3. A distinct change in behaviour is observed at $t \approx 1\text{–}2t_{p,s}$ in accordance with the inferences made in §4.4.2, marking the end of the settling phase. However, a downward trend of $\langle y_c \rangle_{\mathbb{T}}$ is still observed even for $t \gg t_{p,s}$ which indicates that the particle transport has not yet entered the quasi-steady phase. In fact, the sustained downward migration trend perceived for all cases under investigation suggests a phase were contaminant particles are entrained into the sediment bed. For the cases without sediment bedforms, this trend appears to be rather monotonous, and a more pronounced entrainment is observed for G10-D6-LR compared to Ga30-D10-LR, which may be due to the difference in particle size. In the presence of ripples, the mean vertical position oscillates in time while still maintaining a net downward movement. Albeit the amplitude of these oscillations is more pronounced for G10-D6-RP than for G30-D10-RP, the effectiveness of entrainment process does not appear to differ considerably between the two parameter points as opposed to the observations which have just been made for macroscopically flat beds. The local solid volume fraction defined in eq. (3.40) averaged over the time interval $t \in [50t_b, t_{obs}]$ is shown in fig. 4.19(b). The results regarding the cases with $Ga \approx 30$ are unsurprising since their parameter points are matched with those of the sediment particles, and accordingly the evolution of the solid volume fraction approximately overlap with that for the sediment. For $Ga \approx 10$, the profile extends significantly further into the bulk of the channel, which is expected due to the higher value of the relative turbulence intensity. A comparison with the trend predicted by the Rouse formula (not depicted) suggests that the spatial evolution is not significantly altered at a distance from the interface. Yet, due to the vagueness of the virtual origin to be chosen, it is unclear whether the larger sediment particles affect the profile close to the interface where the curve largely overlaps with that for the sediment.

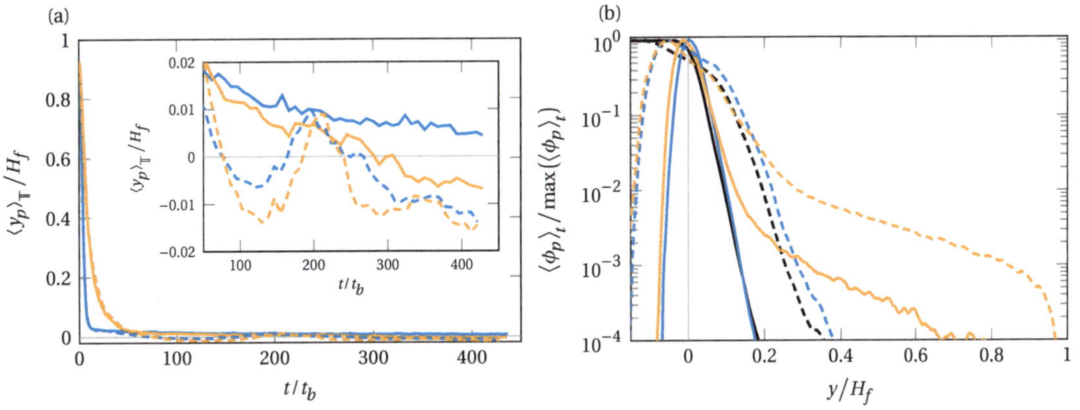

Figure 4.19: (a) Mean vertical position of the contaminants as a function of time, similar to fig. 4.1 (a), but displayed for a considerably longer observation time which exceeds the initial transient. The inset shows a zoom into the quasi-steady regime. (b) Local solid volume fraction as defined in eq. (3.40) evaluated exclusively for contaminant particles (coloured lines) or sediment particles (black lines). A temporal average is shown over the time interval $t \in [50t_b, t_{obs}]$ which excludes the initial settling phase. Regarding the sediment particles, the volume fraction is shown in absence of contaminants, as the modification was found to be insignificant. Legend: G30-D10-LR (—), G10-D6-LR (—), G30-D10-RP (- - -), and G10-D6-RP (- - -).

In regard to the propagation of contaminants, it is interesting to compare the streamwise particle flux in the initial settling phase to that due to bedload transport. For this purpose, the overall particle flux is defined as

$$Q_{p,x}(t) := \frac{1}{H_f} \int_{-H_b}^{H_f} q_{p,x} \, \mathrm{d}y, \tag{4.15}$$

where $q_{p,x}(y,t)$ denotes the particle flux density introduced in eq. (3.41). As fig. 4.20(a) suggests, the particle flux is up to two orders of magnitude higher in the settling phase than in the bedload transport phase, and decreases steadily as the contaminants approach the bed. This is easily explained by the mean streamwise fluid velocity experienced by the contaminants which decreases with increasing progress of the transient. During the settling phase, the influence of bedforms on the transport rate appears to be marginal. The same is true for the monodisperse parameter points ($Ga \approx 30$) in the bedload phase, yet, a significantly enhanced particle flux is observed for bidisperse contaminant transport ($Ga \approx 10$) in the presence of ripples. Although the particle flux oscillates similar to what has been observed for the vertical particle position, the rate is on average two times as high as that over the macroscopically flat bed. When looking at the vertically-resolved contributions to the streamwise particle flux displayed in fig. 4.20(b), it can be seen that the presence of ripple features significantly shifts the distribution of streamwise particle momentum towards greater heights despite the low local volume fractions previously observed away from the bed. This suggests that rather few particles significantly contribute to the streamwise propagation of the contamination, namely those which are located within the sediment transport layer or above, and that the ramp-up effect of ripples might help eject contaminants with a higher Shields number into the wake region of the flow where significantly higher streamwise momentum can be attained. In the monodisperse case, the contaminant transport is equivalent to that of the sediment particles as might have been anticipated, and the observed shift in the peak location is merely caused by the increased number of particles present.

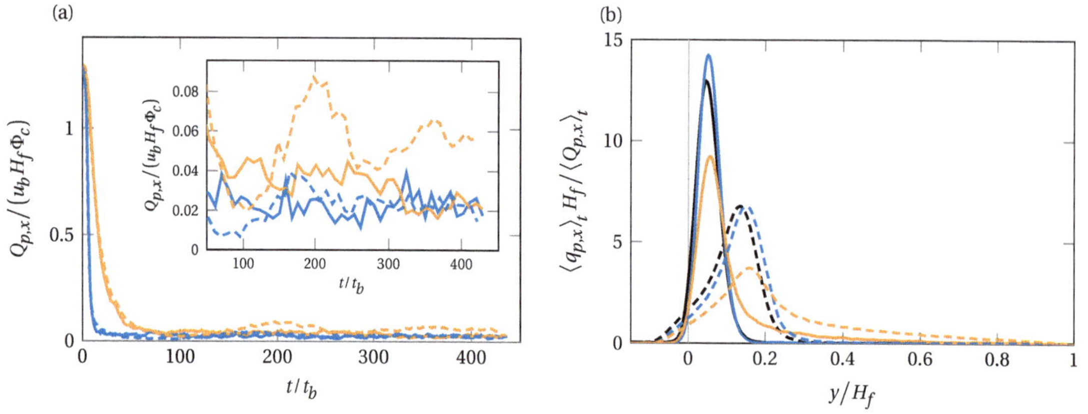

Figure 4.20: (a) Streamwise contaminant flux as defined in eq. (4.15) as a function of time. The flux is scaled in terms of outer units and the global solid volume fraction which are consistent among the cases displayed and allow for an immediate comparison. The inset shows a zoom into the quasi-steady regime. (b) Flux density of the streamwise particle flux evaluated exclusively for contaminant particles (coloured lines) or sediment particles (black lines). A temporal average is shown over the time interval $t \in [50t_b, t_{obs}]$ which excludes the initial settling phase. For the sediment particles, the flux density is shown in absence of contaminants, as the modification was found to be insignificant. Legend: G30-D10-LR (—), G10-D6-LR (—), G30-D10-RP (- - ·), and G10-D6-RP (- -).

4.7 Summary and concluding remarks

In this chapter, the phenomenology of the transient settling of contaminant particles in a sediment-laden turbulent open-channel flow is presented in regard to the statistics of the solid phase. The simulations are conducted either over a macroscopically flat bed (LR,HR) or in the presence of ripple features (RP) at three distinct values of the friction Reynolds number, namely $Re_\tau = \{241, 340, 838\}$ for LR, RP and HR, respectively. In total, eight distinct parameter points for the contaminants have been investigated and compared in regard to their settling velocity and streamwise transport behaviour.

The instantaneous velocity of the contaminants is investigated using an averaging procedure based on crossings of a control plane which avoids an averaging bias towards slowly settling particles a binned averaging procedure would exhibit. It is found that the instantaneous settling velocity is generally higher than its counterpart in a isolated quiescent environment. The excess in settling velocity correlates well with the standard deviation of the vertical fluid velocity for all cases, and the effectiveness of this mechanism is itself dependent on the relative turbulence intensity. No significant dependency on the flow Reynolds number or bedforms could be detected. Regarding the streamwise component of particle velocity, a lead was found in regard to the corresponding mean fluid velocity. In contrast to the settling velocity, the magnitude of this excess is not solely a function of relative turbulence intensity, but also at least Reynolds number dependent. A combined quadrant analysis of fluctuations in fluid and particle velocities suggests that the enhancement in particle velocity is driven by sweep events. The relative turbulence intensity divides the (u', v')-space into regions of upward and downward crossings, and this distinction alone partly explains the prevalence of positive fluctuations in u_p' provided that fluid and particle velocities are instantaneously and locally correlated, which is yet to be demonstrated. Trustworthiness of the results of the numerical simulations is assured by a comparison with the akin experiments of Cuthbertson and Ervine (2007) which yielded a good agreement. Furthermore, as these experiments were conducted at a much higher value of the Reynolds number, confidence is gained that the excess in instantaneous settling velocity is indeed independent of the Reynolds number.

By assessing the number of crossings which occur during the transient through a given control plane, it is shown to be possible to estimate the sufficiency of the observation time of the numerical experiments.

An evaluation of the correlation between cross-through time and instantaneous settling velocity yields that, although the transient is incompletely captured for some parameter points, the analysis on the instantaneous particle velocity stands as is. Moreover, it is found that the distribution of cross-through times follows a generalised Gamma distribution and this knowledge alone is sufficient to estimate the observation time required to capture a given percentage of the full transient. It also enables the reconstruction of quantities which would otherwise be biased by a truncated transient, first and foremost the mean of the cross-through time distribution.

Using the combination of the local distance to the plane of initialization and the reconstructed mean cross-through time, the group velocity of the contaminants is defined which allows a comparison of the propagation speed of the ensemble of particles to its ambient reference. This comparison yields that for $I_\tau \gtrapprox 1$ the group velocity is enhanced, albeit to a much smaller degree than the instantaneous velocity. The same is true for the temporal peak of the ensemble-averaged contaminant settling velocity at a given time, which can also be understood as a measure of the global vertical contaminant flux. The group velocity enables a spatially-resolved view on the net settling velocity, whereas the ensemble-averaged velocity disregards any inhomogeneities in the instantaneous vertical distribution of contaminants. It was ascertained that the presence of bedforms does not perceptibly alter the settling velocity if the induced modification in the wall shear stress is accounted for.

The transport of contaminants beyond the initial settling phase has been investigated for $Ga \approx \{10, 30\}$ in the presence and absence of ripples, as for these cases substantially longer time series are available. Within approximately $500 t_b$, no equilibrium behaviour was achieved, as the contaminants appear to get progressively entrained into the sediment bed. In the absence of ripples, the entrainment is monotonous and its rate depends on the contaminant parameter point. In the presence of ripples, the mean vertical particle position exhibits cycles of downward and upward movement, and the average rate of entrainment does not vary substantially between the two cases investigated. The streamwise flux of contaminant particles is similar to the sediment flux for all cases with the notable exception of the bidisperse transport over ripples (G10-D6-RP), where the contaminant flux oscillates in time similar to the mean vertical position and attains approximately twice the rate of sediment transport.

To conclude, this chapter has shown that many phenomena associated to contaminant settling appear to be driven by fluid-particle interactions, and thus, the subsequent chapters will focus on different aspects of this interaction. At first, it will be necessary to establish a direct correlation between fluid and particle velocity fluctuations, since so far such a relationship is only hinted at by the overlap of fluctuation statistics of both phases. Another aspect to be further investigated is the distinction of u_c' and v_c' in terms of Reynolds number dependence. Moreover, the interaction between contaminant and sediment particles during the entrainment phase deserves further attention, especially in regard to the cyclic behaviour exhibited in the presence of ripples.

5 Preferential sampling of the fluid flow

In the preceding chapter it has been assessed that the contaminant particles' vertical velocity varies substantially depending on the relative turbulence intensity. The reported scaling behaviours, and the fact that fluctuations of the particle velocity are presumably linked to fluid velocity fluctuations and in particular sweep events, suggest that an enhancement of the settling velocity is caused by interactions with turbulent structures. The purpose of the current chapter is therefore to examine these interactions, and more specifically, to assess the role of the preferential sweeping mechanism during the settling phase, cf. §2.3.3. Moreover, since the lead in streamwise velocity by the contaminants cannot solely be explained by a statistical bias, the effect of preferential sampling of intense high-momentum regions is addressed. As the impact of bedforms on the settling velocity has been deemed insignificant, the following analysis will predominantly focus on the settling over macroscopically flat beds, i.e. on the LR and HR background flows.

5.1 Secondary motion and particle velocity

To begin with, the analysis from §4.3 concerning the localised particle velocity based on crossings of wall-parallel planes is revisited. This time, however, the plane is further tessellated in the spanwise direction such that the number of crossings and the spatial-averaged velocity are now also functions of the spanwise position, i.e. $N_{c,\mathbb{X}} = N_{c,\mathbb{X}}(y,z)$ and $\langle \mathbf{u}_c \rangle_{\mathbb{X}} = \langle \mathbf{u}_c \rangle_{\mathbb{X}}(y,z)$ for the remainder of this section. Furthermore, the mean secondary motion of the channel flow is introduced and defined as the streamwise-temporal average of the cross-sectional fluid velocity, viz. $\langle v_f \rangle_{xt}(y,z)$ and $\langle w_f \rangle_{xt}(y,z)$. Although both $\langle v_f \rangle_{xt}$ and $\langle w_f \rangle_{xt}$ vanish for sufficiently long domains or sufficiently long observation times, it is a well-established phenomena that in finite domains and observation times secondary flow patterns can be observed (e.g. Scherer (2022)). Moreover, regions for which $\langle v_f \rangle_{xt} < 0$ are associated to so-called *high-momentum pathways* exhibiting $\langle u_f' \rangle_{xt}(y,z) > 0$, which are complemented by *low-momentum pathways* for which the opposite is true. Momentum pathways can be understood as the statistical footprint of the large-scale sweeps and ejections described in §2.2.3, whereas the secondary flow is associated to the roller structures connecting them. In the following, the time interval for averaging is taken to be the observation time of the settling simulations, which is typically of $\mathcal{O}(10t_b)$, and thus sufficiently short to observe the secondary motion induced by large streamwise-aligned coherent structures. For the sake of brevity, the analysis focuses on the case G10-D10-HR, yet, all observations are qualitatively similar for the other cases.

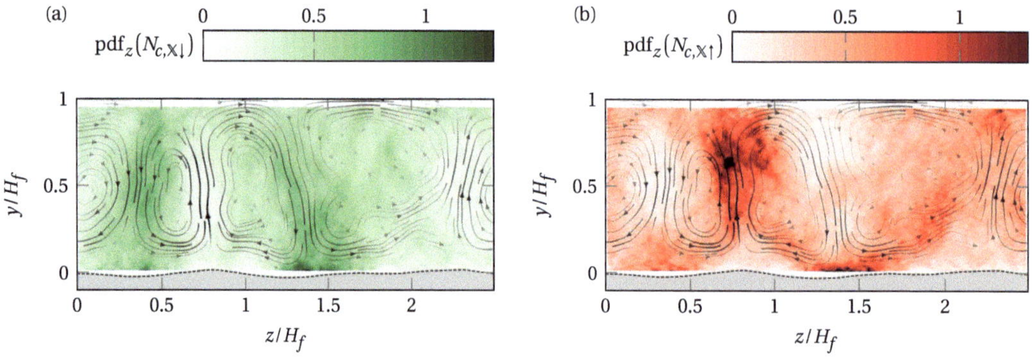

Figure 5.1: Histograms of the number of crossings per spanwise unit, presented in the form of one probability density function per crossing plane. The secondary fluid motion in the yz-plane, averaged over the entire observation time and streamwise direction, is shown by arrows whose thickness and opacity is proportional to the local magnitude of the secondary motion, viz. $\sqrt{\langle v_f\rangle_{xt}^2 + \langle w_f\rangle_{xt}^2}$. The grey area at the bottom of the domain shows the spanwise-resolved average height of the sediment bed, $\langle h_b\rangle_{xt}$. The case G10-D10-HR is shown. (a) Only downward crossings. (b) Only upward crossings.

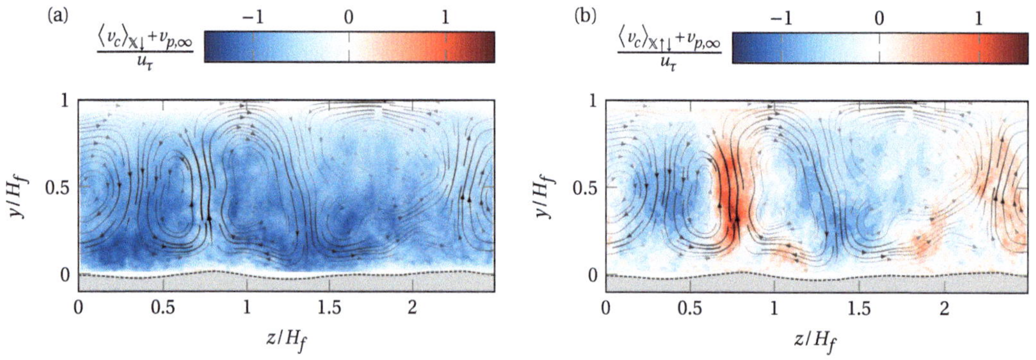

Figure 5.2: Contaminant settling velocity averaged over all crossings of the corresponding y-normal plane within spanwise bins. Secondary fluid motion is indicated by arrows as described in fig. 5.1. The case G10-D10-HR is shown. (a) Averaged over all local downward crossings. (b) Averaged over all local up- and downward crossings.

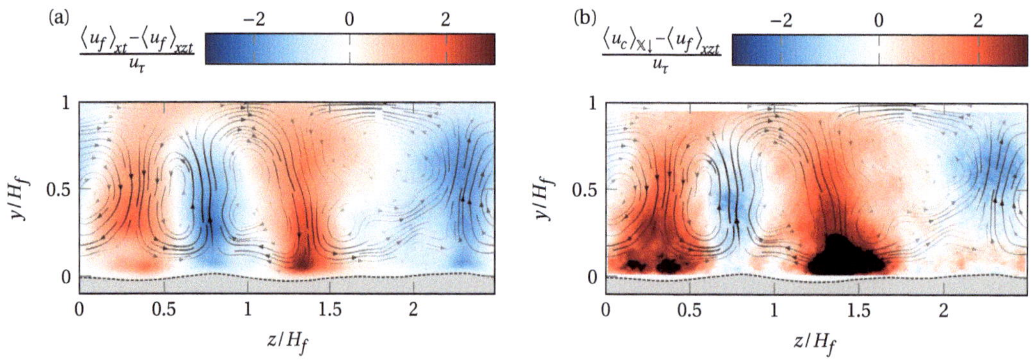

Figure 5.3: Streamwise component of (a) fluid velocity and (b) contaminant velocity at crossing averaged within spanwise bins and presented as differences from the overall mean streamwise fluid velocity. Secondary fluid motion is indicated by arrows as described in fig. 5.1. The case G10-D10-HR is shown.

The spanwise-resolved number of crossings is presented in form of the probability density of the number of crossings for each crossing plane, viz.

$$\mathrm{pdf}_z(N_{c,\mathbb{X}}) := \frac{N_{c,\mathbb{X}}(y,z)\,H_f}{\int_0^{L_z} N_{c,\mathbb{X}}(y,z)\,\mathrm{d}z}, \tag{5.1}$$

which is again a function of vertical and streamwise position and effectively normalises $N_{c,\mathbb{X}}(y,z)$ by the total number of crossings detected for each vertical position, cf. fig. 4.4. A juxtaposition of this probability density and the secondary motion is given in fig. 5.1 for both downward and upward crossings, separately. Under both conditions, the probability of crossings is inhomogeneous in lateral direction. Furthermore, the probable regions of downward or upward crossings are complementary, i.e. a region of high probability of downward crossings is at the same time a region of low probability of upward crossings, and vice versa. The only exception to this observation is found close to the sediment bed, where both regions predominantly overlap.

When comparing the crossing probabilities to the secondary motion, a clear correlation can be detected: whenever the average cross-sectional fluid velocity is pointing upwards, fewer downward crossings occur and the probability of upward crossings is increased. The opposite is true for regions of downward fluid motion. While it is certainly conceivable that the secondary motion is actually induced by the settling contaminants themselves, this mechanism can quickly be discarded by comparing the secondary flow pattern to a numerical simulation of the same setup in absence of contaminants, which differs only little quantitatively (figure omitted). As a consequence, it is likely that the lateral organization of contaminants has its root in the same phenomenon which induces the secondary flow, and thus, it can be hypothesised that particles preferentially sweep turbulent structures on a large scale.

The correlation between the secondary flow and contaminant distribution also affects the local average particle velocity. Figure 5.2 displays the spanwise-resolved excess in settling velocity of the contaminants averaged over all downward crossings, as well as averaged over all up- and downward crossings. For the downward-conditioned average, a strong enhancement in settling velocity is typically observed in fluid downdrafts, whereas in updrafts the settling velocity is significantly slower in comparison. It should be noted, however, that the settling velocity is still enhanced on average even in regions of mean upward fluid motion, which indicates that the secondary flow is not sufficient to explain the enhancement. This conclusion is supported by the results obtained for the average over all up- and downward crossings. While here the correlation between the settling velocity and secondary motion is especially obvious, as due to the inclusion of upward moving particles the average vertical velocity in updrafts is significantly shifted towards high positive particle velocities, the magnitude of enhancement is inexplicable by the large-scale motion alone since the variation in excess settling velocity is at least a factor two higher than that of the mean vertical fluid velocity in terms of range as well as standard deviation (not shown in the figure). Potential causes for this deviation could be inertial or collective effects for the contaminant particles, but also the disregard of fluid motion on smaller—and thus shorter—scales.

In terms of the streamwise component, the discrepancy in velocity magnitude between the mean fluid and contaminant velocity is much less significant for the most part, as shown in fig. 5.3. Here, the velocities are in decent agreement with the only notable exception being the lower quarter of the clear channel, where the solid phase generally possesses higher streamwise velocities than the fluid. The conformity of this region with the occurrence of increased mean shear indicates that this discrepancy may be explained by a delayed response of the contaminants to the rapidly changing mean streamwise fluid momentum, cf. fig.3.6(b).

Although the detailed mechanisms still need to be evaluated, both fig. 5.2 and fig. 5.3 suggest that the enhancement of the contaminants' vertical and streamwise components of momentum is indeed predominantly a result of preferential sampling of specific flow regions, as has been first hypothesised in §4.3.1.

5.2 Particle-conditioned flow field

The juxtaposition of the secondary flow and the particle velocity gave a first hint on how the contaminants are redistributed within the flow and which mechanisms could potentially explain the enhancement in contaminant velocity. However, this perspective does not take into account the instantaneous and largely local nature of particle-fluid interactions, but rather correlates two quantities which have been independently collapsed in space and time. Thus direct evidence for a connection between the local fluid surrounding and the particle velocity is yet to be provided.

5.2.1 Particle-conditioned averaging operator

A new conditional averaging procedure is now defined to evaluate the three-dimensional flow field in the vicinity of the contaminants averaged at the time of crossing a given wall-parallel plane. In order to achieve this, the field is first translated such that the origin of the new coordinate system is centred at the centroid of the particle to be sampled, and subsequently the spatial-averaging procedure introduced in §4.2 is applied. Hence, the particle-conditioned averaging operator reads

$$\langle u_{f,\alpha} \rangle_{\mathbb{PX}}(\tilde{\mathbf{x}}, y) := \frac{\langle u_{f,\alpha}(\mathbf{x}-\mathbf{x}_c, t)\, \chi_f(\mathbf{x}-\mathbf{x}_c, t) \rangle_{\mathbb{X}}}{\langle \chi_f(\mathbf{x}-\mathbf{x}_c, t) \rangle_{\mathbb{X}}}, \tag{5.2}$$

where $\chi_f := 1 - \chi_p$ is the indicator function of the fluid phase which has been introduced in order to disregard the solid phase from the average. This operator maps a three-dimensional and time-dependent input onto an output which is three-dimensional with respect to the transformed coordinates, $\tilde{\mathbf{x}} := \mathbf{x} - \mathbf{x}_c$, but also depends on the y-coordinate of the crossing plane under investigation.[1] By definition, the origin of the transformed coordinate system coincides with the centroid of the particle, and thus, in simple terms, this operation averages the fluid field relative to the particle's location whenever a sample particle crosses a given xz-plane.

Whereas the definition of the particle-conditioned averaging operation is straightforward, its evaluation for fluid data poses difficulties. For attributes intrinsic to the particles, that is, those which are stored alongside the trajectories, the spatial-averaging operator $\langle \bullet \rangle_{\mathbb{X}}$ can readily be evaluated by interpolating the finely sampled data onto the time of crossing in the same manner as in chapter §4. In principle, the same procedure could be used for the fluid as well, however, in practice the sampling frequency of fluid samples in time is orders of magnitude scarcer than that of attributes associated to the particle trajectories because of the vastly larger demand on storage. This circumstance in conjunction with the problematic nature of temporal interpolation of a highly nonlinear chaotic system demands the search for an alternative way to evaluate the spatial-averaging operator which is more practical for fluid data.

[1] As was stated before in §4.2, the y-coordinate of the crossing plane may be regarded as a time-like coordinate for the transient, and hence, the time-dependency of the input is not discarded, but rather substituted such that the averaging is conducted at the particle-dependent time where the fluid conditions for all samples are statistically homogeneous, i.e. at the time where the particles are at the same wall-normal position.

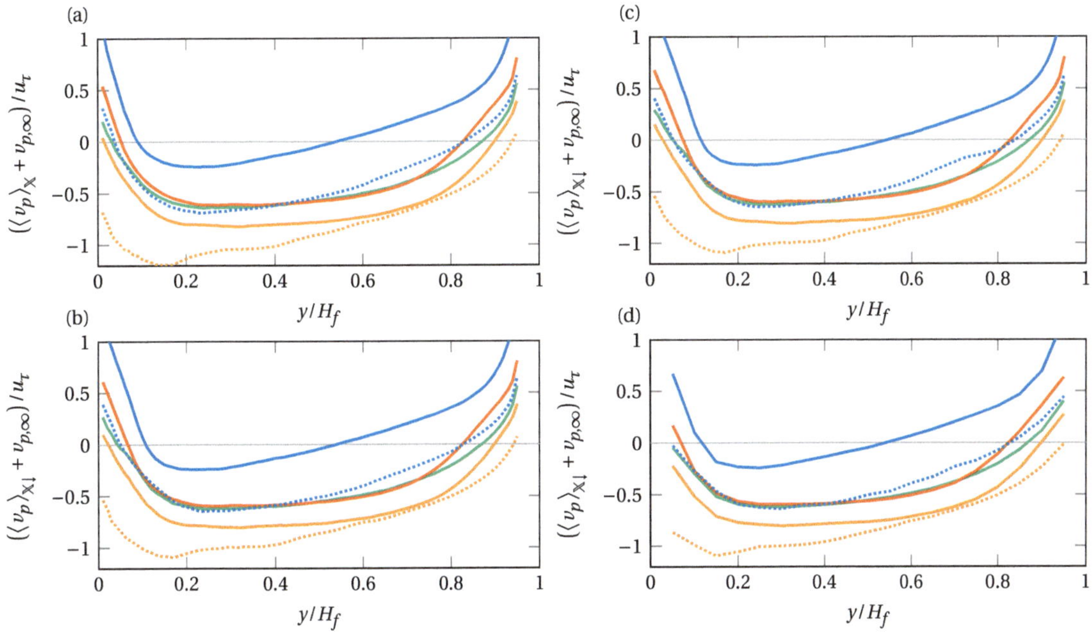

Figure 5.4: Assessment of the differences between the weighted-bin approximation of the spatial-averaging operator given in eq. (5.3) and the corresponding evaluation using (4.5) and interpolation of finely sampled trajectory data. As an example the vertical particle velocity is shown, cf. fig. 4.7. (a) Using the original definition (4.5) and considering only the first crossing. (b) Same, but evaluated for all downward crossings. (c) Using the approximation (5.3) with trajectory data at full temporal resolution and a bin width and spacing of $\varepsilon = 0.01$. (d) Same, but for bin width and spacing of $\varepsilon = 0.05$ and temporal samples only considered where fluid data is available, i.e. the exact setup as used for the fluid data in this work.

As demonstrated in §A in the appendix, the spatial-averaging operator defined in eq. (4.5) can be approximated by a weighted-binning method, viz.

$$\langle \varphi \rangle_{\mathbb{X}}(y) \approx \frac{\int_0^{t_{obs}} \sum_i^{N_{c\Sigma}} \Delta_{\mathbb{X}}\big(y_c^{(i)}(t), v_c^{(i)}(t), y\big) \left| v_c^{(i)}(t) \right| \varphi^{(i)}(t) \, \mathrm{d}t}{\int_0^{t_{obs}} \sum_i^{N_{c\Sigma}} \Delta_{\mathbb{X}}\big(y_c^{(i)}(t), v_c^{(i)}(t), y\big) \left| v_c^{(i)}(t) \right| \mathrm{d}t}. \tag{5.3}$$

Here, $\Delta_{\mathbb{X}}$ is a function which determines whether or not a particle is to be sampled. Whereas the definition of a binning function which approximates exclusively the first crossing of a particle is an intricate task, the definition of a function which is specific to downward crossings is straightforward and reads

$$\Delta_{\mathbb{X}\downarrow}(y_p, v_p, y) := \begin{cases} 1 & \text{if} \quad y_p \in \left[y - \frac{\delta y}{2}, y + \frac{\delta y}{2}\right) \text{ and } v_p < 0 \\ 0 & \text{otherwise} \end{cases}, \tag{5.4}$$

and a variant specific to upward crossings can be constructed analogously. The free parameter δy denotes the width of the bin and its value is set to $\delta y := 0.05 H_f$ in the following. The spacing of the bins is chosen as small as possible under the condition that the bins do not overlap, and hence, it is equivalent to the bin width.

As a consequence, the approximation employed in the current chapter differs from the spatial-averaging procedure employed in chapter §4 in two ways. First, uncertainty in the wall-normal position of samples is introduced as a result of finite-width bins. Second, the nature of samples is slightly different as all downward moving particles are now included as samples by default as opposed to merely the first crossing of a particle. Fortunately, it is straightforward to assess the impact of these differences by applying the approximation to any intrinsic particle attribute where a reference value is known from trajectory interpolation. Such an assessment can be found in terms of the contaminant settling velocity in fig. 5.4. As can be seen, only minor differences in the results arise from the distinct ways of evaluating the operator, and therefore, results obtained in this chapter can without reservation be compared to the results obtained in the previous chapter.

Finally, the approximation of the particle-conditioned averaged fluid velocity as employed in the following is given by

$$\langle u_{f,\alpha}\rangle_{\mathbb{P}\mathbb{X}\downarrow}(\tilde{\mathbf{x}},y) \approx \frac{\int_0^{t_{obs}} \sum_i^{N_{c\Sigma}} \Delta_{\mathbb{X}\downarrow}\left(y_c^{(i)}(t), v_c^{(i)}(t), y\right) \left|v_c^{(i)}(t)\right| u_{f,\alpha}(\mathbf{x}-\mathbf{x}_c^{(i)}(t),t)\, \chi_f(\mathbf{x}-\mathbf{x}_c^{(i)}(t),t)\, \mathrm{d}t}{\int_0^{t_{obs}} \sum_i^{N_{c\Sigma}} \Delta_{\mathbb{X}\downarrow}\left(y_c^{(i)}(t), v_c^{(i)}(t), y\right) \left|v_c^{(i)}(t)\right| \chi_f(\mathbf{x}-\mathbf{x}_c^{(i)}(t),t)\, \mathrm{d}t} \tag{5.5}$$

for the variant conditioned to downward crossing particles. For upward crossings, the procedure differs merely with regards to the binning function employed.

5.2.2 Conditioned fluid velocity in the crossing plane

The averaging procedure previously described is now employed to examine the conditioned fluid velocity. Figure 5.5 displays this quantity within the wall-parallel plane located at $y = 0.3H_f$. At this position, the contaminants' vertical and streamwise velocities are generally enhanced, cf. fig. 4.7 and fig. 4.8. This enhancement is clearly reflected in the nearby fluid environment, as the conditioned vertical(streamwise) fluid velocity fluctuations are biased towards negative(positive) values, thus establishing a direct link between the velocities of both phases.[2] This observation holds for both parameter points shown.

At this point, it has not yet been finally resolved whether the nearby fluid velocity is induced by the contaminants themselves or has its origin in a preferential sampling bias, though the analysis regarding secondary motion hinted towards the latter. The preferential sampling hypothesis is further reinforced for the streamwise velocity component by comparing the two cases shown, which mainly differ in their relative turbulence intensity and particle mobility. If the excess in streamwise velocity was to be induced solely by the finite response time of the contaminants at this position, one would expect G30-D15-HR to exhibit higher values of $\langle u_f' \rangle_{\mathbb{P}\mathbb{X}\downarrow}$ than G10-D10-HR, as these contaminants carry more streamwise momentum per particle, settle faster in absolute terms and adapt more slowly to changes in the fluid environment. This is not the case. In contrast, the short response time and higher relative turbulence intensity of G10-D10-HR likely enables these contaminants to perform preferential sampling more effectively, which is in agreement with the actual trend observed for the two cases. However, this inference is limited to the bulk of the fluid, for which fig. 5.5 is an adequate representative. It was noted already in §4.3.1 that the near-wall behaviour of the streamwise contaminant velocity differs from that in the bulk, and an analysis thereof in terms of the local fluid environment is subject of upcoming §5.2.3.

Regarding the vertical fluid motion, an assessment of the cause of the biased downward fluid velocity is more intricate. This is mainly due to the intrinsic relative velocity between the two phases in the direction of gravity whose value is difficult to quantify as it may deviate from its ambient value in either direction due to turbulence interaction or collective effects. The direct effect of this velocity difference is observable in fig. 5.5 at distances of $\mathcal{O}(10^{-1}d_p)$ from the particle's surface where the conditioned fluid

[2] To be precise, the term "fluctuation" is used in the current context to denote deviations from the classical spatio-temporal mean of the fluid flow, i.e. $\mathbf{u}_f - \langle \mathbf{u}_f \rangle_{xzt}$, and not with respect to deviations from the particle-conditioned average which was just introduced. This implies that spatio-temporal mean of these fluctuations is zero by definition, whereas the particle-conditioned mean may be nonzero if a sampling bias exists.

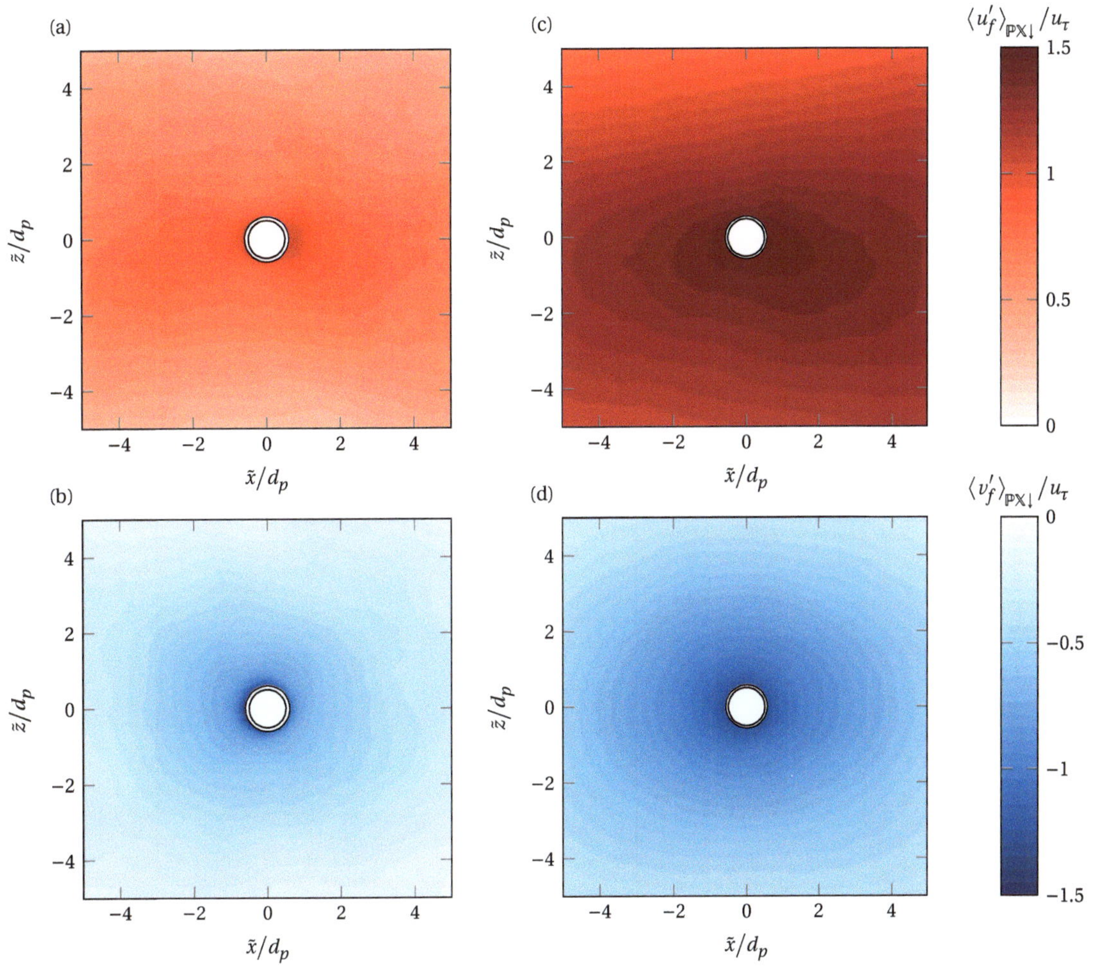

Figure 5.5: Particle-conditioned average of the fluid velocity in the vicinity of the contaminant particles for the cases (a,b) G30-D15-HR and (c,d) G10-D10-HR. The xz-plane is located at $y = 0.3H_f$. The top row (a,c) shows the streamwise velocity fluctuations $\langle u_f' \rangle_{\mathbb{P}\mathbb{X}\downarrow}$, whereas the bottom row (b,d) depicts the vertical component $\langle v_f \rangle_{\mathbb{P}\mathbb{X}\downarrow}$.

velocity has its highest downward values. The difficulty lies, however, in determining the radial distance at which this effect becomes negligible, as this distance could in principle be of the order of multiple particle diameters if the contaminants frequently form clusters. As will be shown in upcoming chapter §6, clustering indeed occurs more frequently for G10-D10-HR than for G30-D15-HR, and hence, this phenomenon may serve as an explanation for the slower radial decay of the vertical velocity observed at this parameter point. Nonetheless, it can equally be argued that the difference in radial decay is caused by the preferential sampling mechanism: due to the higher mobility for G10-D10-HR, these contaminants likely sample downdrafts more effectively which potentially leads to a similar pattern in the conditional average if the downdraft regions are of the size of multiple particle diameters.[3] There is, however, an argument to be made in favour of the preferential sampling mechanism. For both cases, the conditioned vertical fluid velocities are $\mathcal{O}(u_\tau)$. While for G30-D15-HR this information does not allow discrimination between particle-induced and turbulence-induced motion as $v_{p,\infty} \approx u_\tau$, these velocity scales are clearly separated for G10-D10-HR where $v_{p,\infty} \approx 0.35u_\tau$. Hence, a cluster of particles would be required to settle

[3] The assumption that preferentially sampled downdrafts have a length scale of at minimum the particle diameter appears obvious, as any fluctuations significantly smaller than the particle are experienced as noise as they are uncorrelated on the particle scale, and thus, are unlikely to produce the directional effect required for preferential sampling.

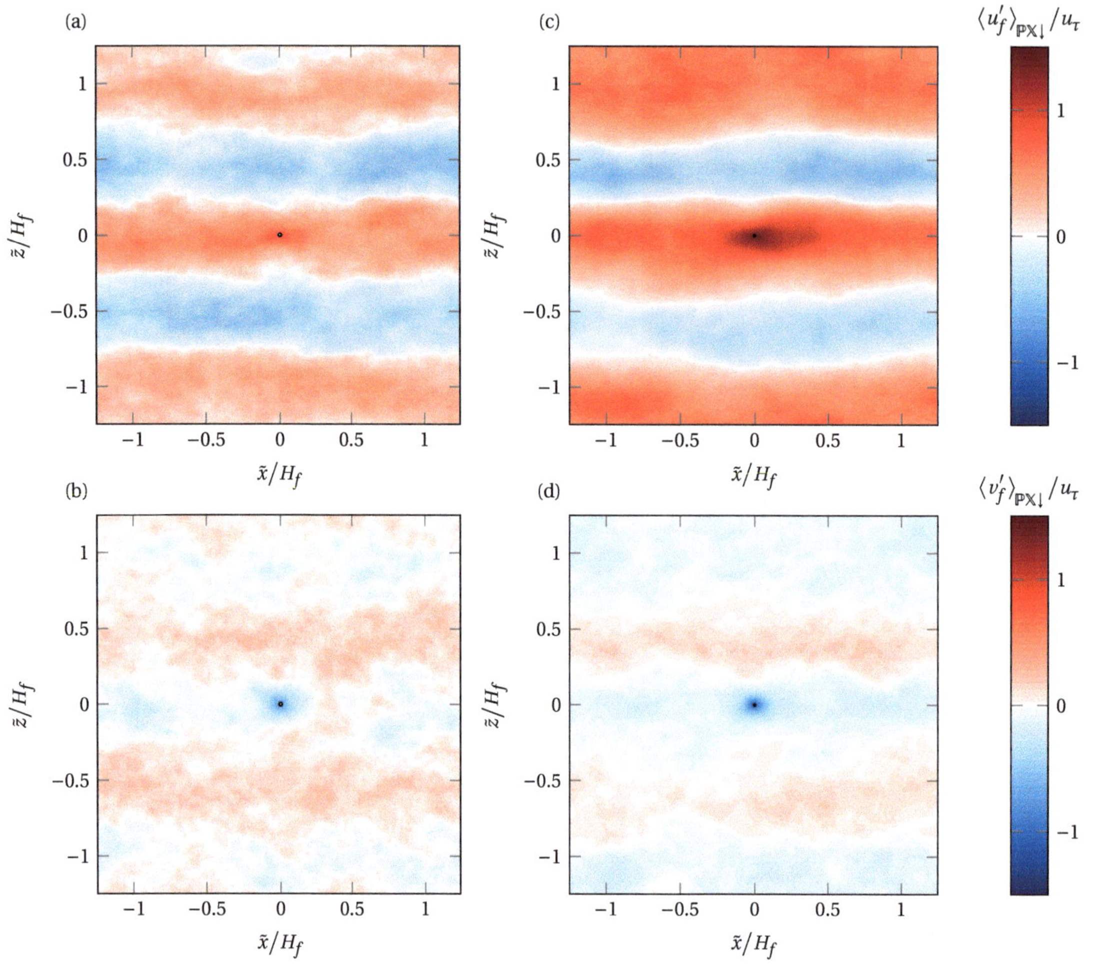

Figure 5.6: Particle-conditioned average of the fluid velocity as in fig. 5.5, but now showing the entire streamwise/spanwise extent of the computational domain. The xz-plane is again located at $y = 0.3H_f$. The left column shows (a) the streamwise and (b) the vertical component of the fluctuating fluid velocity for G30-D15-HR. The right column shows the same quantities for G10-D10-HR, i.e. (c) $\langle u'_f \rangle_{\mathbb{P}\mathbb{X}\downarrow}$ and (d) $\langle v'_f \rangle_{\mathbb{P}\mathbb{X}\downarrow}$.

at around three times the speed of the isolated terminal velocity of an individual particle in order to be the sole explanation of the observed velocity magnitude, a value which is well above those typically observed for wake-induced clustering, see e.g. Uhlmann and Doychev (2014). Furthermore, the values of the Galileo numbers at the parameter points currently discussed are—from the current understanding of this phenomenon—insufficient to generate wake-induced particle clusters. A potential formation of clusters is consequently more likely a byproduct from preferential sampling, and thus might correlate with an enhancement in particle velocity without being the principal cause.

The insights which can be gained from the particle-conditioned fluid velocity are not restricted to the near vicinity of the particle. In fact, as was already observed in §5.1, the contaminants interact with large-scale flow structures with characteristic sizes of the order of the clear fluid height, and this behaviour is also reflected in the conditionally-averaged velocity field. Figure 5.6 presents the same data as previously shown, but at a vastly larger scale. Here, the entire cross-section of the computational domain is shown rather than just the immediate proximity of the particle. For both the streamwise and the vertical component of the fluid velocity, alternating streamwise-aligned stripes of positive and negative fluctuations are conspicuous. These stripes have an infinite extent in streamwise direction—they connect to themselves across the periodic domain—and a spanwise spacing of the order of the

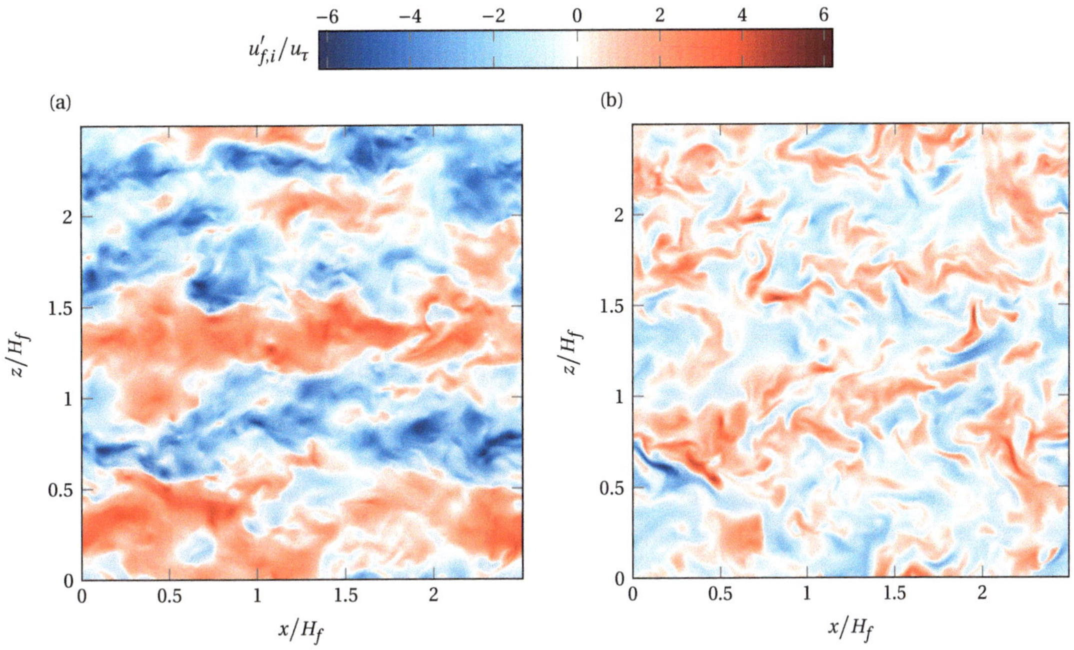

Figure 5.7: Exemplary visualization of an instantaneous state of the background flow at $y = 0.3H_f$ (HR, $Re_\tau = 838$). In this particular numerical experiment depicted, no contaminant particles are present in the flow. (a) Streamwise fluctuations of the fluid velocity u'_f. (b) the corresponding wall-normal fluctuations v'_f.

fluid height. It can be readily acknowledged that the conditionally-averaged field is complementary to the observations from §5.1, as there the contaminants were seen to accumulate in regions of positive $\langle u'_f \rangle_{xt}$ and negative $\langle v_f \rangle_{xt}$, which in the particle-conditioned perspective must be reflected in the form of streamwise-aligned stripes.[4] The advantage of the particle-conditioned average is, however, that it is also resolved in streamwise direction and that the temporal average is weighted by the number of crossings within a given time interval such that it more adequately reflects the flow field at the time at which contaminants are actually affected by the flow.

Regarding the streamwise fluid velocity, the inferences differ only little from the ones made in §5.1. The primary reason for this is that the large-scale streaks which leave their footprint in the streamwise mean are indeed rather homogeneous in the streamwise direction, and their typical lifetime is significantly longer than the observation time of the settling experiments. Therefore, the increased spatial resolution and temporal accuracy of the particle-conditioned averaging procedure is not required for their detection. An example for the instantaneous state of a large-scale streak in the current setup is provided in fig. 5.7(a). The structure exhibits only little variation in the streamwise direction even instantaneously, and its streamwise extents fills the entire length of the computational domain. Indeed, an argument can be made the the numerical domain is too short to capture the dynamics of these large-scale structures accurately—a circumstance which originates from the artificial suppression of the ripple-instability in the sediment bed. A discussion on the implications of the domain size with regards to contaminant settling is subject of §5.5.

The instantaneous vertical fluid velocity is significantly more intermittent in streamwise direction than u_f, see fig. 5.7(b), albeit large-scale coherence may also be detected to some degree. In fact, it is this

[4] Please keep in mind that if the contaminants were to sample the crossing plane randomly with respect to the (x, z)-position and time, the particle-conditioned fluid fluctuations would be nil, as this would directly correspond to the classical spatio-temporal fluid average. The obvious exception to this statement is the local flow modification caused by the particle itself.

streamwise-coherent part which manifests itself in the secondary flow and as stripes in the particle-conditioned average. With regards to $\langle v_f \rangle_{\mathbb{PX}\downarrow}$ it may be observed that the bias in fluid velocity is noticeably stronger in the near vicinity of the particle compared to the large-scale coherent motion at greater distances, which is much less the case for $\langle u_f \rangle_{\mathbb{PX}\downarrow}$. This nearby zone of influence is exactly the one displayed in detail in fig. 5.5(b,d), and for the reasons stated in the discussion thereof, is presumably caused by a preferential sampling of downward fluid motion. Recalling the findings up to this point that

1. the magnitude of $\langle v_f \rangle_{xt}$ is significantly smaller than $\langle v_c \rangle_{\mathbb{PX}\downarrow}$ at the same (y,z)-position
2. the nearby fluid velocity is unlikely to be induced by the contaminants themselves
3. the vertical fluid velocity is highly intermittent in streamwise direction

it can be hypothesised that the enhancement of the settling velocity is primarily caused by the preferential sampling of turbulent fluctuations of the vertical velocity in negative direction which occur at a scale larger than the particle size, but considerably smaller than the coherent turbulent motion at the largest scales. This hypothesis will be subject of further investigations.

5.2.3 Fluid velocity sampled by the contaminants

While the particle-conditioned averaging operator defined in eq. (5.2) proved to be useful for the assessment of the average flow environment experienced by the contaminants, a similar quantity of lower dimensionality would be more expedient to investigate trends between the various cases and at varying wall-normal positions. For this reason, a virtual sphere of radius r is defined in the particle-attached coordinate system whose centre is located at $\tilde{\mathbf{x}} = (0,0,0)^T$. A reduction in dimensionality is then achieved by averaging the values of the particle-conditioned field on the surface of the virtual sphere S_v with radius r, viz.

$$\langle u_{f,\alpha} \rangle_{\mathbb{SX}\downarrow}(r,y) := \frac{1}{4\pi r^2} \oint_{S_v(r)} \langle u_{f,\alpha} \rangle_{\mathbb{PX}\downarrow}(\tilde{\mathbf{x}},y)\, \mathrm{d}S_v. \tag{5.6}$$

This sphere-average $\langle \bullet \rangle_{\mathbb{SX}\downarrow}$ depends on the vertical coordinate of the crossing plane under investigation and its three-dimensionality in the particle-attached coordinate system is now reduced into a single dimension—the radius of the virtual sphere. The idea behind this approach is to define a characteristic velocity of the background turbulence which is experienced by the contaminants at the time of crossing. In order to achieve this, the radius has to be chosen sufficiently large so that the self-induced velocity fluctuations by the particle are negligible, but as small as possible such that the value of the average accurately reflects the local conditions. Similar approaches have been used in literature by e.g. Kidanemariam et al. (2013) (on a segmented sphere) or Uhlmann and Doychev (2014) (on a full sphere) for the same purpose, and in both studies, the radius has been set to $r = 1.5d_p$. This value will be used in the following as well. Please note that the choice of this parameter does indeed influence the absolute values of the obtained velocities, as one might readily assume by assessing fig. 5.5. However, the relative trend among cases and vertical positions was found to be quite robust with respect to r within a reasonable range, such that the particular choice is not decisive for the inferences made in the following.

In the following, an attempt is made to explain the finding of enhanced contaminant velocities from chapter §4 in terms of the fluid velocity in the vicinity of the contaminants. To begin with, the focus is placed on the vertical velocity component. Figure 5.8(a) shows the sphere-averaged vertical fluid velocity for all cases at varying vertical positions in the same scaling as the corresponding fig. 4.7(a) for the

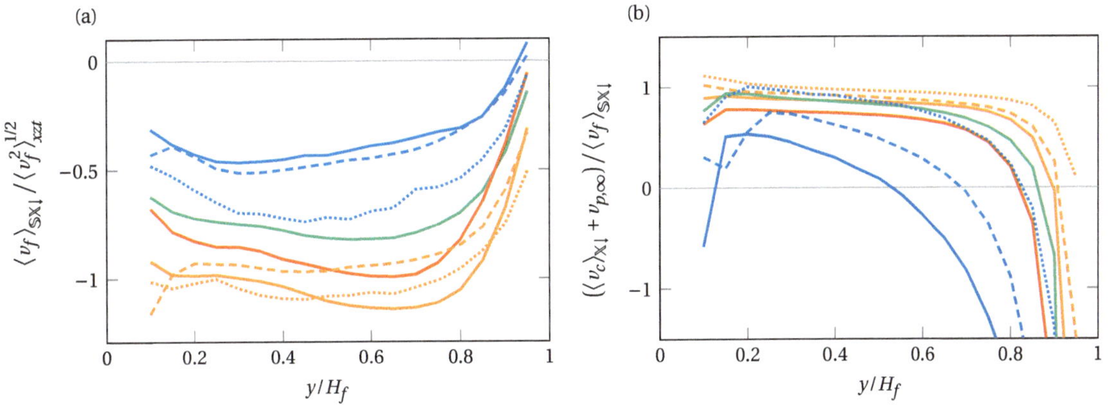

Figure 5.8: Sphere-averaged vertical fluid velocity experienced by the contaminants at the time of crossing of a given xz-plane. (a) Normalised as in fig. 4.7(a) using the local unconditioned value of the standard deviation. (b) As normalisation of the excess in contaminant settling velocity. Legend: G30-D15-HR (······), G10-D10-HR (······), G30-D10-LR (——), G10-D6-LR (——), G15-D10-LR (——) and G15-D6-LR (——).

particle velocity. A comparison of the two figures suggests that the excess in settling velocity is correlated to the surrounding fluid velocity, which is to be expected if the enhancement is indeed a result of the preferential sweeping mechanism. Interestingly, for some cases the velocity sampled by the contaminants appears to decrease with decreasing height when compared to the standard deviation of the local vertical fluid velocity after a peak at the end of the initial transient phase. If the particles were to sample fluctuations of the same length scale consistently, this observation may be related to the variation in turbulent length scales, i.e. small turbulent scales increase their contribution to $\langle v_f' \rangle_{xzt}$ closer to the lower boundary, but this contribution may be too small to be sampled by the contaminants. In fact, one has to keep in mind that averaging over the surface of a sphere acts as an implicit filter for any variation whose length scales are smaller than the diameter of the virtual sampling sphere. A discussion on the wall-normal dependence of the scales of turbulent fluctuations in the flow configuration of current interest is provided in §5.5.

When the excess settling velocity is scaled by $\langle v_f \rangle_{\mathbb{S}\mathbb{X}\downarrow}$, an approximate plateau is reached—or at least approached—by all cases, see fig. 5.8(b). In fact, the convergence towards the plateau exhibits similarities to the initial acceleration in the ambient settling experiments, in particular, the distance travelled until the particles reach an equilibrium of all acting forces as displayed in fig. 3.8(c). Perhaps, the relaxation observed in the turbulent environment can be interpreted in a similar way, in the sense that some distance needs to be travelled until the interaction between the background turbulence and the contaminants is approximately quasi-steady.[5] If so, fig. 5.8(b) provides a potential explanation on why G30-D10-LR exhibits a behaviour deviant from the other cases on some statistics—the available settling distance may be too short for those contaminants to reach their equilibrium settling state. This realization should be kept in mind whenever an anomaly concerning G30-D10-LR is to be interpreted.

Turning to the streamwise velocity, the wall-normal trend of the sphere-averaged fluid velocity portrayed in fig. 5.9(a) appears to explain the corresponding evolution of the particle velocity (fig. 4.8(a)) reasonably well in the bulk of the channel. Interestingly enough, the correlation gets lost below $y \lessapprox 0.3H_f$ where the surplus in streamwise particle velocity with respect to the local fluid mean has its pronounced peak. It was hypothesised in §4.3.1 that this peak is not due to preferential sampling, but rather

[5] Certainly, no steady state in the proper sense can be reached during the transient as the background flow is inhomogeneous in the direction of settling.

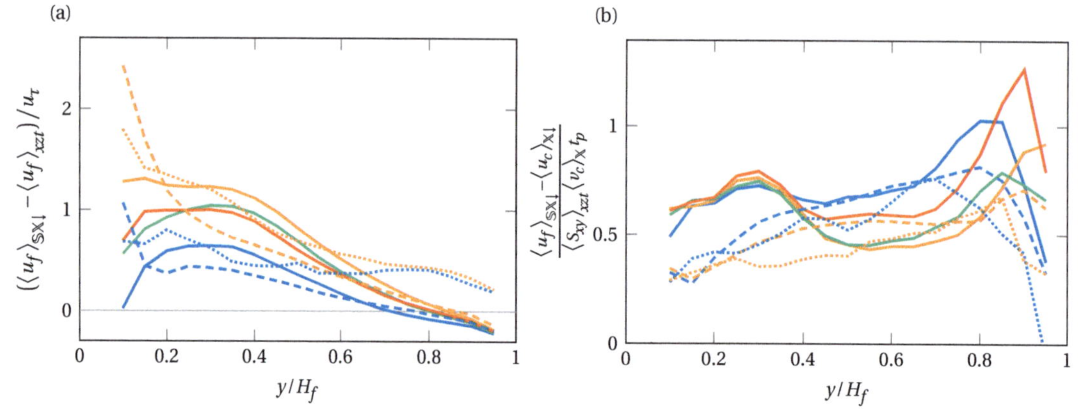

Figure 5.9: Sphere-averaged streamwise fluid velocity experienced by the contaminants at the time of crossing of a given wall-normal plane. (a) Normalised as in fig. 4.8(a) using the friction velocity. (b) Difference between the velocities of the solid and fluid phase normalised by the product of mean shear in the channel, local settling velocity and particle response time. Legend: G30-D10-LR (—), G10-D6-LR (—), G15-D10-LR (—), G15-D6-LR (—), G30-D10-RP (- - -), G10-D6-RP (- - -), G30-D15-HR (⋯⋯), and G10-D10-HR (⋯⋯).

a result of the inability of the contaminants to adapt to the mean fluid velocity which is rapidly decreasing towards the sediment bed. In order to reexamine this hypothesis, the surplus of the particle velocity with respect to the sphere-average fluid velocity is shown in fig. 5.9(b) scaled by the three factors this phenomenon ought to be proportional to—the mean shear ($\langle S_{xy} \rangle_{xzt}$), the velocity with which the contaminants travel along this gradient ($\langle v_c \rangle_{\mathbb{X}\downarrow}$) and the particle response time (t_p). The surplus may be interpreted as a particle velocity which adjusted to disregard preferential sampling effects, and it indeed collapses reasonably well for a given background flow at around $y \lessapprox 0.3H_f$. As a consequence, one can conclude that the mean fluid shear needs to be taken into account in the lower quarter of the channel in order to explain the surplus in $\langle u_c \rangle_{\mathbb{X}\downarrow}$ with respect to $\langle u_f \rangle_{xzt}$.

5.2.4 Quadrant analysis of conditioned fluid velocity

Now that then enhancement in mean streamwise and vertical particle velocities was successfully explained by means of the particle-conditioned average fluid velocities, the more detailed quadrant analysis of §4.3.2 is revisited. As a reminder, the premise of the heuristic proposed in that analysis was that the fluctuations in particle velocities directly reflect the turbulent fluctuations in close proximity to the contaminants. This conjecture is the subject of this paragraph.

In order to construct a histogram of the fluid fluctuations in the vicinity of the contaminants at a given vertical position, eq. (5.6) has to be generalised in the sense that the subset of instantaneous velocities which gives rise to the sphere average needs to be defined. In a first step, the time-dependent set of fluid velocities sampled by the i-th contaminant particle along its trajectory is defined as

$$\left[u_{f,\alpha}\right]_{\mathbb{S}}^{(i)}(r,t) := \left\{u_{f,\alpha}(\tilde{\mathbf{x}}^{(i)},t) : \tilde{\mathbf{x}}^{(i)} \in S_v(r) \wedge \chi_f(\tilde{\mathbf{x}}^{(i)},t) = 1\right\}. \tag{5.7}$$

Consequentially, the complete set of fluid velocities sampled for $\langle u_{f,\alpha} \rangle_{\mathbb{S}\mathbb{X}\downarrow}$ is given by the union of $\left[u_{f,\alpha}\right]_{\mathbb{S}}^{(i)}$ evaluated at all instances in time where downward crossings at y occur for all contaminant particles, viz.

$$\left[u_{f,\alpha}\right]_{\mathbb{S}\mathbb{X}\downarrow}(r,y) := \bigcup_{j=1}^{N_{c,\mathbb{X}\downarrow}} \left[u_{f,\alpha}\right]_{\mathbb{S}}^{(i)}\left(r, t_{c,\mathbb{X}\downarrow}^{(j)}(y)\right) \tag{5.8}$$

where i is the index of the particle to which the j-th downward crossing of the plane is attributed. The arithmetic mean of $[u_{f,\alpha}]_{\mathbb{S}\mathbb{X}\downarrow}$ is equivalent to $\langle u_{f,\alpha} \rangle_{\mathbb{S}\mathbb{X}\downarrow}$. A variant which considers only upward crossings can be constructed analogously.

The primary quantity of interest for the current analysis is the joint probability density of the conditioned streamwise and vertical fluid velocities. However, because of the scarce temporal resolution of the samples, cf. §5.2.1, it is not feasible to construct the set defined in eq. (5.8) explicitly. Instead, the probability distribution is approximated using a weighted histogram in a manner similar to the approximation of the particle-conditioned average, cf. eq. (5.5). This weighted histogram is given by

$$\mathrm{hist}\Big([u_f]_{\mathbb{S}\mathbb{X}\downarrow}, [v_f]_{\mathbb{S}\mathbb{X}\downarrow}, y\Big) \approx \frac{\int_0^{t_{obs}} \sum_i^{N_{c\Sigma}} \Delta_{\mathbb{X}\downarrow}\big(y_c^{(i)}(t), v_c^{(i)}(t), y\big) \left|v_c^{(i)}(t)\right| \mathrm{hist}\big([u_f]_{\mathbb{S}}^{(i)}(t), [v_f]_{\mathbb{S}}^{(i)}(t)\big)\,\mathrm{d}t}{\int_0^{t_{obs}} \sum_i^{N_{c\Sigma}} \Delta_{\mathbb{X}\downarrow}\big(y_c^{(i)}(t), v_c^{(i)}(t), y\big) \left|v_c^{(i)}(t)\right| \mathrm{d}t} \tag{5.9}$$

and normalisation yields the corresponding joint probability density function. Please note that the dependency on the radius of the sampling sphere has been disregarded in this equation as it is not varied in this study.

The resulting joint probability functions of the (u_f, v_f)-fluctuations experienced by downward moving particles displayed in fig. 5.10 are strikingly similar to those of the corresponding particle velocity fluctuations (fig. 4.9). This reinforces the conjecture that these quantities are closely related. Moreover, it can be observed that the (u_f, v_f)-space also exhibits an approximate upper bounded dictated by the relative turbulence intensity as virtually no fluctuations faster than the terminal velocity of the particle are sampled by the contaminants—at least for the vertical position under investigation. This observation is remarkable as the analogous (rigorous) property with respect to the particle velocity fluctuations plays a decisive role for the reasoning of §4.3.2, and many of the inferences made relied on the supposition that the fluid velocity fluctuations behave equally, which indeed seems to be adequate assumption.

Interestingly enough, the joint PDFs of fluid fluctuations conditioned to upward crossings (fig. 5.11) suggest that there is a fundamental difference between downward and upward particle movement. Here, the fluctuations sampled by the contaminants show little to no indication of boundedness by the relative turbulence intensity. One potential interpretation of this observation is that downward moving particles are more effective at sampling regions of the flow which are favourable for their direction of movement than upward moving particles.

As may be inferred from the conditioned mean statistics, the sampling bias towards specific quadrants depends on the wall-normal position. For this reason, the contribution of the individual quadrants to the (u_f, v_f)-space sampled by the contaminants is examined in the following. The probability of an instantaneous fluctuation to belong to i-th quadrant, which will be denoted as X_{Qi}, is given by

$$\begin{aligned} &\underbrace{\int_0^{\infty}\int_0^{\infty} \mathrm{jpdf}(u_f', v_f)\,\mathrm{d}u_f'\,\mathrm{d}v_f}_{:=X_{Q1}} + \underbrace{\int_0^{\infty}\int_{-\infty}^{0} \mathrm{jpdf}(u_f', v_f)\,\mathrm{d}u_f'\,\mathrm{d}v_f}_{:=X_{Q2}} + \\ &\underbrace{\int_{-\infty}^{0}\int_{-\infty}^{0} \mathrm{jpdf}(u_f', v_f)\,\mathrm{d}u_f'\,\mathrm{d}v_f}_{:=X_{Q3}} + \underbrace{\int_{-\infty}^{0}\int_0^{\infty} \mathrm{jpdf}(u_f', v_f)\,\mathrm{d}u_f'\,\mathrm{d}v_f}_{:=X_{Q4}} = 1, \end{aligned} \tag{5.10}$$

respectively. In this definition, u_f and v_f may refer to either the conditioned and unconditioned sample velocities alike, although in both cases u_f' is to be understood as the difference from the unconditioned fluid mean $\langle u_f \rangle_{xzt}$.

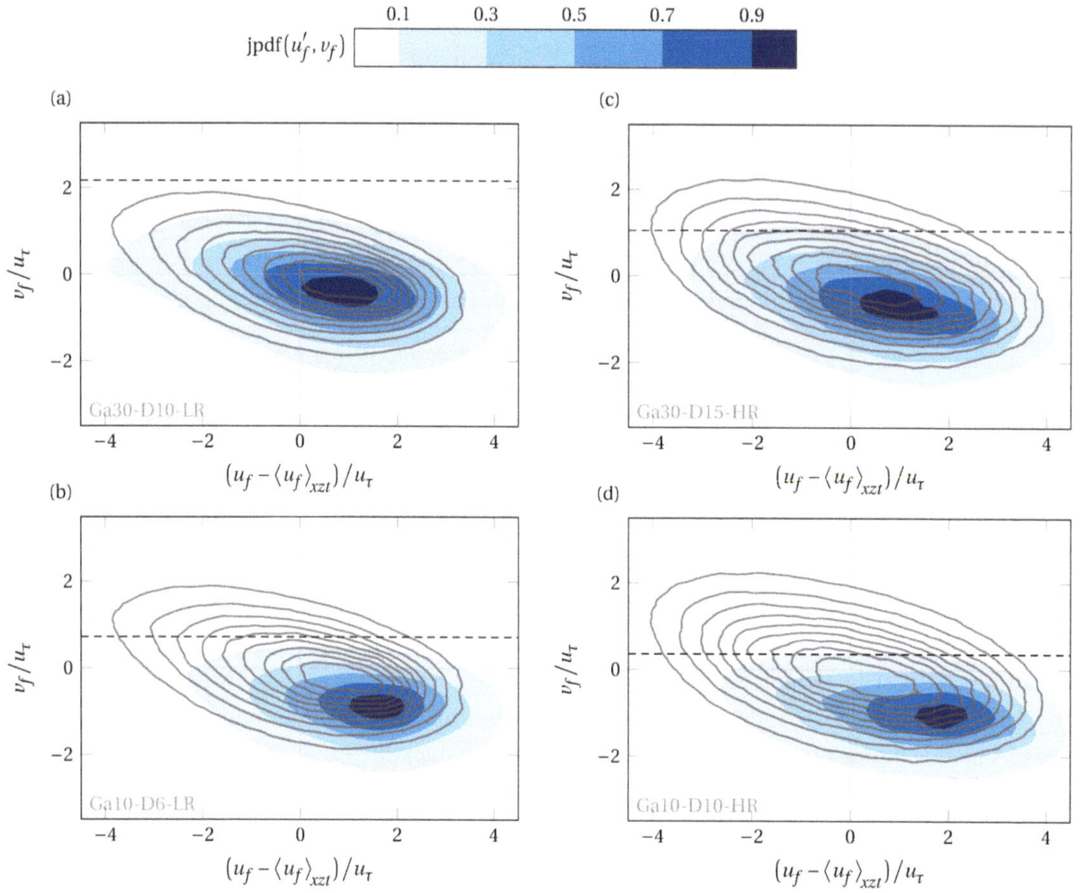

Figure 5.10: Joint probability density function of the streamwise and vertical components of the fluid velocity conditioned to the contaminants, jpdf$([u_f']_{\mathbb{S}\mathbb{X}\downarrow}, [v_f]_{\mathbb{S}\mathbb{X}\downarrow})$ (shaded regions) and the unconditioned fluid velocity, jpdf(u_f', v_f), (contour lines) at $y = 0.3H_f$ for all downward crossings. The data is presented as a function of the velocity fluctuations, which means that the mean unconditioned streamwise velocity at the position shown is subtracted from the conditioned and unconditioned streamwise velocities alike. The colouring corresponds to $\{0.1, 0.3, 0.5, 0.7, 0.9\}$ times the respective maximum value of the distribution. An indicator for the inverse of the relative turbulence intensity for each case is given by the dashed line which is located at $v_f = v_{p,\infty}$ (- - -). The following cases are arranged in column-major order: (a) G30-D10-LR, (b) G10-D6-LR, (c) G30-D15-HR and (d) G30-D10-HR.

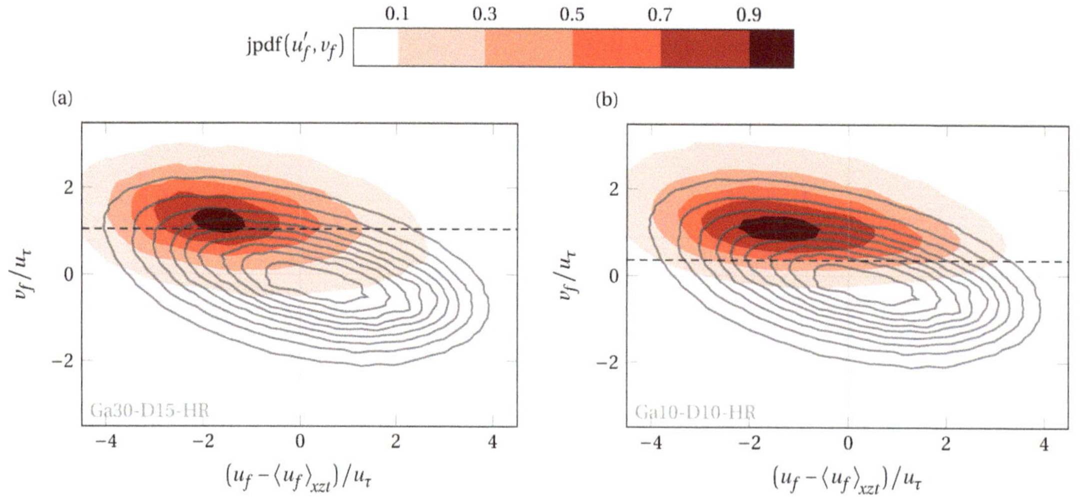

Figure 5.11: Same as fig. 5.10, but the conditioned fluid velocity is now considered for all upward crossings, i.e. the shaded regions correspond to jpdf$([u_f']_{\mathbb{S}\mathbb{X}\uparrow}, [v_f]_{\mathbb{S}\mathbb{X}\uparrow})$. Only the cases at $Re_\tau = 838$ are shown: (a) G30-D15-HR and (b) G30-D10-HR.

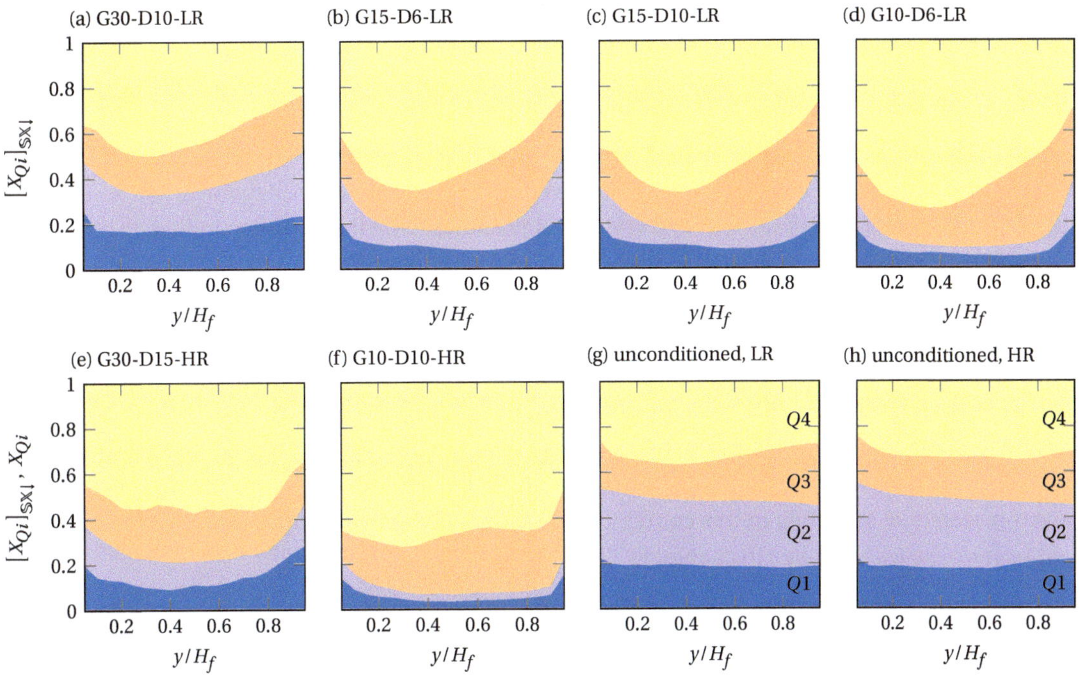

Figure 5.12: Probability of a fluid fluctuation sampled by the contaminants to belong to the i-th quadrant as a function of vertical position in the channel. The data is presented in form of stacked area curves to reflect the property $\sum_i X_{Qi} = 1$. Therefore, X_{Q1} (■), X_{Q2} (■), X_{Q3} (■) and X_{Q4} (■) correspond to the vertical extent of the areas in the respective colour at each y-position. The conditioned probabilities for the four cases at $Re_\tau = 241$ are shown in the upper row (a-d), whereas those for the two simulations at $Re_\tau = 838$ are shown in the lower row (e,f) together with the values obtained for the unconditioned fluid velocity for the (g) LR and (h) HR background flows. The conditioned probabilities are ordered in increasing relative turbulence intensity for each flow Reynolds number.

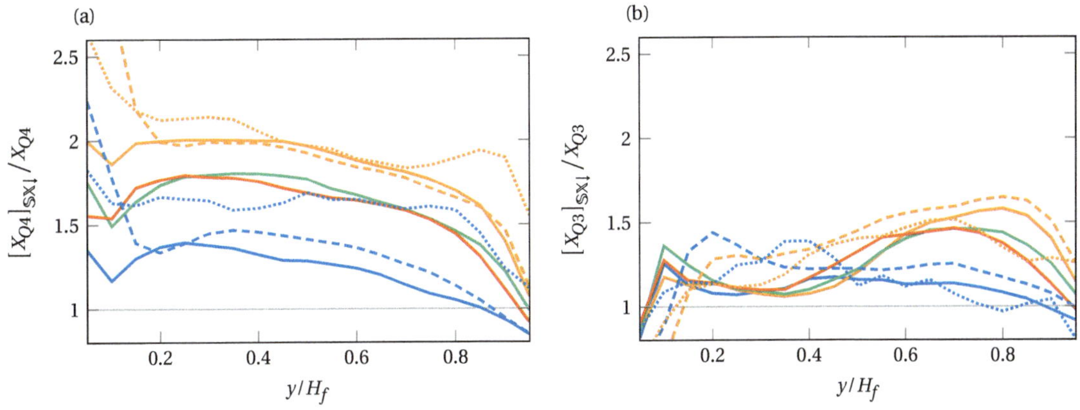

Figure 5.13: Ratio between conditioned and unconditioned probability of a fluid fluctuation to belong to (a) the fourth quadrant where $u_f' > 0, v_f < 0$, and (b) the third quadrant where $u_f' < 0, v_f < 0$. Legend: G30-D10-LR (—), G10-D6-LR (—), G15-D10-LR (—), G15-D6-LR (—), G30-D10-RP (- - -), G10-D6-RP (- - -), G30-D15-HR (⋯⋯), and G10-D10-HR (⋯⋯).

The resulting probabilities are shown in fig. 5.12. When comparing the conditioned values to the unconditioned ones, it becomes apparent that sweep-events (Q4) dominate, and that the prevalence increases with increasing relative turbulence intensity. This increase largely happens at the expense of Q1- and Q2-events, i.e. those for which $v_f > 0$, whereas Q3-events are still likely, but are not as prevalent as Q4. The unconditioned probabilities suggest that the distribution among quadrants slightly varies with respect to the wall-normal coordinate, and this variation is also reflected by the conditioned probabilities—at times even in an amplified manner such as the excursion observed for Q4 at $y \approx 0.35$ for LR. In order to gauge this effect and to display the relative shift in probabilities, fig. 5.13 provides the conditioned probabilities of Q3- and Q4-events normalised by their corresponding unconditioned values. While for a short initial period—the wall-normal position serves as a time-like coordinate of the transient in this picture—the organization into preferentially sampled regions is rather irrespective of the sign of u_f', sweep events are clearly preferred at later times in most cases. A notable exception to this observation is G30-D10-RP for which the probability of sampling either quadrant stays comparable until close to the sediment bed. A major contributing factor for the preference is presumably the accumulation of contaminants in large-scale velocity streaks, which was observed and described in §5.1 and §5.2.2. In order to understand the connection between these observations, the mechanism which causes preferential sampling in this configuration needs to be further understood, and an attempt at gaining more insights is taken in chapter 7.

5.3 Particle wakes

The focus is now shifted towards the flow at the scale of the individual contaminant particles. In particular, the questions of how the wake of the particles interacts with the background flow and how it is modified with respect to a suitable laminar reference state are raised.

5.3.1 Instantaneous fluid motion near the contaminants

To begin with, the instantaneous fluid motion in the vicinity of the contaminants is inspected on a sample basis. For this purpose, two orthogonal slices oriented normal to the spanwise and streamwise direction, respectively, and centred at the centroid of the contaminant particle of interest are extracted from a snapshot of the fluid field at a time where the settling process is fully developed. Four distinctive samples are chosen under the consideration that their settling velocities at the time of inspection preferably span the full extent of the range at this particular instance in time. The fluid motion is then inspected from two perspectives. First, the vertical velocity is examined from the perspective of a steady observer, i.e. a observer who is not moving with respect to the channel's walls, which enables a straightforward assessment of the background turbulence in the vicinity of the particle. Second, the velocity is inspected in a frame of reference translating with the particle of interest and particular attention is drawn to vertical velocities of the order of the terminal velocity of the contaminants. This perspective is especially advantageous for the detection of particle-induced fluid motion as caused by a particle's boundary layer or wake.

It should perhaps be noted that in the range of Galileo numbers under investigation in this thesis ($Ga \leq 30$), no wake in the proper sense is to be expected, which means that no zone of recirculation exists in the undisturbed flow around such particles. Nonetheless, the particle induced motion exhibits a pronounced front-rear-asymmetry at all parameter points, and it is the velocity deficit in the rear of

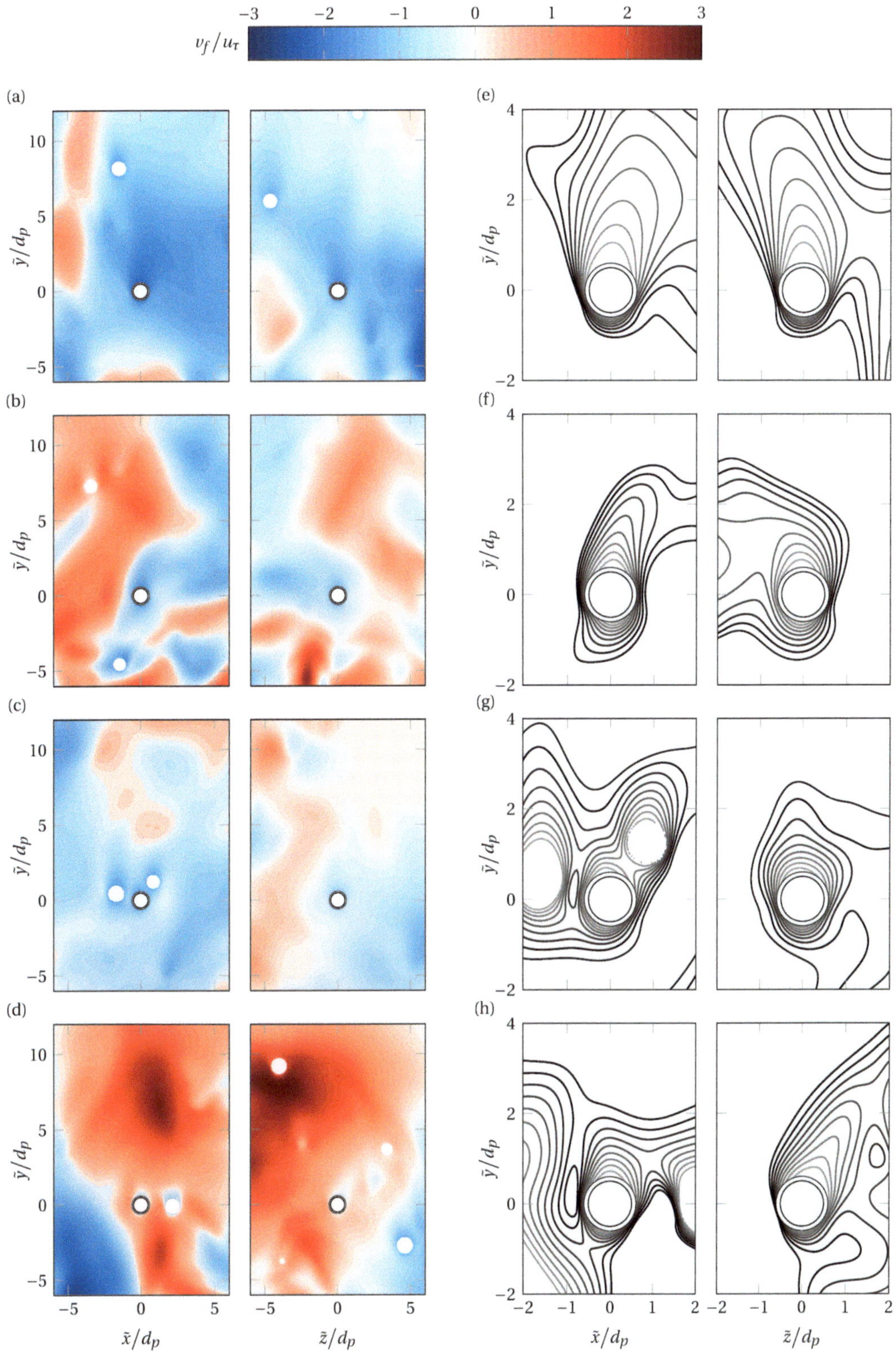

Figure 5.14: Instantaneous flow field around four selected contaminant particles for G30-D15-HR at $t = 7.9t_b$. (a-d) Vertical fluid velocity scaled by the friction velocity in a stationary reference frame centred at the particle of interest. (e-h) Vertical fluid velocity scaled by the particles terminal velocity in a reference frame moving with and centred at the particle of interest. The lines correspond to $v_f = [-0.9:0.1:-0.1]\,v_{p,\infty}$. Per subfigure, a cross-sectional xy/zy-plane is shown on the left/right, respectively. The particles' instantaneous vertical positions and velocities are (a,e) $0.48H_f$, $-2.0v_{p,\infty}$, (b,f) $0.36H_f$, $-1.3v_{p,\infty}$, (c,g) $0.48H_f$, $-1.1v_{p,\infty}$, (d,h) $0.48H_f$, $-0.2v_{p,\infty}$, respectively.

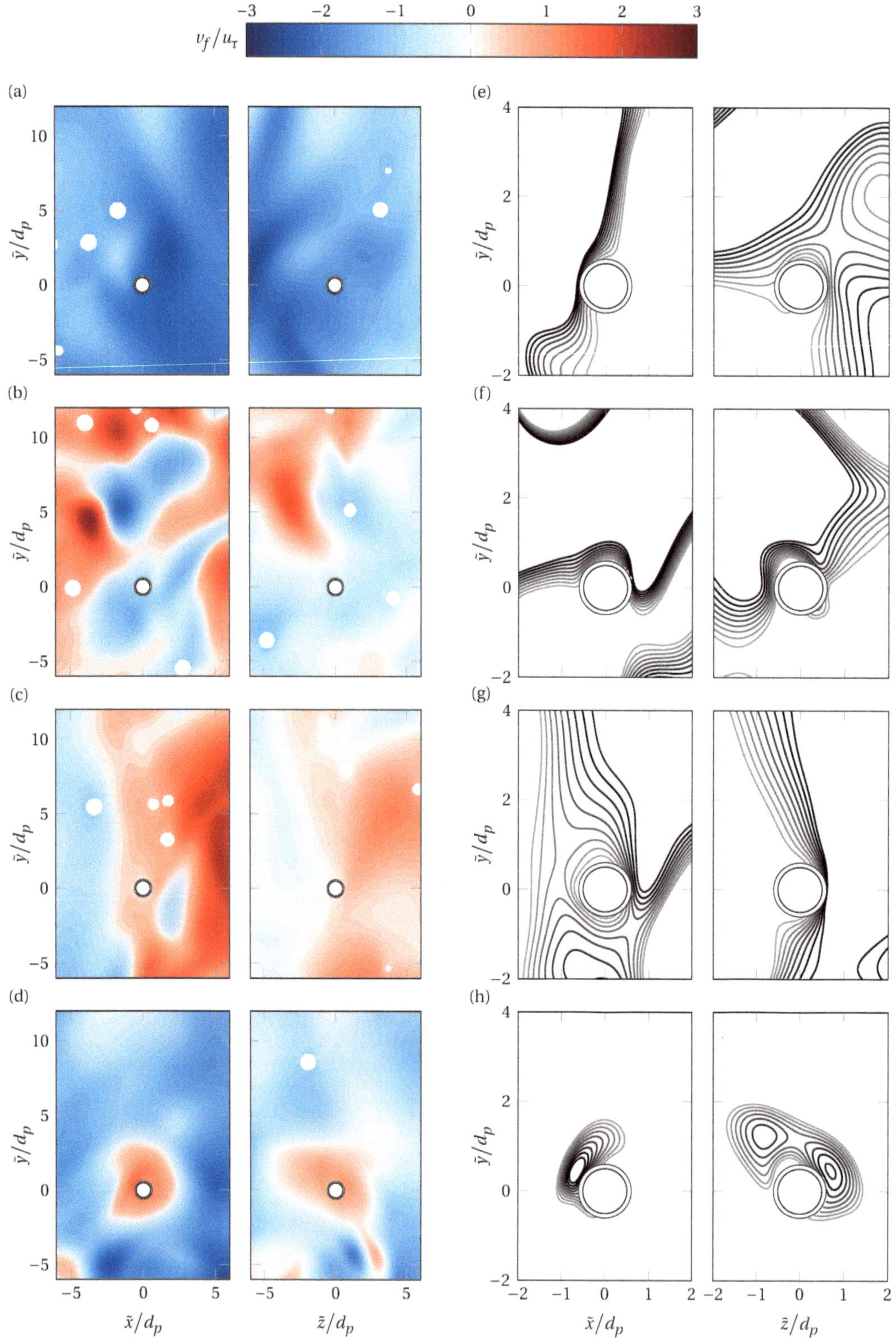

Figure 5.15: Same as fig. 5.14, but for G10-D10-HR at $t = 13.1\,t_b$. The particles' instantaneous vertical positions and velocities are (a,e) $0.39H_f$, $-5.8v_{p,\infty}$, (b,f) $0.25H_f$, $-1.7v_{p,\infty}$, (c,g) $0.43H_f$, $-0.3v_{p,\infty}$, (d,h) $0.24H_f$, $+2.7v_{p,\infty}$, respectively.

the particle which is referred to as wake in the following. Visualizations of the undisturbed wakes at $Ga = \{27.1, 10.8\}$ are provided at a later point in fig. 5.16(a) and fig. 5.17(a), respectively.

Figure 5.14 shows the instantaneous fluid motion in the vicinity of selected contaminant particles for G30-D15-HR. From the point-of-view of a steady observer, it is obvious that the fluid environment experienced by the contaminants during settling is intermittent, as for all samples shown, patches of negative and positive vertical fluid velocity coexist within only few diameters distance from the particle. Regarding the relative velocity between the solid and fluid phase, it can be noticed that the contaminants induce a fluid motion which resembles a wake, although these structures are visibly deformed by the background flow, especially at greater distances. A decent example is given by fig. 5.14(f) where a distinct velocity deficit can be observed in the rear of the sample particle.

The observations for G10-D10-HR are similar to those at the higher value of the Galileo number with regards to the fluid environment experienced by the contaminants, see fig. 5.15. However, there is a disparity regarding the relative fluid motion in the near vicinity of the particle. Whereas for $Ga \approx 30$ structures resembling a particle boundary layer are still perceptible, it seems that for $Ga \approx 10$ and the corresponding higher relative turbulence intensity, fluctuations of the order of the terminal velocity are mostly driven by—or at least severely distorted by—the background turbulence. This suggests that high values of I_T might reduce the difference in relative velocity between the two phases below the ambient terminal velocity, or that the time scale at which a distortion by turbulent fluctuations occurs simply becomes much shorter than the characteristic time scale of the particle-induced flow.

5.3.2 Comparison with undisturbed wake

The particle-conditioned average introduced in §5.2.1 is now used to investigate the vertical velocity field in close proximity of the particle in order to understand how the boundary layer is modified by the background turbulence. Of particular interest in this investigation are variations of the order of the ambient terminal velocity in a reference frame moving with the particle, as fluid velocities of this order of magnitude can be induced by the intrinsic relative velocity between the particle and its environment.

In order to have a reference to which the wake under the influence of turbulence can be compared to, the undisturbed flow around a particle at the corresponding value of the Galileo number has to be examined first. For this reason, the converged state of the reference simulations previously presented in §3.7 is visualised in fig. 5.16(a) for $Ga = 27.1$. As previously stated, the boundary layer is steady and axisymmetric and no recirculation occurs in the rear of the particle. However, a strong asymmetry of the velocity deficit in the direction of movement can be observed with significant disturbances being still noticeable at distances of the order of ten particle diameters from the rear.

In contrast, the velocity deficit decays within a considerably shorter distance in the averaged turbulent case, see fig. 5.16(b-e). This result is consistent with e.g. Bagchi and Balachandar (2004) or Homann et al. (2013), although only few data points at comparably low(high) values of the Reynolds number(relative turbulence intensity), respectively, are available in literature. More interestingly, the general shape of the boundary layer from the quiescent reference is somewhat maintained in the turbulent mean for the vertical positions shown in the sense that it is approximately axisymmetric and slightly tapered at the sides of the particle.

For $Ga = 10.8$, the undisturbed boundary layer is notably shorter and less confined in the horizontal direction than at higher Galileo numbers, see fig. 5.17(a). In the presence of background turbulence

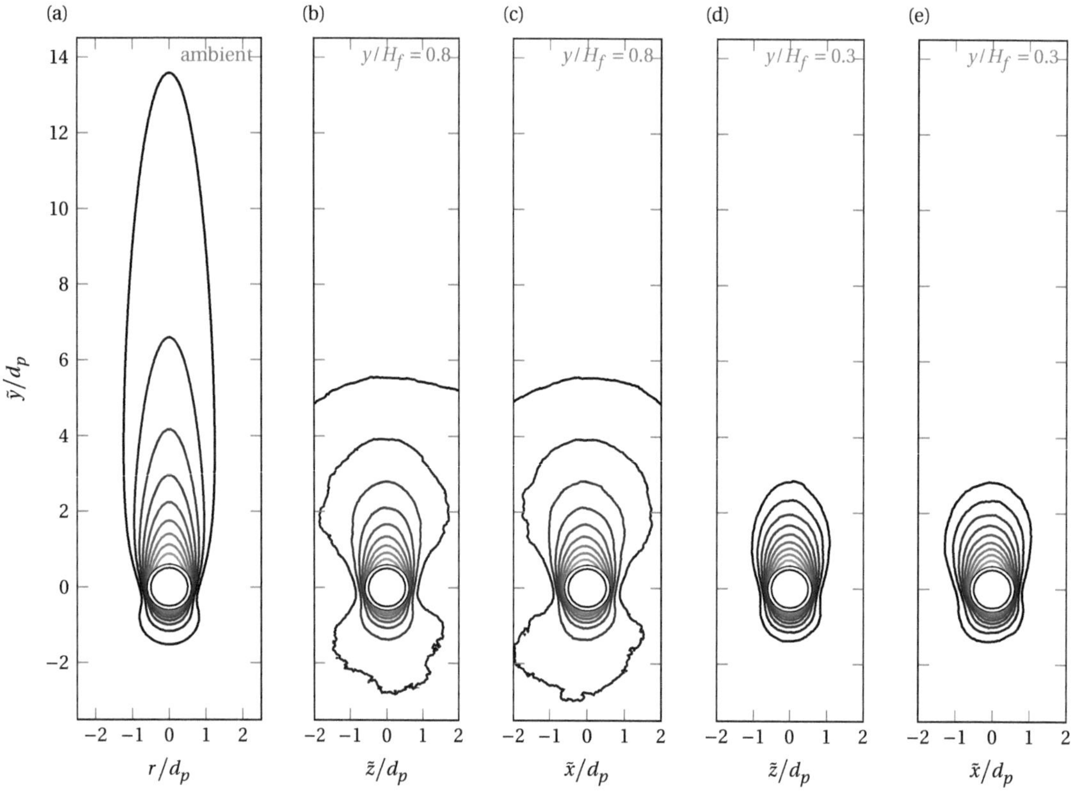

Figure 5.16: Vertical component of particle-conditioned average fluid velocity $\langle v_f \rangle_{\mathbb{P}\mathbb{X}\downarrow}$ in the vicinity of the contaminant particle for G30-D10-HR. The lines indicate isolines of velocity with values at $[-0.1:-0.1:-0.9]$ times the ambient terminal velocity. (a) The undisturbed axisymmetric boundary layer for $Ga = 27.1$ as a reference. (b) The averaged boundary layer from the turbulent simulations at $y = 0.8H_f$ in the streamwise-normal plane. (c) Same as before, but in the spanwise-normal plane. (d) The averaged boundary layer from the turbulent simulations at $y = 0.3H_f$ in the streamwise-normal plane. (e) Same as before, but in the spanwise-normal plane.

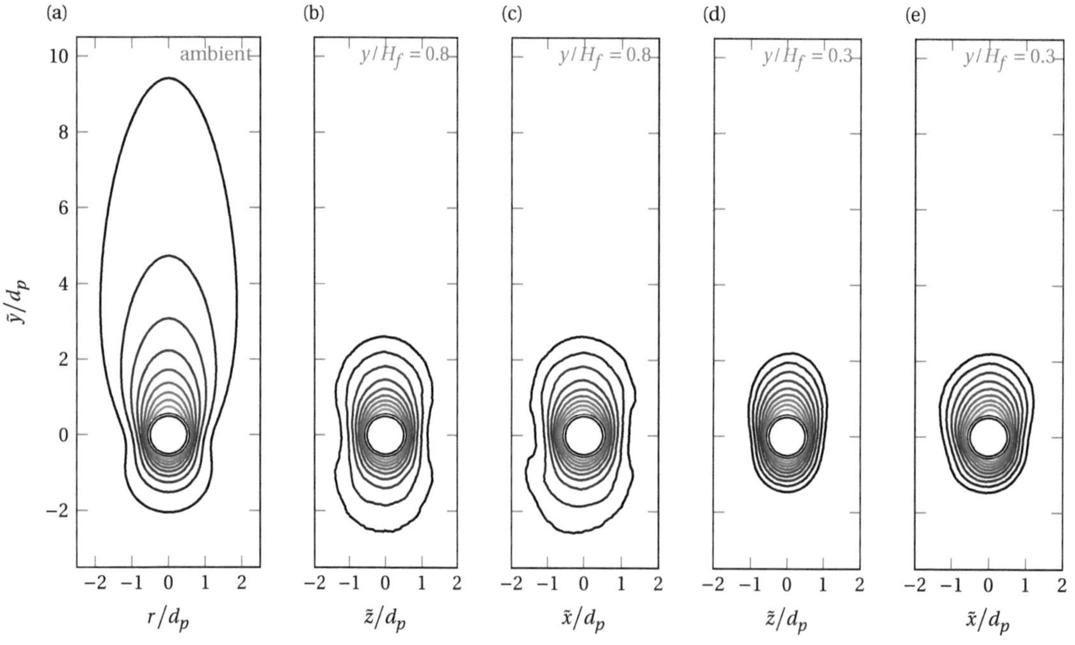

Figure 5.17: Same as fig. 5.16, but for G10-D10-HR.

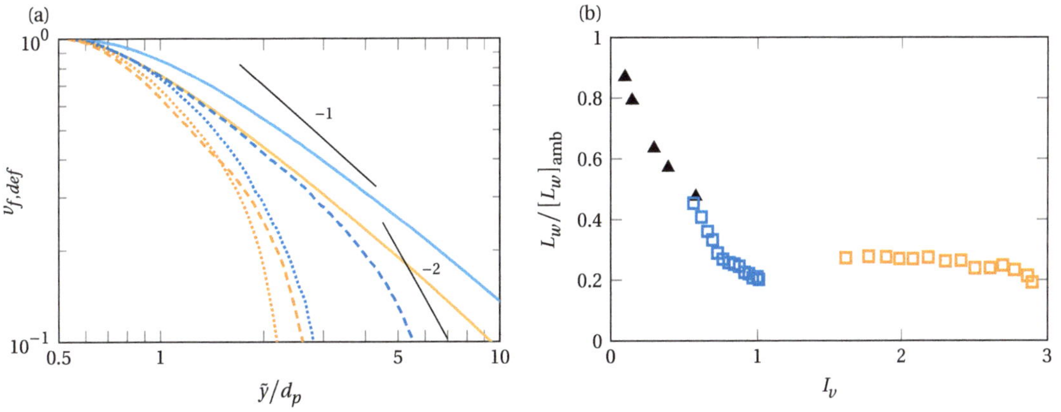

Figure 5.18: (a) Velocity deficit as a function of distance from the particles' centroid for the cases G30-D15-HR (- - ·, ······) and G10-D10-HR (- - ·, ······) at $y = 0.8H_f$ (- - ·) and $y = 0.3H_f$ (······). Also shown are the curves obtained for the converged states of the respective ambient settling simulations at $Ga = 27.1$ (—) and $Ga = 10.8$ (—) as well as black lines indicating the slope of $\tilde{y}^{-1}$ (laminar decay) and $\tilde{y}^{-2}$ (decay in presence of background turbulence). (b) Length of the wake as defined in the main text as a function of local relative turbulence intensity for the cases G30-D15-HR (□) and G10-D10-HR (□). The samples are taken at vertical positions in the range $y \in [0.2, 0.85]\,H_f$ and the values are normalised by the corresponding value for an undisturbed wake at the same value of the Galileo number. The black triangles (▲) indicate the relative length of the wake obtained by Homann et al. (2013) for a sphere towed in homogeneous isotropic turbulence at $Re_{p,\infty} = 20$. Please note that the reference data of Homann et al. (2013) is evaluated for a velocity deficit of 0.5 as opposed to 0.1 for the simulations presented here.

(b-e), most of the defining features are lost, which is unsurprising given the virtual absence of particle-induced motion in the instantaneous picture.

Both fig. 5.16 and fig. 5.17 depict the mean wake at two positions within the channel, and it can be noted that the wake considerably differs between those positions. In order to quantify the length and decay of the wake in the rear of the particles, the velocity deficit is defined as

$$v_{f,def}(\tilde{y}, y) := 1 - \frac{\langle v_f \rangle_{\mathbb{P}\mathbb{X}_\downarrow}(\tilde{x} = 0, \tilde{y}, \tilde{z} = 0, y) - \langle v_c \rangle_{\mathbb{X}_\downarrow}(y)}{v_{p,\infty}}. \tag{5.11}$$

Please note that this deficit is defined with regards to the ambient terminal velocity—similar to the wakes previously shown—and not with respect to a zero mean and also not with respect to an actual slip velocity. The former is due to the fact that $\langle v_f \rangle_{\mathbb{P}\mathbb{X}_\downarrow}$ does not decay to zero anywhere near the particles due to preferential sampling, and the latter is avoided since the definition of an actual slip velocity is ambiguous as will be discussed in §5.4.1.

Figure 5.18(a) shows the decay of the velocity deficit for the two undisturbed and four mean wakes displayed in fig. 5.16 and fig. 5.17. Whereas the undisturbed wakes follow the decay expected for a laminar wake proportional to $\tilde{y}^{-1}$ (Wu and Faeth 1993) reasonably well, this regime can merely be observed over a considerable distance for G30-D10-HR at $y = 0.8H_f$. In the presence of turbulence, the velocity deficit decays significantly quicker, and especially at high relative turbulence intensities no clear power law such as the quadratic decay proposed by Wu and Faeth (1994), Legendre et al. (2006) or Amoura et al. (2010) for the wake of a sphere in a turbulent background could be identified over a significant extent.

A possible explanation for the distinct wake behaviour at a given Galileo number, but at different positions in the channel is given by the varying turbulence intensity experienced by the contaminants as they move vertically through the channel. In particular, the intensity of the fluctuations in the direction of gravity varies by approximately a factor of two for the two positions shown, cf. fig 3.6(c). For this reason, a local relative turbulence intensity is defined in terms of the standard deviation of the vertical fluid velocity, viz.

$$I_v(y) := \frac{\langle v_f^2 \rangle_{xzt}^{1/2}(y)}{v_{p,\infty}}. \tag{5.12}$$

In order to examine the evolution of the velocity deficit as a function of I_v, the length of the wake is defined as the smallest vertical distance on the centre axis originating from the particle's centroid at which the particle-conditioned mean flow is still disturbed by 10% of the value of the ambient terminal velocity in the relative reference frame. In other words, $L_w(y)$ is the distance in the particle-attached coordinate system for which $v_{f,def}(\tilde{y}, y) = 0.1$. The threshold is motivated by the varying decay rates for different parameter points and the observation that for the selected threshold all wakes examined decay proportionally to $\tilde{y}^{-2}$.

Figure 5.18(b) shows the length of the wake normalised by its ambient counterpart as a function of the local relative turbulence intensity, i.e. at different vertical positions in the channel, for the two Galileo numbers in the HR background flow. A consistent trend can be observed for G30-D15-HR, where for decreasing I_v the length of the wake approaches its ambient value as one would expect. Furthermore, the ratios which are observed for $Ga = 27.1$ ($Re_{p,\infty} = 17.9$) agree well with the results of Homann et al. (2013) for a sphere ($Re_{p,\infty} = 20$) towed through homogeneous isotropic turbulence in terms of the general trend as well as the quantitative value within the short parameter range in which they overlap. When examining higher relative turbulence intensities, a saturation in wake length seems to occur around $I_v \approx 1$ where the vertical velocity fluctuations are of the same order as the ambient terminal velocity. The nearly constant wake length observed for G10-D10-HR for which $I_v \gg 1$ at all positions supports this observation. This might correspond to what Amoura et al. (2010) describe as "a wake controlled by the incident turbulence" for which the far wake results "from the distortion of the incident turbulence by the sphere". They state that the length of the attached wake in this regime is approximately $1.5d_p$ and that as soon as the wake is fully controlled by the incident turbulence, it ought to decay proportional to $\tilde{y}^{-2}$. Both of these predictions agree well with what is shown in fig. 5.18.

5.3.3 Effect of shear on settling velocity

The wakes previously presented are all axisymmetric on average to a good estimation. This is generally not the case when vertical positions closer to sediment bed are investigated where the fluid shear is considerably higher in the space-time average, cf. fig 3.6(b). For this reason, the role of significant shear on the averaged particle wakes will now be subject of discussion.

As the comparison with an undisturbed wake in §5.3.2 proved to be expedient, an extension of this comparison with regards to shear may aid in the interpretation of the mean turbulent wakes as well. For this reason, the ambient settling simulations of §3.7 are modified such that the particles now settle in the presence of a linear velocity profile whose gradient is coaligned with the direction of gravity. To achieve this, the free-slip upper boundary is substituted by a no-slip boundary which moves with a constant speed of U_w in x-direction, and the background flow is initialised by its laminar profile, viz.

$$u_f(y) = \frac{y}{L_{y,c}} U_w, \tag{5.13}$$

where $L_{y,c}$ denotes the vertical extent of the simulation domain. This setup corresponds to plane Couette flow and a sketch and further information on the computational domain are provided in fig. 5.19(a).

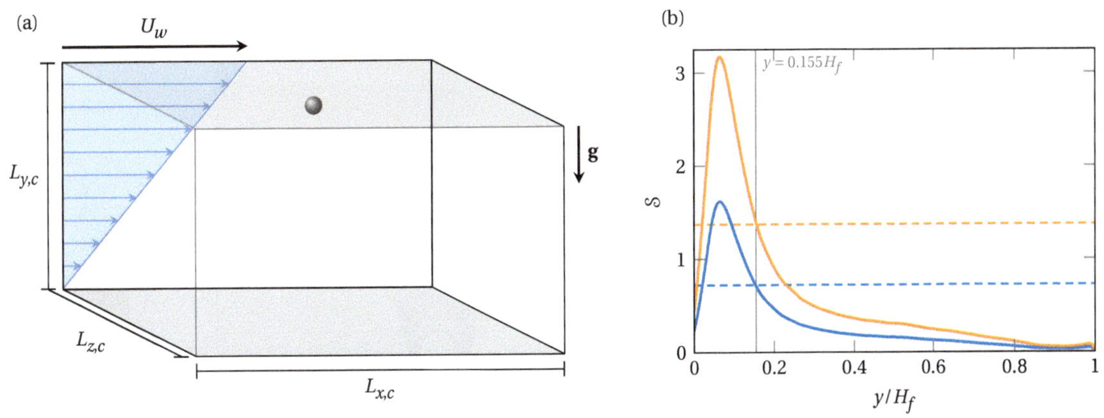

Figure 5.19: (a) Sketch of the computational setup used for the reference simulations to investigate the effects of shear on the particle wakes. The setup is similar to that of the ambient settling reference cases described in §3.7. However, a constant background shear of $U_w/L_{y,c}$ is achieved throughout the domain by means of introducing a no-slip wall as the upper boundary which moves at velocity U_w in x-direction. The numerical parameter are identical to those of the ambient settling simulations of G30-D10-LR for $Ga \approx 30$ and G10-D6-LR for $Ga \approx 10$ except for a substantially larger domain size of $L_{x,c} \times L_{z,c} = 5L_{x,a} \times 2L_{z,a}$. (b) The value of the shear parameter based on the mean shear as a function of vertical position for the turbulent cases G30-D15-HR (—) and G10-D10-HR (—). The constant values of $\mathcal{S}$ for the reference simulations in laminar Couette flow are indicated by dashed lines: $Ga \approx 30$ (- - -) and $Ga \approx 10$ (- - -).

It should be noted that while laminar plane Couette flow is stable with regards to infinitesimal perturbations (Davey 1973, Romanov 1973) independent of the Reynolds number, finite-amplitude perturbations as introduced by the presence of the moving particle may trigger the transition to a turbulent state. The value of the Reynolds number for the reference simulations reads

$$Re_c := \frac{U_w L_{y,c}}{\nu_f} = 4.5 \cdot 10^3, \tag{5.14}$$

and is indeed sufficiently large such that turbulence may be triggered by three-dimensional finite-amplitude perturbations (Orszag and Kells 1980). However, during the time of observation, no significant deviations from the laminar state apart from the wake of the particle were observed, indicating that either the perturbation is too insignificant to trigger the transition or that its time-scale is significantly longer than the observation time.

The presence of background shear introduces another particle-specific parameter which needs to be matched between the reference and the turbulent simulation, see e.g. Bagchi and Balachandar (2002) for the parameter space of a similar problem. Therefore, the shear parameter is introduced as

$$\mathcal{S} := \frac{2\langle S_{xy}\rangle_{xzt}\, d_c}{v_{p,\infty}}. \tag{5.15}$$

Please note that this parameter is typically defined with regards to the relative velocity between the two phases, but in the spirit of the previous analysis it is assumed that the ambient terminal velocity is a sufficiently good approximation instead. Figure 5.19(b) displays the local value of the shear parameter for the cases of interest, G30-D15-HR and G10-D10-HR, as well as the corresponding values for the reference simulations. As indicated, the reference setup corresponds approximately to the conditions that are experienced on average by the contaminants[6] at $y \approx 0.15H_f$.

[6] As the shear parameter is based on the unconditioned spatio-temporal mean of the fluid shear, it is not taken into account that due to preferential sampling the shear experienced by the contaminants might be different.

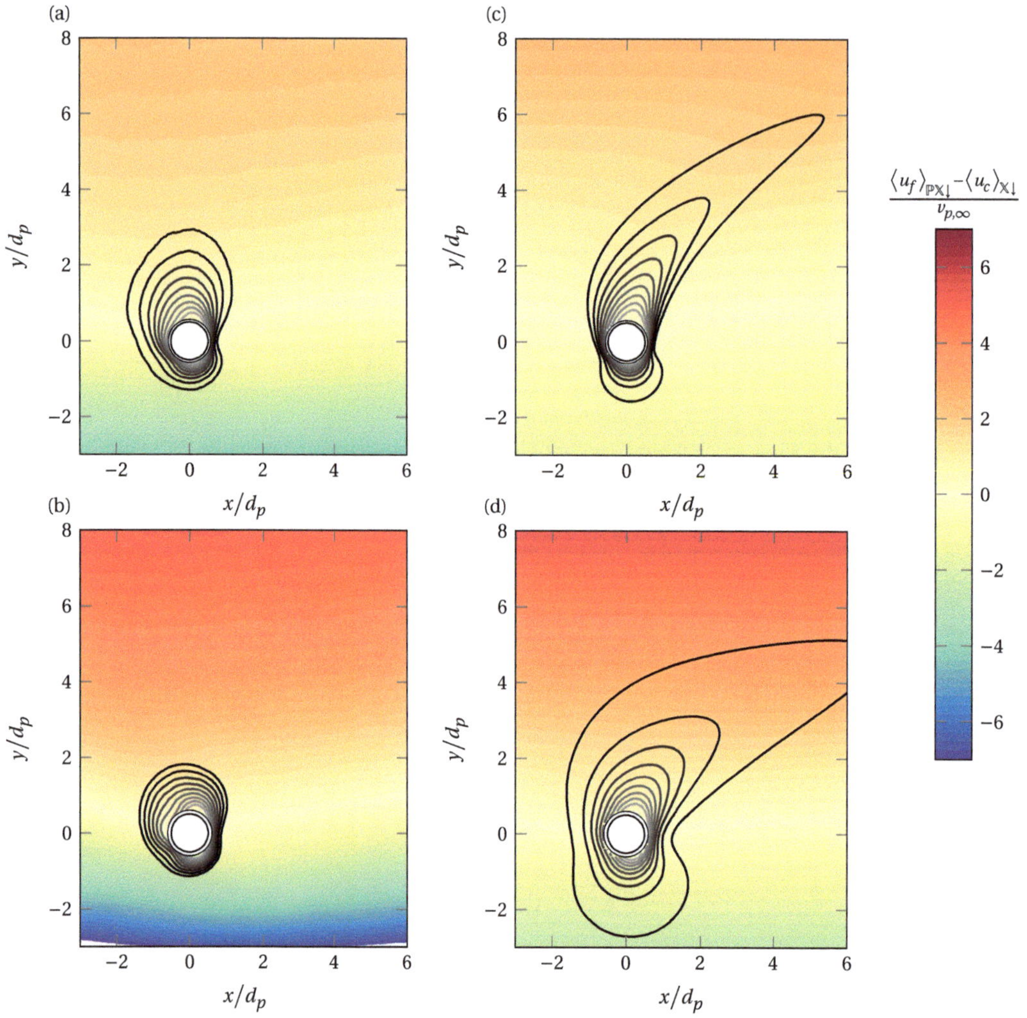

Figure 5.20: Particle wakes in the presence of significant mean shear. The solid lines indicate isovalues of the vertical fluid velocity $\langle v_f \rangle_{\mathbb{P}\mathbb{X}\downarrow}$ for the values $[-0.1:-0.1:-0.9]$ times the ambient terminal velocity in the particle-attached frame of reference. The coloured background shows the streamwise fluid velocity $\langle u_f \rangle_{\mathbb{P}\mathbb{X}\downarrow}$ in the relative frame. (a) Mean wake for G30-D15-HR. (b) Mean wake for G10-D10-HR. (c) Same parameter point as G30-D15-HR, but for an isolated particle settling in laminar Couette flow. (d) Same parameter point as G10-D10-HR, but for an isolated particle settling in laminar Couette flow. The value of the shear parameter reads $\mathcal{S} = 0.71$ for (a,c) and $\mathcal{S} = 1.37$ for (b,d).

Figure 5.20 displays the mean particle wakes from the turbulent simulations and the corresponding reference cases in laminar Couette flow. The normalised fluid shear in the rear of the particles is in good agreement between the turbulent and laminar cases, however, a steeper gradient in the front can be observed for the turbulent mean shear due to its nonlinear trend. Qualitatively, the wakes of the laminar reference and the turbulent mean are rather similar in terms of their shear-induced deformation. Their primary distinction is the considerably shorter wake in the turbulent case, which is in accordance with the observations made in §5.3.2. Nonetheless, the near-particle region exhibits congenial modifications caused by the shear, which suggests that the laminar reference might be a suitable proxy to assess its effect. When evaluating the settling velocity under linear shear conditions, it is found that its value remains unchanged for $Ga \approx 30, \mathcal{S} = 0.71$ and is reduced by approximately 4% for $Ga \approx 10, \mathcal{S} = 1.37$. Under the conjecture that these inferences are transferable to the turbulent case, it can be concluded that the effect of mean shear on the settling velocity is negligible in the current context.

5.4 Relative velocity on a microscale

The preceding analysis regarding particle wakes was conducted under the supposition the relative velocity between the fluid and solid phase is of the order of the ambient terminal velocity. In the following, an attempt is made to quantify the actual slip velocity in order to justify this assumption and to gain insights on the impact of wake modifications on the settling velocity under the influence of turbulence on a microscale.

5.4.1 Slip velocity

Figure 5.8(b) suggests that the ratio between the settling velocity of the contaminants and the fluid velocity experienced by them is approximately constant over a large vertical range for the majority of parameter points under investigation. By subtracting a measure for the instantaneous fluid velocity experienced by the particle from the particle velocity, the effect of preferential sampling is effectively disregarded, and the flow conditions in the vicinity of the particles can be investigated on a microscale. For this purpose, an instantaneous variant of the sphere-average is introduced, which reads

$$\langle u_{f,\alpha}\rangle_{\mathbb{S}}^{(i)}(r,t) := \frac{\oint_{S_v(r)} u_{f,\alpha}(\tilde{\mathbf{x}}^{(i)},t)\,\chi_f(\tilde{\mathbf{x}}^{(i)},t)\,\mathrm{d}S_v}{\oint_{S_v(r)} \chi_f(\tilde{\mathbf{x}}^{(i)},t)\,\mathrm{d}S_v} \tag{5.16}$$

for the i-th contaminant particle. Here, the instantaneous fluid velocity field $u_{f,\alpha}(\tilde{\mathbf{x}}^{(i)},t)$ is translated such that the origin of the coordinate system coincides with the centroid of the i-th particle, and again, the surface of the virtual sampling sphere $S_v(r)$ is defined in this coordinate system and its radius is set to the same value as earlier.

Under the assumption that the interaction between the particle and the nearby fluid velocity—nearby meaning of the order of the radius of the virtual sampling sphere—is invariant with regards to the flow conditions at a larger scale, i.e. preferential sampling, the difference in $u_{c,\alpha}$ and $\langle u_{f,\alpha}\rangle_{\mathbb{S}}$ is not subject to the bias described in §4.2.2 anymore, and hence, the spatial-averaging procedure can be replaced by an unweighted binning approach for investigations at the microscale.[7] Therefore, the instantaneous slip velocity $u_{c,\alpha} - \langle u_{f,\alpha}\rangle_{\mathbb{S}}$ is averaged within a wall-parallel slab using a standard binned-averaging operation, viz.

$$\langle u_{c,\alpha} - \langle u_{f,\alpha}\rangle_{\mathbb{S}}\rangle_{\Delta}(y_0,y_1) := \frac{\int_0^{t_{obs}} \sum_i^{N_{c\Sigma}} \Delta\big(y_c^{(i)}(t),y_0,y_1\big)\,\big(u_{c,\alpha}^{(i)}(t) - \langle u_{f,\alpha}\rangle_{\mathbb{S}}^{(i)}(t)\big)\,\mathrm{d}t}{\int_0^{t_{obs}} \sum_i^{N_{c\Sigma}} \Delta\big(y_c^{(i)}(t),y_0,y_1\big)\,\mathrm{d}t}. \tag{5.17}$$

with the binning kernel function

$$\Delta(y_p,y_0,y_1) := \begin{cases} 1 & \text{if} \quad y_p \in [y_0,y_1) \\ 0 & \text{otherwise} \end{cases}, \tag{5.18}$$

where y_0, y_1 are the extents of the bin to be specified.

Figure 5.21 shows this slip velocity as a function of the relative turbulence intensity averaged in the region $y/H_f \in [0.3, 0.8]$ where $\langle v_p\rangle_{\mathbb{X}\downarrow} / \langle v_f\rangle_{\mathbb{S}\mathbb{X}\downarrow}$ only varies little for the majority of cases. Its value is found

[7] This assumption appears to be justified if the length scale of the regions which are preferentially sampled is considerably larger than the particle diameter.

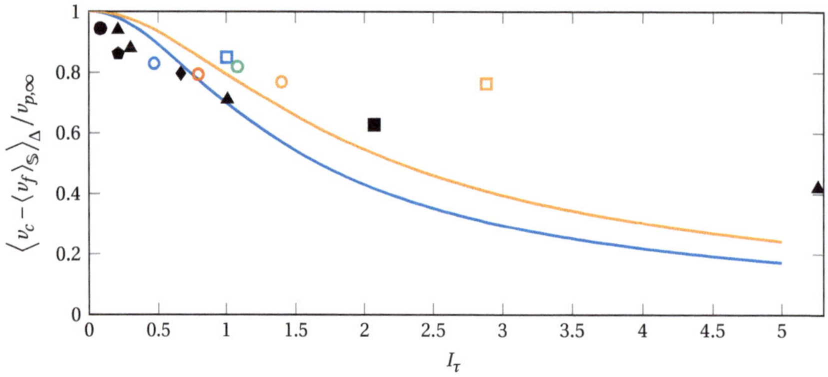

Figure 5.21: Average relative particle velocity of the contaminants at the microscale as a function of relative turbulence intensity and scaled by its ambient value. Reference data by Fornari et al. (2016) in sustained homogeneous isotropic turbulence (▲) is shown for $Ga = \{200, 145, 60, 19\}$ (from left to right) at density ratios of $\rho_p/\rho_f \in [1.0004, 1.04]$ and a particle diameter of twelve Kolmogorov length scales, where the relative turbulence intensity is defined as the ratio of the root-mean-square of the turbulent velocity fluctuations to the ambient terminal velocity of the particles. Further reference points are available from vertical channel flow evaluated on the centre line, for which the parameters in terms of the triplet $(\rho_p/\rho_f, Ga, d_p^+)$ are given by ●: (2.2, 115.4, 11.3), García-Villalba et al. (2012). ■: (2, 23, 19), Zhu et al. (2020). ♦: (2, 40, 18), Zhu et al. (2020). ⬟: (2, 99.5, 21), Yu et al. (2021). The solid lines indicate the reduction in settling velocity expected from the non-linear drag model by Homann et al. (2013), evaluated for $Ga = 30$ (—) and $Ga = 10$ (—). Legend: G30-D15-HR (□), G10-D10-HR (□), G30-D10-LR (○), G10-D6-LR (○), G15-D10-LR (○) and G15-D6-LR (○).

to be generally reduced in comparison to the ambient value. Keeping in mind the arbitrariness of the definition of a local fluid velocity and the resulting uncertainty of the obtained relative velocity, the general magnitude and trend of this observation conform reasonably well to the result of Fornari et al. (2016) who found a reduction in the settling velocities of particles in sustained homogeneous isotropic turbulence in a similar range of relative turbulence intensities, as well as data available from vertical channels.

One potential explanation for the reduction of the settling velocity in a turbulent environment is the nonlinear dependence of the drag coefficient on the particle Reynolds number (based on the slip velocity) for $Re_p \gg 1$, where a temporal variation in the slip velocity may result in an increased mean drag force, cf. §2.3.3. Homann et al. (2013) derived an expression for this increase assuming a nonlinear drag law and Gaussian fluctuations in the relative velocity. The resulting drag coefficient depends on the relative turbulence intensity and is given by[8]

$$C_d^{HBG}(Re_p, I_\tau) := \left(1 + \frac{0.1897 Re_p^{0.687}}{1 + 0.15 Re_p^{0.687}} I_\tau^2\right) C_d^{SN}(Re_p). \tag{5.19}$$

Here, C_d^{SN} is the ambient drag coefficient as obtained from the empirical drag law of Schiller and Naumann (1933), viz.

$$C_d^{SN}(Re_p) = \frac{24}{Re_p}\left(1 + 0.15 Re_p^{0.687}\right). \tag{5.20}$$

It should be noted that the considerations of Homann et al. (2013) were made for a sphere translating at a fixed velocity. Extending their argument towards freely moving particles, a variation in the drag force naturally leads to a variation in the relative velocity, resulting in a feedback loop between these

[8] Please note that the prefactor of the term involving I_τ differs from the one stated in Homann et al. (2013) as a numerical evaluation of the underlying integral yielded a value differing from the one stated in the publication (private communication, Uhlmann, 2015).

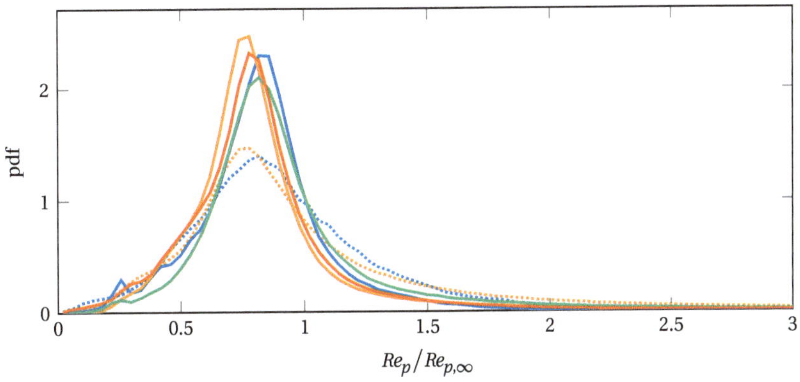

Figure 5.22: Probability density function of the instantaneous values of the particle Reynolds number based on the three-dimensional slip velocity in the region $y/H_f \in [0.3, 0.8]$. Legend: G30-D15-HR (······), G10-D10-HR (······), G30-D10-LR (—), G10-D6-LR (—), G15-D10-LR (—) and G15-D6-LR (—).

two quantities. If all other unsteady effects as well as lateral movement are neglected, C_d^{HBG} can be substituted into eq. (2.49) in order to obtain a solution for the modified slip velocity by solving

$$\tfrac{4}{3}Ga^2 = C_d^{HBG}(Re_p, I_\tau)Re_p^2 \tag{5.21}$$

given the Galileo number and the relative turbulence intensity. In fig. 5.21, two families of solutions are shown for $Ga = \{10, 30\}$, respectively. While the predictions for $I_\tau \approx 1$ are in the same range as the numerical data, this simple model considerably overpredicts the reduction in relative velocity for $I_\tau \gg 1$.

5.4.2 Particle Reynolds number

In the previous discussion on the slip velocity, the lateral components of the relative velocity were disregarded in order to simplify the argument. For some applications, and especially for the accurate modelling of particle-laden flows using unresolved methods, it might, however, be of interest to be aware of the range of particle Reynolds numbers which are reached instantaneously. For this purpose, the probability density distributions of the particle Reynolds number based on the three-dimensional slip velocity,

$$Re_p = \frac{\left|\mathbf{u}_c - \langle \mathbf{u}_f \rangle_{\mathbb{S}}\right| d_p}{\nu_f}, \tag{5.22}$$

are provided in fig. 5.22. While the instantaneous Reynolds number is most likely lower than that based on the terminal velocity—an observation in accordance with the findings regarding only the vertical component—it can occasionally be found to be increased by a factor of three.

In general, it is found that the particle velocity is severely correlated to the fluid velocity sampled on the virtual sphere. Table 5.1 provides the Pearson correlation coefficients for each velocity component and various parameter points. The most notable deviations from unity are observed for $Ga \approx 30$, and decorrelation is generally higher for the vertical and spanwise component than for the streamwise component. It can be assumed that a high correlation between $u_{c,\alpha}$ and $\langle u_{f,\alpha} \rangle_{\mathbb{S}}$ is a prerequisite to obtain similar probability distributions for the particle velocity fluctuations and the particle-conditioned fluid velocities as presented in §4.3.2 and §5.2.4.

Table 5.1: Linear correlation coefficients defined in similar to eq. 4.8 for the individual components of the instantaneous particle velocity and the corresponding sphere-averaged fluid velocity.

	G30-D10-LR	G15-D6-LR	G15-D10-LR	G10-D6-LR	G30-D15-HR	G10-D10-HR
$\varrho\left(u_c, \langle u_f \rangle_{\mathbb{S}}\right)$	0.98	0.99	0.99	1	0.98	1
$\varrho\left(v_c, \langle v_f \rangle_{\mathbb{S}}\right)$	0.95	0.98	0.98	0.99	0.95	0.99
$\varrho\left(w_c, \langle w_f \rangle_{\mathbb{S}}\right)$	0.96	0.99	0.98	0.99	0.95	0.99

5.5 Domain size effects

It was found in §5.2.2 that the preferential arrangement of the settling contaminants likely coincides with the lateral position of large-scale high-speed streaks which are present throughout the outer layer of the turbulent background flow. While the vertical length scale of these turbulent structures is of the order of the clear fluid height, their streamwise extent was shown to be of as long as 3–100 H_f in closed channels (Lozano-Durán and Jiménez 2014a). The lower end of this range corresponds to the structures known as large-scale motions (LSMs) while the upper range is known as very-large scale motions (VLSMs), cf. §2.2.3. These structures are known to contribute substantially to the overall Reynolds stress (Guala et al. 2006, Lozano-Durán et al. 2012b), and since contaminants were shown to preferentially sample Q4-events during settling, it ought to be assumed that an adequate representation of these large-scale structures is required for an accurate description of transient settling.

It is apparent that the streamwise period of the computational domains employed for all background flows considered in this thesis are insufficiently long to adequately capture events at these length scales—a circumstance which arises partly from the desired suppression of bedform instabilities, but is also due to the large computational cost of particle-laden flow simulations. In fact, it was previously noted in §5.2.2 that the velocity streaks of the outer layer self-connect across the periodic boundary, and hence, constitute a coherent region of infinite length. Since no suitable reference data for turbulent flow over a mobile sediment bed exists, the energy content of the various spatial scales of the turbulent motion is in the following compared to an open-channel flow over a smooth wall at a comparable friction Reynolds number in a computational domain with substantially larger streamwise and spanwise periods. For this purpose, the velocity components are expanded as a Fourier series with discrete coefficients $\hat{u}_{f,\alpha}^{(k,l)}(y,t)$ which are defined under the convention that the backtransform reads

$$u_{f,\alpha}(x,y,z,t) = \sum_{k=-\infty}^{\infty} \sum_{l=-\infty}^{\infty} \hat{u}_{f,\alpha}^{(k,l)}(y,t) \exp\left(-\mathrm{i}\left(\kappa_x^{(k)} x + \kappa_z^{(l)} z\right)\right). \tag{5.23}$$

The l-th/k-th harmonics of the periodic system in x/z-direction are indicated by superscripts to the Fourier coefficients, and the discrete wavenumbers are given by

$$\kappa_x^{(k)} := \frac{2\pi k}{L_x} \quad \text{and} \quad \kappa_z^{(l)} := \frac{2\pi l}{L_z}. \tag{5.24}$$

The Fourier series expansion allows the overall energy of turbulent fluctuations to be expressed as sum of the energy contained at each wavenumber (Parseval's theorem), viz.

$$\sum_{k=-\infty}^{\infty} \sum_{l=-\infty}^{\infty} \left|\hat{u}_{f,\alpha}^{(k,l)}\right|^2 = \left\langle u_{f,\alpha}^2 \right\rangle_{xz} \tag{5.25}$$

Therefore, the variation of the absolute square of the Fourier coefficients with respect to the streamwise and spanwise wavenumbers is referred to as the energy spectrum of the particular fluid velocity component. Following common practice, these spectra will be presented in terms of the wavelengths which are related to the corresponding wavenumbers by

$$\lambda_x^{(k)} := \frac{2\pi}{\kappa_x^{(k)}} = \frac{L_x}{k} \quad \text{and} \quad \lambda_z^{(l)} := \frac{2\pi}{\kappa_z^{(l)}} = \frac{L_z}{l}. \tag{5.26}$$

The reference spectra for turbulent open-channel flow over a smooth wall are provided by Bauer et al. (2022). Their domain size has a wall-parallel cross-sectional area of $L_x \times L_z = 12\pi \times 4\pi$ in terms of the fluid height, which is approximately 75 times larger than that of the computational domain employed in the transient settling simulations. As a consequence, the LSMs and VLSMs are resolved to a far greater extent in their simulation. The value of the friction Reynolds number is $Re_\tau = 596$ which closely matches that of HR when the bed is still at rest ($Re_\tau \approx 622$). Indeed, the following discussion will be held exclusively with regards to HR, as its domain is the shortest and LSMs and VLSMs are thought to be more relevant at high Reynolds number due to their different scaling compared to other turbulent structures (del Álamo and Jiménez 2003) and the growing separation between inner and outer turbulence scales.

Figure 5.23(a) shows a comparison between the two-dimensional spectra of the streamwise and spanwise fluid velocities for the HR background flow in the absence of contaminant particles and for the reference smooth open-channel flow at $y = 0.5$. The spectra are premultiplied by the corresponding wavenumbers such that energy contribution over any wavenumber range is obtained from the area under the curve, and thus, the most energetic modes can be identified by the peak locations in this representation (Perry et al. 1986). As was expected, the spectrum of the streamwise velocity is considerably truncated by the period of the domain. However, the resolved portion of the spectrum coincides reasonably well with the reference data despite the notable differences in setup. It should be noted that the contours are presented in absolute terms, i.e. they are scaled by their respective friction velocity and not relative to their total energy $\langle u_f' u_f' \rangle_{xzt}$. Furthermore, the total energy content agrees reasonably well between the HR background flow and the reference data, and is thus not considerably affected by the truncation. This type of behaviour was already described by Álamo et al. (2004) and Lozano-Durán and Jiménez (2014a) for varying box sizes in smooth-wall closed channel flow, and Lozano-Durán and Jiménez (2014a) argued that the truncated unresolved part of the spectrum is essentially captured by the $(k = 0, l > 0)$ harmonics, and that the interaction of these modes with the resolved part of the spectrum is well represented. If this is indeed the case, one might argue that the fluid-particle interaction is likely not considerably different in a truncated domain compared to a flow with fully resolved LSMs and VLSMs. This is because a direct interaction presumably occurs around the scale of the particle size, i.e. within the fully resolved turbulent scales which are captured accurately, and large-scale organization of particles is then a result of the interaction between fluid modes of different sizes which seems to be little affected by the truncation according to the arguments of Lozano-Durán and Jiménez (2014a).

More evidence that the short streamwise length of the domain likely does not substantially alter the observed settling behaviour is provided by the spectra of vertical fluid velocities—the component of velocity which is directly linked to the contaminants' settling velocity and plays a key role in preferential sampling due to the turbulence intensity effect described in §4.3.2 and §5.2.4. As is demonstrated in 5.23(b), the fluctuations of v_f generally occur at smaller scales which are captured for the most part by the computational domain.[9] This disparity in length scale may be explained in terms of the tall attached

[9] The difference in length scale and anisotropy between the streamwise and wall-normal velocity component is already visually evident in the instantaneous visualizations shown previously in fig.5.7

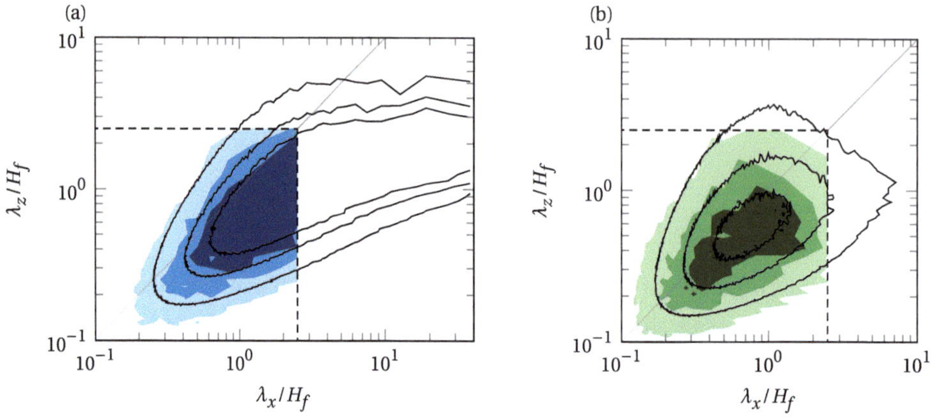

Figure 5.23: Two-dimensional premultiplied energy spectra of the fluid velocity $\kappa_x^{(k)}\kappa_z^{(l)}\langle|\hat{u}_{f,\alpha}^{(k,l)}|^2\rangle_t$ at $y = 0.5H_f$. The shaded regions display the spectra obtained for HR in the absence of contaminant particles. The black lines represent the corresponding spectra for open-channel over a smooth wall at $Re_\tau = 596$ in a considerably larger computational domain (Bauer et al. 2022). The fundamental wavelengths $(\lambda_x^{(1)}, \lambda_z^{(1)})$ of the transient settling domain are indicated by dashed lines (- - -). (a) Streamwise component $\hat{u}_f^{(k,l)}$ with contour levels at $\{0.1, 0.2, 0.3\}\kappa_x^{(1)}\kappa_z^{(1)}u_\tau^2$. (b) Vertical component $\hat{v}_f^{(k,l)}$ with contour levels at $\{0.08, 0.16, 0.024\}\kappa_x^{(1)}\kappa_z^{(1)}u_\tau^2$. An average of all quadrants is shown exploiting the symmetry of positive and negative wavenumbers.

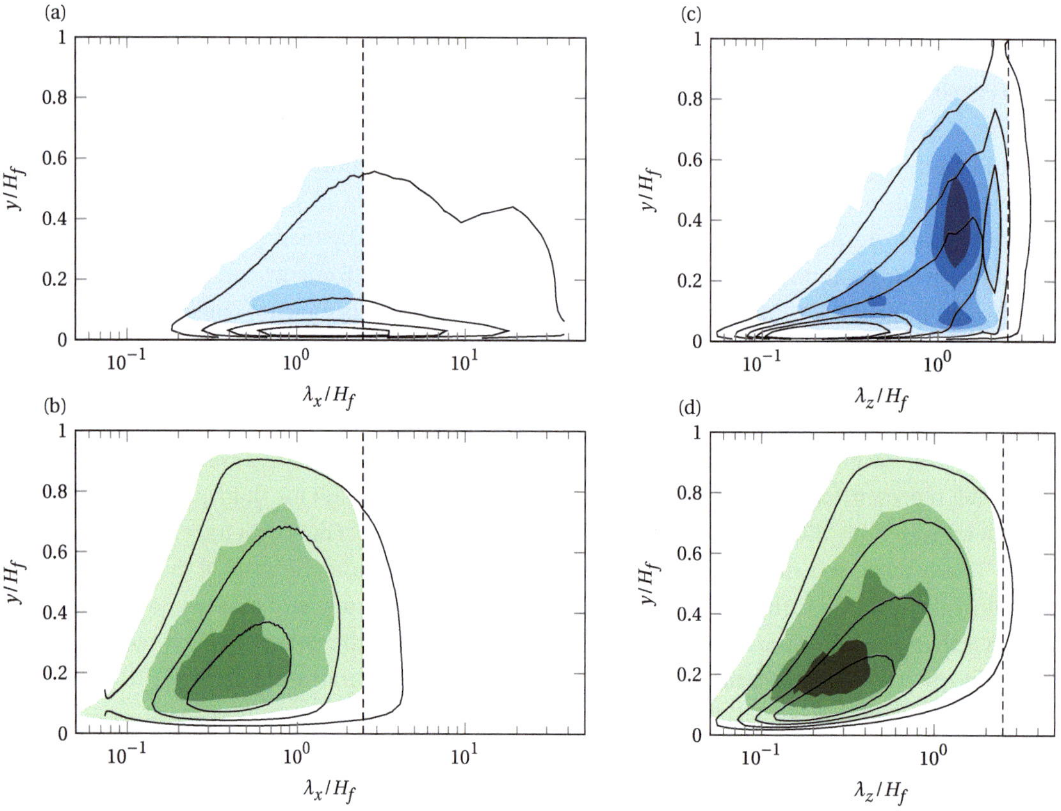

Figure 5.24: Premultiplied spectra (a,b) summed over all spanwise modes $\kappa_z^{(l)}\sum_{k=-\infty}^{\infty}\langle|\hat{u}_{f,\alpha}^{(k,l)}|^2\rangle_t$, and (c,d) summed over all streamwise modes $\kappa_x^{(k)}\sum_{l=-\infty}^{\infty}\langle|\hat{u}_{f,\alpha}^{(k,l)}|^2\rangle_t$. The shaded regions display the spectra obtained for HR in the absence of contaminant particles. The black lines represent the reference data from Bauer et al. (2022). The respective fundamental wavelength ($\lambda_x^{(1)}$ or $\lambda_z^{(1)}$) of the background flow is indicated by a dashed line (- - -). (a,c) Streamwise component with contours at $\{0.2, 0.4, 0.6, 0.8, 1.0\}\kappa_z^{(1)}u_\tau^2$. (b,d) Vertical component with contours at $\{0.05, 0.1, 0.15, 0.2\}\kappa_z^{(1)}u_\tau^2$. An average over positive and negative wavenumbers is shown.

vortex clusters described by del Álamo et al. (2006) where the cluster itself is relatively compact and a contributor to large-scale vertical motion, whereas its "wake" is significantly longer and gives rise to the very long anisotropic motions observed in the spectrum of u_f (Jiménez 2012).

In order to inspect the wall-normal dependence of the energy distribution among the scales, the one-dimensional spectra obtained by means of summation over one homogeneous direction are now considered. Figure 5.24(a,c) displays the one-dimensional spectrum of the streamwise velocity as a function of y and the wavelength in the non-collapsed homogeneous direction. The first observation to be made is the lack of the energy peak at small wavelengths close to the (virtual) wall which is associated to the streaks of the buffer layer, which has already been discussed in §2.3.1. However, it is known that the absence of buffer layer structures has essentially no implications on the statistics of the logarithmic layer (Flores and Jiménez 2006) nor on the properties of attached vortex cluster (Flores et al. 2007), which is reflected by the fact that the remainder of the spectrum coincides reasonably well with the reference data. The second observation to be made concerns the spanwise period of the computational domain. When Flores and Jiménez (2010) investigated the minimal flow units of the logarithmic layer they found that a minimal spanwise period of $L_z \approx 3H_f$ is required in order to have "healthy" unrestricted turbulence. However, according to their analysis, a smaller value of the spanwise period does not lead to the decay of turbulence, but rather restricts the region where the turbulence is described accurately close to the wall. For the HR background turbulence, the spanwise period is $L_z \approx 2.5H_f$ and hence smaller than the minimum width stated by Flores and Jiménez (2010). In terms of the one-dimensional spectra, this manifests itself in terms of slightly compressed spectra of both u_f and v_f, see fig. 5.23(c,d), although the remainder of the spectra does not show conclusive anomalous behaviour. According to the arguments of Flores and Jiménez (2010), a subminimal domain restrict the healthy turbulence to a height of approximately $0.3L_z$, which corresponds to $0.75H_f$ for HR, and thus, one should be keep this uncertainty in mind when interpreting the simulation results.

On top of the restrictive properties of a small computational box on flow quantities, it should also not be disregarded that a single realization of an initial value problem—such as the transient settling discussed—may differ substantially based on the state of the flow during initialization. For example, the large-scale streak in the minimal flow unit of the logarithmic layer is known to burst quasi-periodically (Flores and Jiménez 2010), and hence, the flow on the largest scale considerably varies in time. As a consequence, even if the amount of sample particles is plenty in a single realization, the obtained particle statistics may exhibit a bias if the relevant flow characteristic are variant on a time scale comparable or larger than the settling time. This is less likely to be a problem in a larger domain where different states of the flow can coexist at a single instance in time. With regards to the current study, the LR and RP simulations presumably do not suffer from this problem as multiple initial conditions or a longer domain have been simulated, while for the HR simulations the settling was only realised once because of the high computational costs.

5.6 Summary and concluding remarks

In this chapter, the settling of contaminants was examined with regards to the regions of the background flow which are predominantly visited by the particles. A comparison between the streamwise-time-averaged flow and the spanwise-resolved crossing locations of wall-parallel planes suggests that the contaminants accumulate in high-momentum pathways, and consequently in downdrafts of the secondary flow. By means of particle-conditioned averaging, it was deduced that this corresponds to an accumulation in the high-momentum regions associated to large-scale streaks of the turbulent outer layer in the instantaneous picture, which are thought to be a result of the vortex clusters described by del Álamo et al. (2006).

While the locations of the large-scale streaks and the spanwise arrangement coincide well, this interaction at the length scales of $\mathcal{O}(H_f)$ is deemed insufficient to fully explain the enhancement of settling velocity described in chapter §4. In fact, the particle-conditioned average indicates that the contaminants also preferentially sample downward fluid motion at scales which are significantly smaller than H_f, but must be larger than the particles themselves. Although it is a possibility that velocity fluctuations at this scale are induced by the contaminants themselves, it is considered more likely that these regions originate from the background turbulence as the magnitude of the average nearby fluid velocity decreases with with increasing terminal velocity. However, this argument requires further refinement regarding the analysis of clustering.

Given the importance of large-scale flow structures, the adequacy of the size of the computational domain is discussed regarding the ability to appropriately represent these features. It is argued that while the LSMs are indeed truncated and the VLSMs remain unresolved, no substantial implications on the settling behaviour are to be expected. This is largely due to the fact, that the energy of the unresolved motion is properly captured by spanwise modes which are infinitely long in streamwise direction, and that the interaction between these modes and the resolved portion of the spectrum is well represented according to the arguments of Lozano-Durán and Jiménez (2014a).

A systematic analysis of the immediate surrounding of the contaminants reveals that the enhancement in streamwise and vertical particle velocity can be explained adequately by the fluid velocity sampled by the particles. It is found that the contaminants preferentially sample sweep events, and that the tendency to do so increases with increasing relative turbulence intensity. Moreover, a quadrant analysis of the sampled fluid velocity exhibits a striking similarity to that of the particle velocity fluctuations which reinforces the heuristic inferences made in chapter §4 regarding the role of turbulence intensity.

On a microscale, the averaged wakes of the contaminants retain some similarities to their undisturbed settling states. This similarity is progressively lost at increasing relative turbulence intensity until the average wake is confined into a narrow region of $\mathcal{O}(d_p)$ in the rear of the particle which is consistent with the predictions of Amoura et al. (2010). At low to moderate turbulence intensities, the observed length of the wake is in good agreement with the observations of Homann et al. (2013) in a similar parameter range for a towed sphere. The transition in behaviour is found to occur at a relative turbulence intensity of around unity. Furthermore, the characteristics of the wake at a single value of the Galileo number vary with respect to vertical channel coordinate in a trend which can be explained by a local value of the relative turbulence intensity.

Mean shear plays a subordinate role for the settling in the bulk of the channel, but becomes increasingly important as the contaminants approach the lower boundary. Close to the sediment, the wakes of the particles are visibly sheared, albeit a comparison with a particle settling in a comparable laminar shear flow suggests that the impact on the settling velocity is low. Nonetheless, shear is an important factor in explaining the excess in mean streamwise particle velocity in the lower quarter of the channel where preferential sampling loses its relevance in this regard.

An evaluation of the slip velocity, although not devoid from arbitrariness, indicates that it might be reduced with respect to the ambient terminal velocity in a way which is consistent with the results of Fornari et al. (2016). This suggests that the settling velocity is reduced on a microscale, whereas on a macroscale preferential sampling is prevalent, and thus, overshadows this reduction.

In conclusion, the major observations of chapter §4 regarding the particle statistics are readily explicable in terms of the fluid regions which are preferentially visited by the contaminants during the transient. However, this raises the question of what mechanism promotes the particles to enter these regions and at which scales it is active. Furthermore, the role of potential collective effects has not been discussed extensively yet, and particularly the significance of clustering regarding the settling velocity still needs to be addressed.

6 Pattern formation in the dispersed phase

The analyses of the collective settling of contaminants have indicated that contaminants preferentially sample flow regions of enhanced streamwise and downward fluid momentum, which ought to be reflected in the relative arrangement of the particle positions. An assessment of the structure of the dispersed phase during the settling phase and beyond is subject of this chapter, and inferences will be made regarding the role of self-organisation on the spreading behaviour.

6.1 Clustering

First of all, the contaminant trajectories are analysed with regards to the formation of particle clusters. As discussed in §2.3.2 and §2.3.4, clustering may occur due to turbulent effects such as preferential sweeping or may be self-induced, for instance by wake attraction effects. In the following, the emergence of clusters will be assessed and quantified, especially with respect to its potential connection to the enhancement in particle settling velocity.

6.1.1 Reference distribution

In order to detect and quantify particle clusters, a suitable reference distribution is required in order to distinguish random fluctuations in the local particle density from accumulations which are induced by physical mechanisms. In the study of homogeneous domains, the measured distribution of particles is typically compared to random distributions with uniform probabilities in all directions, cf. e.g. Monchaux et al. (2012). However, concerning the transient investigated in this thesis, the settling particles are initially distributed inhomogeneously in the vertical direction. In fact, their initial distribution is a state of maximum order with respect to y, and consequently, vertical dispersion will generally tend to reduce the orderliness compared to a three-dimension uniform random arrangement with time which complicates the detection of potential particle accumulations. For this reason, special care is to be taken in the choice of a suitable reference distribution in order to quantify the formation of clusters.

For the purpose of familiarising oneself to the phenomenon which is to be quantified, fig. 6.1 displays an example for the emergence of three-dimensional structures of the solid phase shortly after the start of the transient. When comparing the contaminant arrangement to the initially homogeneous arrangement in a single wall-parallel plane, cf. fig. 3.1, it becomes immediately obvious that a mechanism exists which leads to the formation of accumulations and voids. While the portrayed arrangement evidently possesses increased order in streamwise and spanwise directions, dispersion reduces the orderliness in the wall-normal direction, and when the arrangement is compared to a three-dimensional uniform random distribution, these effects counteract each other. Therefore, in order to avoid undesired effects

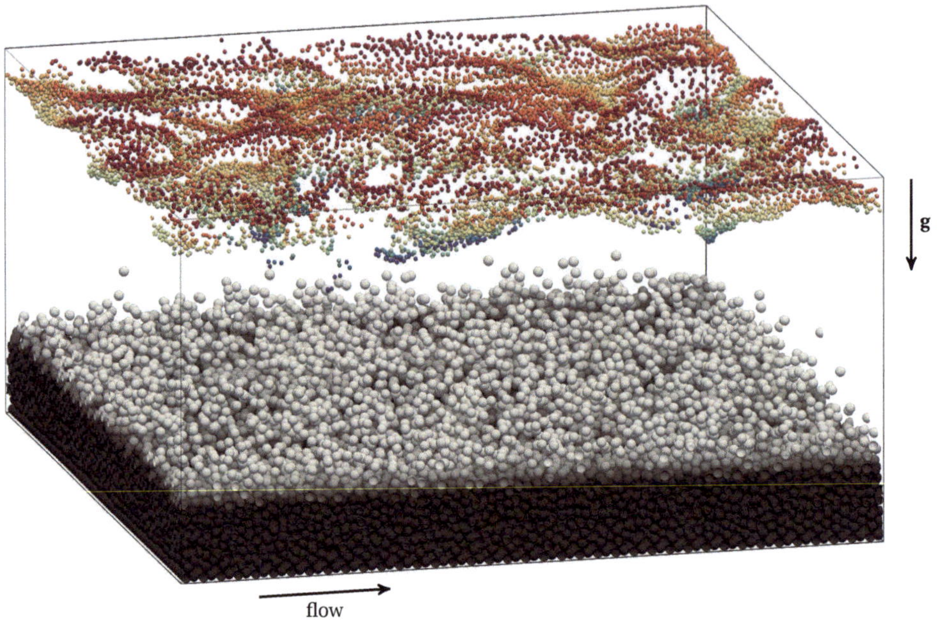

Figure 6.1: Exemplary visualisation of the formation of clusters shortly after the start of the transient settling. The contaminants are coloured by their vertical positions indicating their vertical dispersion. The parameter point shown is G10-D10-HR at $t \approx 2.2 t_b$.

caused by the disorganisation in y, the reference arrangement needs to take into account the localised nature with regards to this direction.

A potential resolution to this difficulty is to disregard the vertical component of the particle position entirely, i.e. to only analyse accumulation effects solely in a projection onto the wall-parallel plane. While this is a practical solution for early times where the vertical localisation is still prominent, it defies the purpose of the clustering analysis with regards to collective effects at later times as the possibility for particles to interact with each other predominantly depends on their instantaneous proximity in three dimensions. The negligence of the vertical position at times where wall-normal localisation is mostly lost does not necessarily quantify this proximity, and thus, complicates the interpretation of the results.

A more suitable solution is to construct a reference distribution which takes into account the confinement of possible particle positions with respect to the vertical direction. In order to achieve this, the probability density for the random distribution cannot be uniform with respect to y, but should rather reflect the progressive dispersion which is observed in the settling simulations. This is accomplished by extracting the time-varying probability distribution of the vertical contaminant positions from the simulation data, which is then randomly sampled to obtain non-uniformly distributed y-positions for the randomly arranged reference particles, whereas uniformity is upheld in the perpendicular directions.

The resulting algorithm then reads as follows. To begin with, the vertical contaminant positions are gathered for all realisations of the parameter point as a function of time passed since the initial release, and from these positions, a time-dependent cumulative density function $\mathrm{cdf}(y_c)$ is constructed. Subsequently, N_c particles are distributed using a random Poisson process under the condition that particle do not overlap, cf. §3.6. Therefore, a candidate for the centre point position is generated by drawing two uniformly distributed pseudorandom numbers from the intervals $[0, L_x)$ and $[0, L_z)$ for the horizontal components, and one pseudorandom number in accordance with $\mathrm{cdf}(y_c)$ for the vertical component.[1]

[1] In practice, the non-uniform distribution is easily achieved by uniformly sampling the inverse of $\mathrm{cdf}(y_c)$.

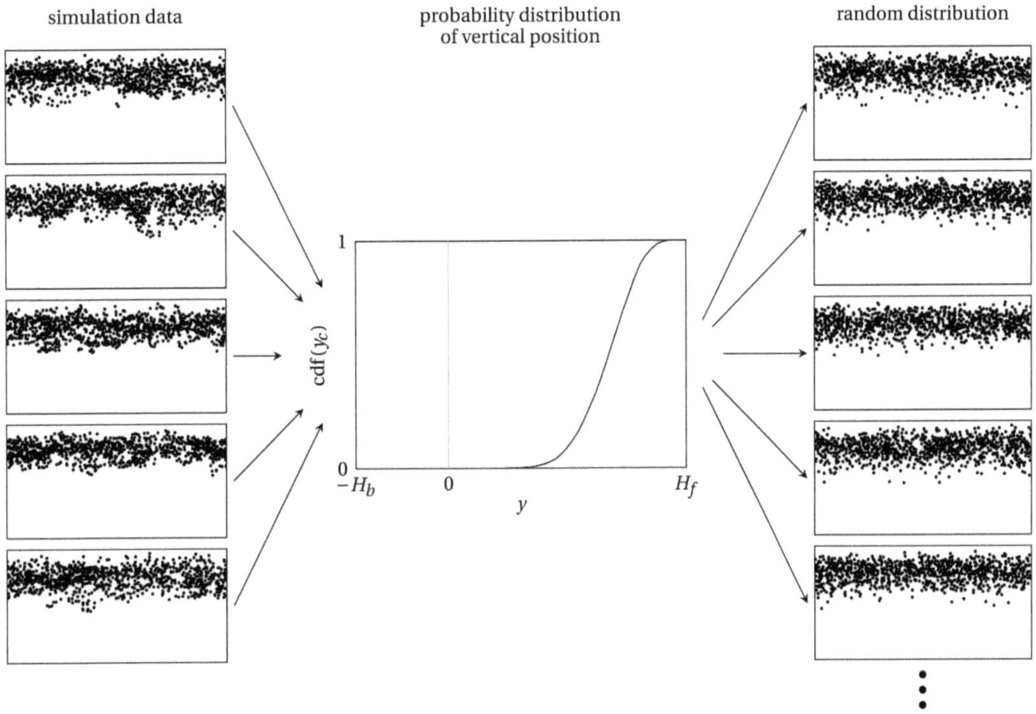

Figure 6.2: Visual sketch describing of the random distribution of contaminant particles which are used as a reference to identify clustering in the following. From the simulation data, the vertical distribution of contaminants is extracted in order to obtain the cumulative density function of y_c as a function of time. The random sample particles are then generated such that they follow a uniform distribution in x and z, whereas the distribution is non-uniform in y following cdf(y_c). The advantage of this method is that the confinement of the wall-normal particle positions is accounted for and vertical dispersion does not conceal the emergence of structures in the wall-parallel directions.

The candidate is examined with regards to overlaps with previously distributed particles. If it does overlap, the process is repeated until a suitable candidate is found and all reference particles have been distributed. In practice, a multitude of such random reference distributions are generated to obtained converged statistics, with the total number of sample particles being of $\mathcal{O}(10^7)$ in the subsequent analyses. A visual sketch of this method of randomly distributing particles is provided in fig. 6.2.

6.1.2 Nearest neighbour analysis

Now that a suitable reference arrangement has been established, an appropriate way of quantifying clustering needs to be defined. A straightforward way of characterising the spatial structure of the dispersed particles is the nearest neighbour analysis proposed by Kajishima (2004b) for this context. In this method, the distance from the i-th particle to its nearest neighbour is computed by

$$R_{NN}^{(i)}(t) := \min\left\{\left|\mathbf{x}_c^{(j)} - \mathbf{x}_c^{(i)}\right| : j \in \{1, 2, \ldots, N_c\} \wedge j \neq i\right\}, \tag{6.1}$$

and a global indicator for clustering can then be deduced by ensemble-averaging $R_{NN}^{(i)}$ and comparing it to the reference arrangement. The choice of this approach was mainly motivated by its capability to provide a meaningful value for the state of clustering for each individual particle in the current context, which is for instance not the case for approaches based on local particle concentrations due to the abrupt decline in this quantity at the front of the settling ensemble which introduces undesired behaviour.

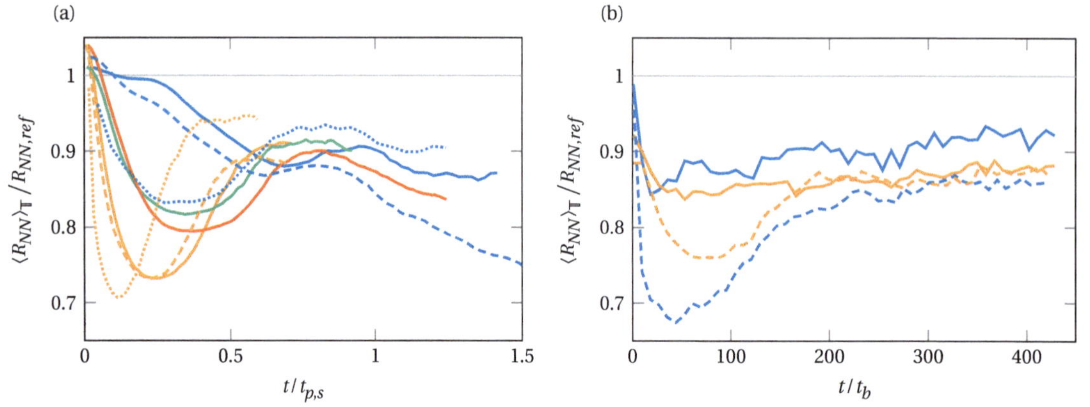

Figure 6.3: Ensemble-averaged distance to the nearest neighbouring contaminant particle normalised by the value obtained for the random arrangement described in §6.1.1 and as a function of time. (a) Only capturing the transient phase and time scaled by the characteristic settling time. (b) Extended temporal range for the cases with long observation times, cf. table 3.3, showcasing the evolution in the bed-interaction phase. The time is scaled in bulk units. Legend: G30-D10-LR (—), G10-D6-LR (—), G15-D10-LR (—), G15-D6-LR (—), G30-D10-RP (- - -), G10-D6-RP (- - -), G30-D15-HR (······), and G10-D10-HR (······).

Figure 6.3 displays the ensemble-averaged distance to the nearest neighbour defined analogously to eq. (4.1). The data is normalised by the corresponding value $R_{NN,ref}(t)$ for the random arrangement described in §6.1.1, and hence, values smaller than unity indicate that the contaminants tend to accumulate more substantially than what is expected for random fluctuations. All parameter points under investigation exhibit signs of clustering. When focusing on the transient settling phase, the time at which clustering is found to be most intense coincides well with the time at which the contaminant settling velocity is most enhanced, cf. fig. 4.1, suggesting a relationship between the two phenomena. Interestingly, it can be noted that the behaviour for G30-D10-LR is qualitatively different from most other parameter points at early times in the regard that clustering is only found to occur for $t \gtrsim 0.25 t_{p,s}$. The corresponding settling over ripple bedforms behaves similarly, albeit to a lesser degree. When inspecting on the bed-interaction phase after the initial settling has concluded, a spatial organisation of contaminants is found to be maintained. In fact, in the presence of ripple features, the contaminants are observed to cluster for a second time around $t \approx 50$–$100 t_b$, and the degree of accumulation is found to be of similar magnitude or even more significant than during the settling phase for both parameters investigated.[2] The time at which the second peak occurs coincides with the minimum of the streamwise contaminant flux observed in fig. 4.20 for each case, and its origin will be subject of investigation of §6.5. In the absence of bedforms or at longer observation times for ripples, the level of organisation stays relatively constant with a slow tendency to approach the random state.

The apparent correlation between $\langle R_{NN} \rangle_{\mathbb{T}}$ and $\langle v_c \rangle_{\mathbb{T}}$ will now be assessed. While there is a known mechanisms available which might give rise to a cause-effect relation between the two quantities—namely the wake attraction discussed in §2.3.4—it was argued in §5.2.2 that both velocity enhancement and clustering are more likely to be caused by preferential sweeping or similar turbulence-driven effects. In order to clarify this, $R_{NN}^{(i)}$ is inspected from the spatially-resolved perspective by introducing the distance to the nearest neighbor at the time of the initial crossing of a wall-parallel plane, viz.

$$R_{NN,\mathbb{X}}^{(i)} := R_{NN}^{(i)}\left(t_{c,\mathbb{X}}^{(i)}\right) \tag{6.2}$$

2 The occurence of the first peak displayed in fig. 6.3(a) is hardly perceptible in fig. 6.3(b) as it occurs at around $5 t_b$ and the resolution of the long-term evolution is of $\mathcal{O}(10 t_b)$.

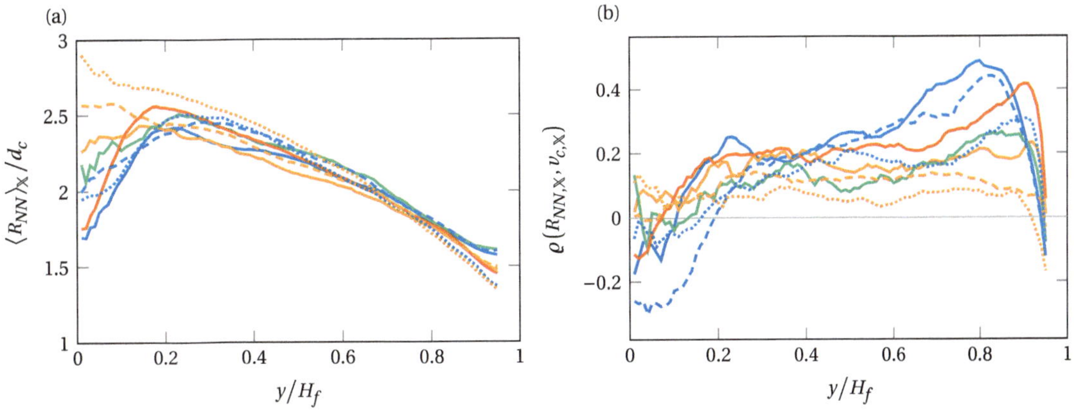

Figure 6.4: (a) Distance to the nearest neighbouring contaminant particle averaged at the time of crossing a given wall-normal plane. The absolute value scaled in multiples of the contaminant particle diameter is shown. (b) Linear correlation coefficient between the nearest neighbour distance and instantaneous settling velocity at the crossing time. Legend: G30-D10-LR (—), G10-D6-LR (—), G15-D10-LR (—), G15-D6-LR (—), G30-D10-RP (- - ·), G10-D6-RP (- - ·), G30-D15-HR (······), and G10-D10-HR (······).

where $t_{c,\mathbb{X}}^{(i)}(y)$ is the time at which the i-th contaminant particle first crosses the wall-parallel plane located at y, cf. §4.2. When the wake attraction mechanism is active, the settling velocity is generally enhanced for accumulations of particles. In fig. 6.4(a), the mean distance to the nearest neighbour is shown as a function of vertical position and in absolute terms, i.e. not scaled with respect to a reference arrangement. As can be observed, the distance between particles generally increases with growing distance from the plane of initialisation which does not align with the hypothesis that the settling velocity enhancement is caused by a collective effect, as the enhancement becomes more significant with growing distance from the initial position, cf. fig. 4.6. A more rigorous and direct evaluation of the correlation between $R_{NN,\mathbb{X}}^{(i)}$ and $v_{c,\mathbb{X}}^{(i)}$ at the time of crossing is achieved using a linear correlation coefficient defined analogously to eq. (4.8). Figure 6.4(b) display the evolution of $\varrho\left(R_{NN,\mathbb{X}}, v_{c,\mathbb{X}}\right)$ with respect to the vertical position. The correlation is rather weak for the most part, and is most significant close to the plane of initialisation for the parameter points which exhibit only little clustering at early times. This phenomenon will be subject of further investigations.

6.1.3 Conditionally averaged particle concentration

In order to examine the relative positioning of particles in clusters, the local solid volume fraction is assessed in the particle-conditioned reference frame. For this purpose, the conditionally averaged solid volume fraction is defined as

$$\langle \phi_c \rangle_{\mathbb{P}\mathbb{X}\varphi}(r, \tilde{y}, y) := \frac{1}{2\pi r} \int_0^{2\pi} \langle \chi_c(r, \varphi, \tilde{y}, t) \rangle_{\mathbb{X}} \, \mathrm{d}\varphi, \tag{6.3}$$

where $\chi_c(r, \varphi, \tilde{y}, t)$ is the solid indicator function applied exclusively to contaminant particles in a cylindrical coordinate system whose axis is aligned with the vertical direction and whose origin coincides with the centroid of the particle considered. Hence, $\langle \bullet \rangle_{\mathbb{P}\mathbb{X}\varphi}$ denotes an azimuthal average in the particle-attached frame of reference conditioned to the time when the contaminant particle first crosses a given wall-parallel plane.

The conditionally averaged particle concentration is displayed in fig. 6.5 for $y = 0.8H_f$ which approximately coincides with the position where the maximum of $\varrho\left(R_{NN,\mathbb{X}}, v_{c,\mathbb{X}}\right)$ occurs. Four parameter points are shown which are selected such that they represent various degrees of correlations. The difference

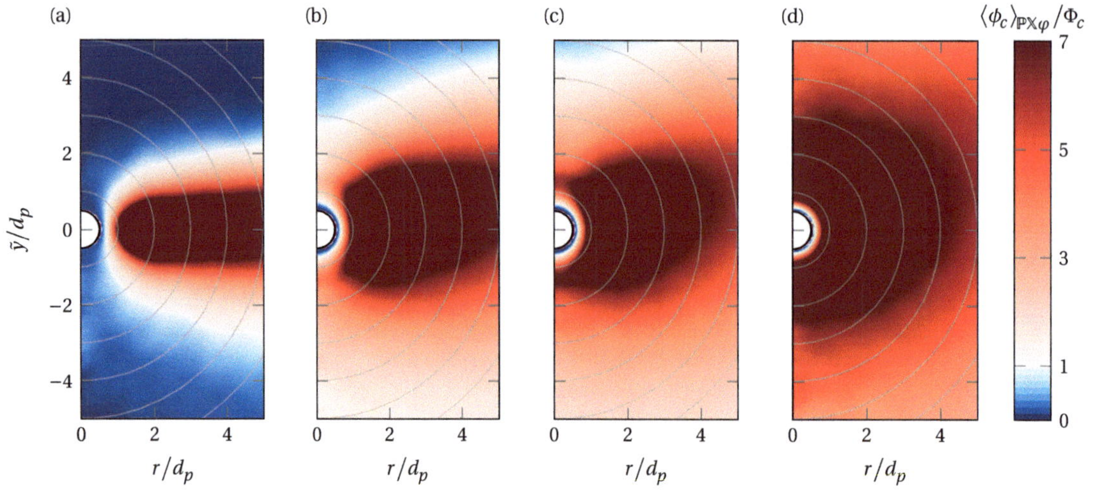

Figure 6.5: Conditionally averaged local solid volume fraction normalised by the total solid volume fraction at $y = 0.8H_f$. (a) G30-D10-LR, (b) G15-D6-LR, (c) G10-D6-LR, and (d) G10-D6-HR. The parameter points are ordered according to the correlation between $\langle R_{NN} \rangle_{\mathbb{T}}$ and $\langle v_c \rangle_{\mathbb{T}}$ at this position of the wall-parallel plane in decreasing order, cf. fig. 6.4(b). The circular grey lines indicate radial distances from the particle centroid in steps of $1d_c$.

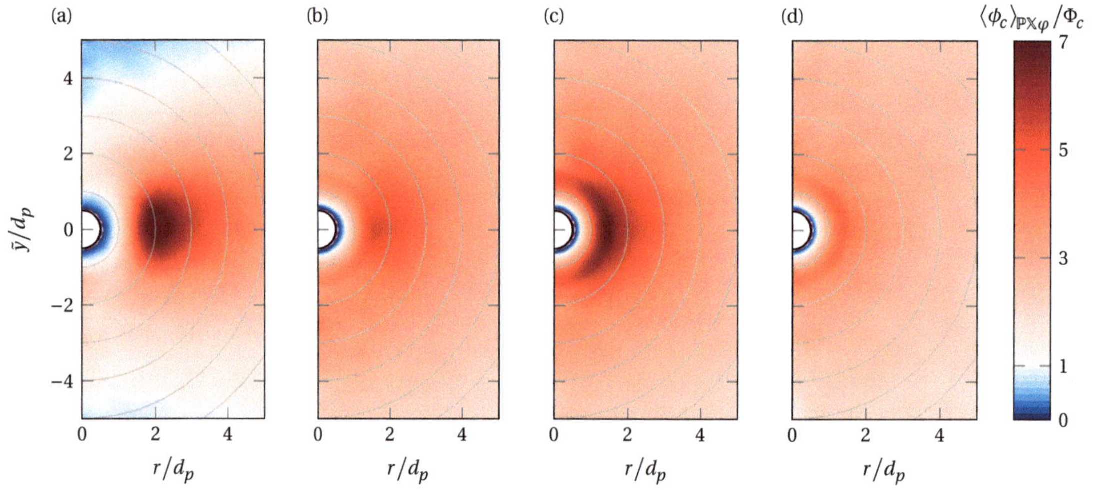

Figure 6.6: Same as fig. 6.5, but for $y = 0.5H_f$. (a) G30-D10-LR, (b) G15-D6-LR, (c) G10-D6-LR, and (d) G10-D6-HR.

in the relative arrangement of contaminant particles is striking: while for G30-D10-LR neighbouring particles are preferentially located side by side, the relative positioning for G10-D6-HR is substantially more homogeneous. The former arrangement resembles the arrangement of particles in the plane of initialisation and is presumably a remnant thereof. Thus, it should be emphasised that this horizontal alignment is not a cluster in the sense of §6.1.2, but rather a result of the lack of vertical dispersion, which is also reflected by fig. 6.3(a). Moreover, it can be hypothesised that a predominantly horizontal arrangement between particles is unfavourable regarding an enhancement of settling velocity as it is likely subject to the hindrance effect described by Richardson and Zaki (1954), cf. §2.3.4. Evidence for this supposition is provided by the trend of $\varrho\left(R_{NN,\mathbb{X}}, v_{c,\mathbb{X}}\right)$ in between the parameter points which suggests that the positive correlation is proportional to the degree of horizontal alignment of $\langle \phi_c \rangle_{\mathbb{P}\mathbb{X}\varphi}$. With advancing progress of the settling phase, all relative particle positionings tend to become more isotropic as can be inferred from fig. 6.6. In fact, a radial peak in the local solid volume fraction indicates that clustering occurs, but no elongated structures are formed in the direction of gravity as would be typical for

Table 6.1: Numerical parameters of the auxiliary simulations conducted to determine the hindrance effect during collective settling in a quiescent or laminar open channel.

case	$[L_{x,h} \times L_{z,h}]/L_{y,h}$	$[N_{x,h} \times N_{y,h} \times N_{z,h}]$	$L_{y,h}/d_p$	$d_p/\Delta x_h$	N_p	Φ_p	$\phi_p(y_{p,0})$
G30-D10-HIN	[1.67×0.67]	[640×384×256]	25.6	15	123	$3.5 \cdot 10^{-3}$	0.13
G15-D10-HIN	[1.67×0.67]	[640×384×256]	25.6	15	123	$3.5 \cdot 10^{-3}$	0.13
G15-D6-HIN	[1.67×0.67]	[640×384×256]	38.4	10	417	$3.5 \cdot 10^{-3}$	0.20
G10-D6-HIN	[1.67×0.67]	[640×384×256]	38.4	10	417	$3.5 \cdot 10^{-3}$	0.20

wake attraction. In conclusion, there is little support for the hypothesis that the formation of clusters is induced by the particles alone.

6.2 Hindrance and self-organisation

To study the consequences of horizontal alignment on the settling of particles, auxiliary simulations regarding the collective settling in an open channel with smooth walls under initially quiescent and laminar flow conditions are conducted. Similar to the shear analysis in §5.3.3, this approach attempts to make qualitative inferences on the role of collective effects on the particle settling velocity by studying proxy problems of reduced complexity for which the difficulties posed by turbulence are absent. The domain within which the auxiliary simulations are conducted has the same dimensions and numerical resolution as the one described in §5.3.3, yet it differs from it in terms of the boundary conditions (no-slip/free-slip) and the number of particles present. In addition to this, the numerical resolution of the particles matches exactly that of the isolated settling simulations described in §3.7. To obtain representative results, N_p is chosen such that the total solid volume fraction is around the same as that for the turbulent simulations. Moreover, the ratio between the vertical extent of the domain $L_{y,h}$ and the particle diameter is kept similar such that the local solid volume fraction in the plane of initialisation is matched. Two flow states are examined for the channel—an initially quiescent environment similar to §3.7, and a laminar flow at the same bulk Reynolds number as the turbulent simulations.[3]

Figure 6.7(a) shows an aggregation of the ensemble-averaged settling velocity, the standard deviation of vertical particle position, and the normalised distance to the nearest neighbour as a function of time for the initially quiescent environment. The occurrence of a hindrance effect is conspicuous. For all parameter points, the settling velocity is found to be significantly reduced compared to the terminal velocity of an isolated particle. While for the cases with the larger particle diameter an approximate plateau is reached and maintained, the cases with smaller particles initially exhibit a plateau with stronger reduction, but subsequently start to accelerate again until a second plateau at a faster velocity is approached. The stronger reduction in case of the latter is in accordance with the finding of Richardson and Zaki (1954) as the local volume fraction at $y_{p,0}$ is higher for the smaller particles. This hindrance is essentially a combined result of fluid displacement, vertical boundedness and incompressibility which demand a fluid flow in the direction opposing the falling direction of the particles. The resulting resistant hydrodynamical force on the particles depends on the arrangement of the ensemble. While the initial homogeneous arrangement in the wall-parallel plane is arguably a regime of high resistance, a reorganisation which yields a vertically stacked arrangement of particles would exhibit a significantly reduced

[3] Despite the obvious possibility for the flow to become turbulent and the perturbations introduced by the settling particles, it has been verified that the flow retains its laminar velocity profile during the relatively short time of observation.

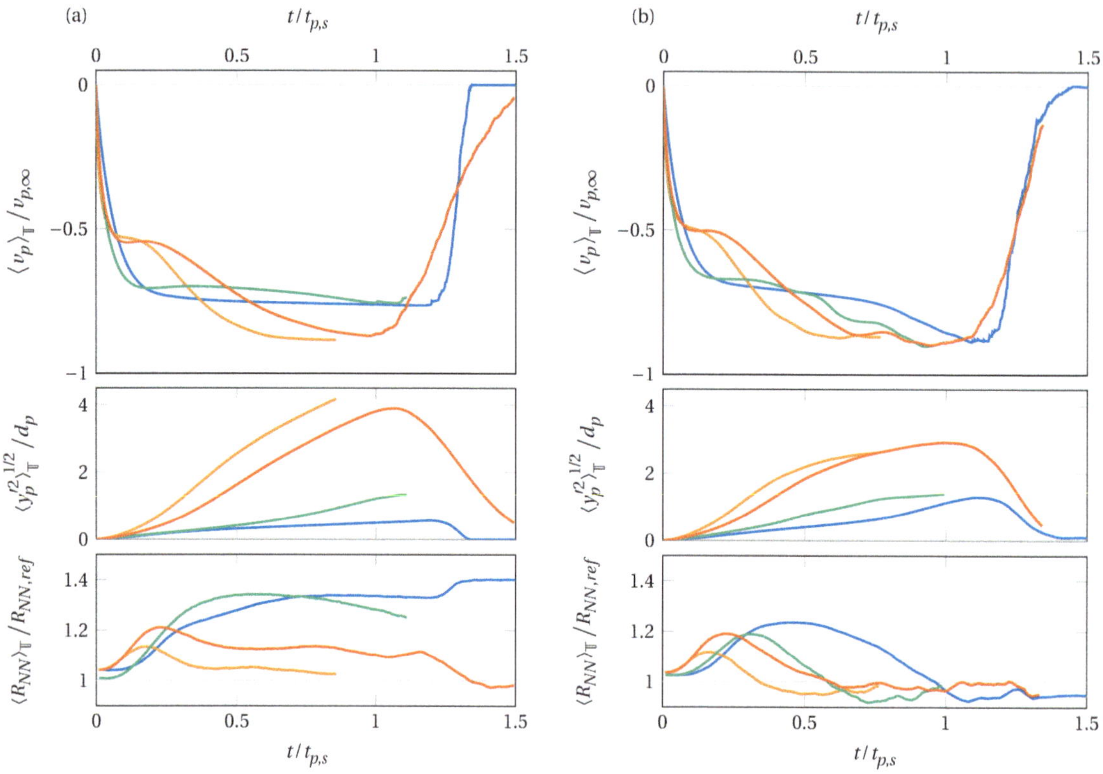

Figure 6.7: Results of the collective settling simulations in (a) a quiescent environment and (b) laminar flow. From top to bottom, the figure shows the settling velocity normalised by the terminal velocity of an isolated particle, the standard deviation of vertical position, and the distance to the nearest neighbour normalised by the value of the corresponding reference distribution. All quantities are given as a function of time scaled by the characteristic settling time. Legend: G30-D10-HIN (—), G10-D6-HIN (—), G15-D10-HIN (—), and G15-D6-HIN (—).

effective frontal area such that it is presumably less affected by the counterflow. Therefore, it can be hypothesised that the second acceleration phase observed for G15-D6-HIN and G10-D6-HIN is associated to the breakup of the initial layer and a rearrangement of the particles. This hypothesis is corroborated by the temporal evolution of the standard deviation of vertical particle position: for the smaller particles the vertical spread increases alongside the particle velocity, whereas for G30-D10-HIN and G15-D10-HIN the standard deviation does not increase substantially beyond one particle diameter indicating that the two-dimensional arrangement stays intact. Similar inferences can be made from the distance to the nearest neighbour normalised by the reference arrangement. Here, it can be observed that the initially randomly distributed particles tend to reorder themselves such that their distance from each other increases.

When comparing the spatial structure of the particle arrangement at initialisation and at a time where the settling velocity has converged for G30-D10-HIN, the organisation into a more regularly spaced arrangement becomes obvious, see fig. 6.8(a), and as a consequence, hindrance essentially exhibits an effect which opposes clustering. It can also be observed that the quasi two-dimensional arrangement eventually develops an instability during which the vertical distribution loses its uniformity. For G30-D10-HIN this loss of uniformity occurs in form of a sinusoidal perturbation in streamwise direction, yet, in a quiescent environment it does not develop rapidly enough to break up the layer within the settling time. However, when the same parameter point is examined in a laminar flow, the layer indeed does disperse vertically which is accompanied by an increase in the settling velocity of the ensemble, see fig. 6.7(b). The flow-induced breakup mechanism appears rather straightforward to explain: any irregularities in the vertical distribution lead to a streamwise motion of the settling particles relative to each other due to the fluid shear. While particles at higher positions get pushed forward, the ones with negative fluctuations y_p' from the mean position are driven backwards which leads to the formation of the spanwise-aligned structures displayed in fig. 6.8(b). This new arrangement considerably reduces the projected frontal area of the ensemble, and the consequential reduction in resistance manifests itself in an increased settling velocity. The breakup mechanism notably resembles the Kelvin–Helmholtz instability at the interface of two sheared fluids, and it may be assumed that they are similar in their nature.

The Kelvin–Helmholtz-like instability is indeed not the only breakup mechanism which arises. In fact, as stated earlier, the settling ensembles with the smaller particle sizes disintegrate already in the absence of a background flow, and the average settling velocity is not substantially altered when the settling occurs in a laminar flow, see fig. 6.7. When examining the spatial structure shortly before the breakup of the first plateau for G10-D6-HIN, it can be perceived that fluctuations in y_p' emerge in compact spots both in absence and presence of a background flow, see fig. 6.9. This behaviour is reminiscent of the Rayleigh–Taylor instability which occurs at the unstable interface of two fluids with different densities under the action of gravity. With time, the initial disturbances increase in amplitude and the layer disperses vertically with little organization in the wall-parallel plane, even if it is sheared. Indeed, a closer inspection of fig. 6.9(a) and a comparison between the initial condition and the state at $t = 0.19t_{p,s}$ yields that the arrangement stays almost unaltered and that the fluctuations in vertical position coincide with the random clusters occuring in the initialisation. The particles which are initially located in a random cluster appear to settle slightly faster, and the resulting dispersion is sufficient to break up the layer before the opposing flow is capable of maximizing the distance between the particles. Thus, this breakup mechanism is fundamentally different from the Kelvin–Helmholtz-like instability in the sense that no substantially ordered state is achieved before the breakup, and that no shear is required for it to be active. Furthermore, the nearest neighbour analysis presented in fig. 6.7 might indicate that the particles which undergo the shear-driven instability show a stronger tendency to cluster after the disintegration of the layer than those which undergo the random-cluster instability. If this is indeed the case, it is likely the result of the tendency of shear to vertically stack particles creating intermediate voids in the xz-plane, whereas the cluster-induced instability does not lead to nor require significant horizontal movement, and hence, retains its random horizontal arrangement.

A substantial recognition from the examination of the proxy settling problems is that a strong reduction in settling velocity goes hand in hand with an increased order in the internal structure of the settling ensemble of particles. In fact, when no effective breakup mechanism for the layer is available, the induced resistant flow will actively maximise the horizontal spacing between the particles which essentially counteracts the formation of clusters as indicated by the nearest neighbour analysis. In some sense, clustering is antagonistic to the hindrance phenomenon as any reorganisation in the wall-parallel plane towards spatial intermittency reduces the effective frontal area of the ensemble and may lead to a bifurcation in settling behaviour through an induced breakup mechanism. As a consequence, the relevance of hindrance on the settling velocity in the turbulent case is likely tied to the tendency of particles to form clusters.

For the majority of the turbulent parameter points under investigation clusters are formed immediately at the beginning of the transient, cf. fig. 6.3, and it is thus likely that the hindrance effect is minimal in these cases. However, this is not the case for G30-D10-LR and G30-D10-RP where an initial reluctance to form cluster is observed. Based on the observations made in the current section in conjunction with the findings of §6.1.3, it stands to reason that hindrance is an important effect to be considered for these cases at early times. This is reinforced by the temporal evolution of the standard deviation of the vertical particle position for the turbulent cases, which is provided in fig. 6.10 and indicates especially

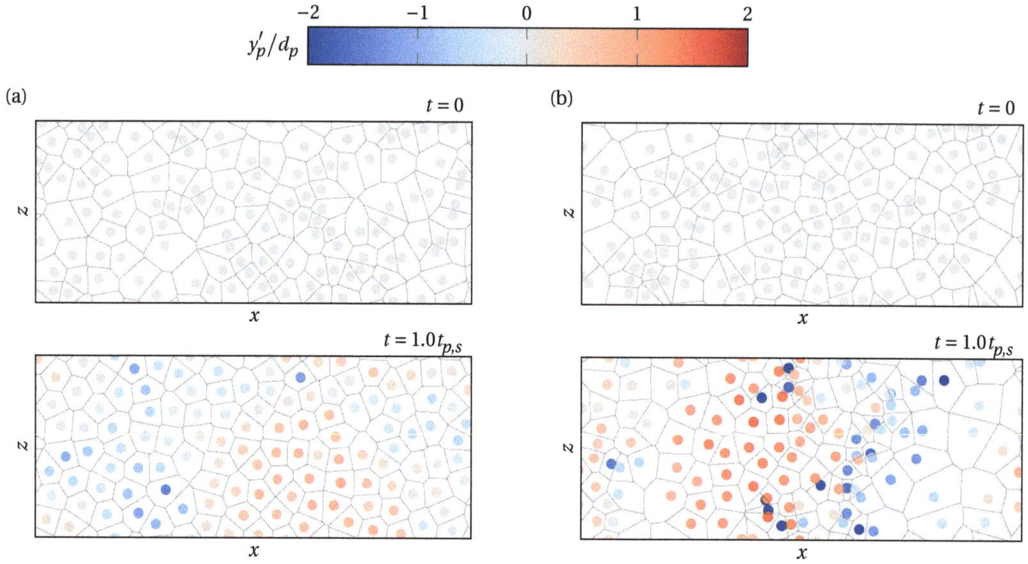

Figure 6.8: Spatial arrangement of particles collectively settling in (a) an initially quiescent environment and (b) in laminar open channel flow. Two stages in time are shown and time increases from top to bottom. The colouring of the particles indicate the deviations from the mean particle position in vertical direction. Cells outline the regions within the plane of projection which are closest to the particle which is embedded into them, i.e. large cell sizes indicate voids whereas small cell sizes show accumulations. The parameter point shown is G30-D10-HIN.

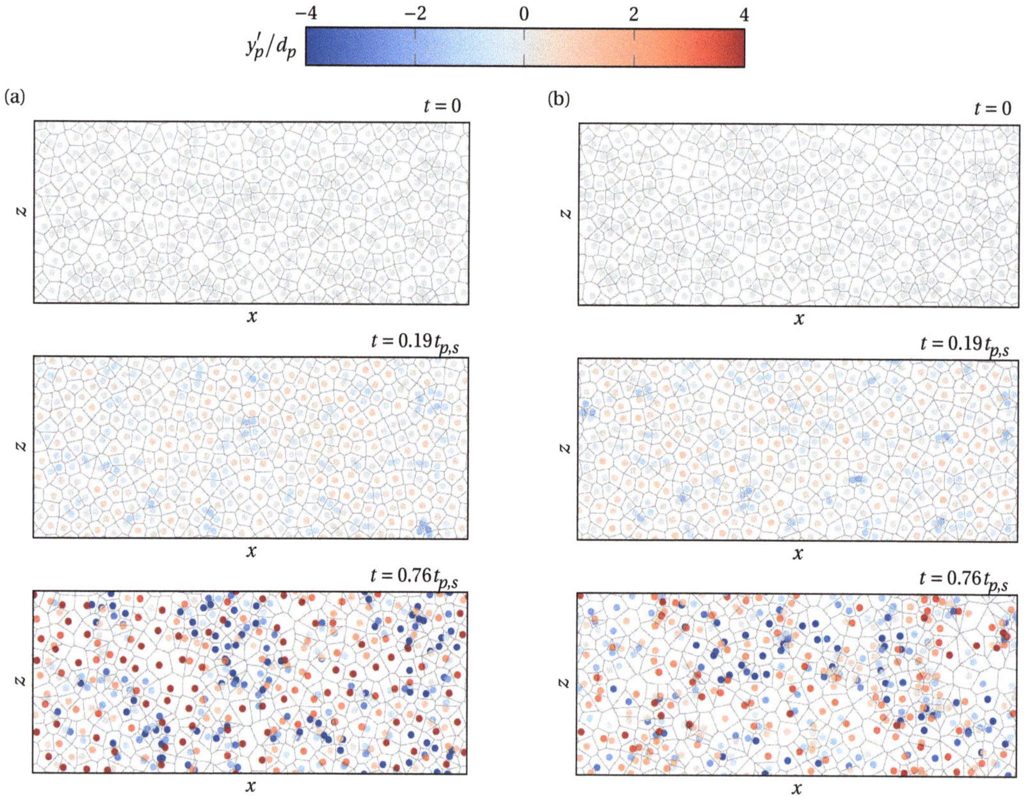

Figure 6.9: Same as fig. 6.8, but for the parameter point G10-D6-HIN and at different times. (a) Quiescent environment, (b) laminar open channel flow.

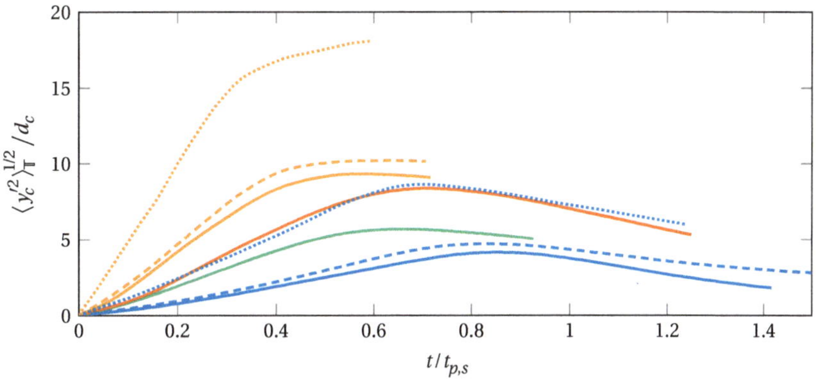

Figure 6.10: Standard deviation of vertical contaminant position as a function of time for the turbulent simulations. Legend: G30-D10-LR (—), G10-D6-LR (—), G15-D10-LR (—), G15-D6-LR (—), G30-D10-RP (- - -), G10-D6-RP (- - -), G30-D15-HR (······), and G10-D10-HR (······).

small vertical dispersion for these cases. Consequentially, it is to be expected that a reduction in settling velocity and a prolongation of the initial acceleration phase occurs for G30-D10-LR and G30-D10-RP due to this collective effect. This may serve as an explanation for the deviant behaviour of these cases observed in fig. 5.8. Moreover, the antagonistic relationship between order-inducing hindrance and intermittency-inducing clustering may unveil part of the reason for the correlation between $R_{NN,\mathbb{X}}$ and $v_{c,\mathbb{X}}$ demonstrated in fig. 6.4(b)—particles which fall in clusters exhibit less resistance due to hindrance. This is especially supported by the fact that the correlation is the highest for parameter points where hindrance is expected to be most significant, and the fact that the correlation decreases with increasing distance from the plane of initialisation.

6.3 Preferential concentration in the crossing plane

The previous analyses lead to the conclusion that clusters of contaminants are likely not self-induced, but rather a result of turbulent interactions. Clustering in turbulence is often associated to preferential concentration. In particular, this connotes that clusters emerge due to a spatially inhomogeneous distribution of the probability density of presence for a particle which dependents on the local flow conditions in the simplest case, or on non-local effects in the more general case, cf. §2.3.2. Thus, in contrast to the analysis of clustering, the analysis of preferential concentration does not necessarily require an evaluation of the instantaneous relative arrangement of particles at a given instance in time. Instead, the crossing locations of settling particles can be sampled over a finite time interval in order to obtain a probability distribution which can be linked to flow features under the assumption that this sampling time interval is sufficiently short compared to the intrinsic time scales of the structures which gives rise to this effect.

6.3.1 Compensation for crossing time

In the following, the quantities of interest will be the streamwise and spanwise positions at which contaminants cross a given wall-parallel plane. In general, these crossings do not take place simultaneously, but instead it was found in §4.4 that the cross-through times are distributed according to a generalised

Gamma probability distribution. Since the particles possess a significant streamwise velocity component on average, the streamwise position at which particles cross a given wall-parallel plane may significantly differ between particles which interact with the same structure, but pass at different times if viewed from the point of view of a steady observer. An appropriate frame of reference would be required to move with the propagation velocity of the structure of interest which is generally not known or may not even be defined unambiguously. A viable alternative is to investigate the crossing locations in a reference frame moving at a time-invariant velocity equal to the local value of the mean streamwise contaminant velocity $\langle u_c \rangle_{\mathbb{X}}$. Letting $x^{(i)}_{c,\mathbb{X}}$ denote the streamwise crossing location of the i-th particle in the steady frame of reference, the crossing location in the moving frame of reference is hence given by

$$\tilde{x}^{(i)}_{c,\mathbb{X}}(y) := x^{(i)}_{c,\mathbb{X}} - \langle u_c \rangle_{\mathbb{X}}\, t^{(i)}_{c,\mathbb{X}}, \tag{6.4}$$

while accounting for the periodicity of the domain during the shift, i.e. $\tilde{x}^{(i)}_{c,\mathbb{X}} \in [0, L_x)$. It is important to note that if $x^{(i)}_{c,\mathbb{X}}$ was distributed randomly and independent of $t^{(i)}_{c,\mathbb{X}}$, the surrogate $\tilde{x}^{(i)}_{c,\mathbb{X}}$ would be randomly distributed as well. This increases the robustness of the procedure and ensures that patterns which are detected in $\tilde{x}^{(i)}_{c,\mathbb{X}}$ are not a result of the choice in propagation velocity of the moving frame, but rather an outcome of spatial and/or temporal correlations in the contaminant dynamics. Furthermore, this recognition provides a criterion for the evaluation of the appropriateness of the frame of reference—if patterns emerge within the plane, the choice of a suitable propagation velocity should roughly maximise their perceptibility.[4] As a matter of fact, part of the motivation behind selecting $\langle u_c \rangle_{\mathbb{X}}$ was that the resulting preferential concentration is more apparent than for other reasonable choices such as $\langle u_f \rangle_{xzt}$ or u_b. In accordance with chapter §4, the wall-parallel crossing location $(\tilde{x}^{(i)}_{c,\mathbb{X}}, z^{(i)}_{c,\mathbb{X}})$ is solely evaluated for the first downward crossing of each particle in the following.

Figure 6.11 shows the time-compensated crossing locations through wall-parallel planes at various heights for Ga10-D10-HR, the case which exhibits the strongest clustering according to the nearest neighbour analysis of §6.1.2. While the initial condition only exhibits random fluctuations as expected, intense organisation can already be observed at a small distance from the plane of initialisation at $y = 0.9H_f$, i.e. when the particles have merely travelled less than four times their diameter downwards. Following the argument made in §6.2, this arrangement likely minimises the hindrance effect at an early stage. The shape of the arrangement strongly resembles the expirimental measurements of Fessler et al. (1994) for small glass particles at the centreline of a vertical channel, and the simulation results of Narayanan and Lakehal (2003) for point-particles near the free-surface of an open channel flow. Indeed, it will be subject of §6.3.3 to demonstrate that the formation of this pattern is the outcome of the interaction between the contaminants and near-surface vortices. While accumulations in the initial arrangement are of rather small length scales with little preferential alignment in neither the streamwise nor spanwise direction, the arrangement observed in the lower parts of the channel are considerably different in their nature. In the lower half, the contaminants tend to align in long streamwise-aligned stripes which resemble the large-scale streaks discussed in §5.2.2 which have already been proven to play a primordial role in the enhancement of streamwise particle momentum.

The crossing locations for Ga30-D15-HR are qualitatively similar to the observations just described for the other parameter point, see fig. 6.12. As both cases are started from the exact same initial flow field,

[4] The concept of a single propagation velocity is owed to the demand of a preferably simple method of analysis. In general, different structures at the same vertical position may propagate at distinct speeds, and as has been hinted at before, a rigorous definition of such a speed is often impossible as eddies are not purely advected, but also continuously evolve in shape due to diffusion and non-linear interactions.

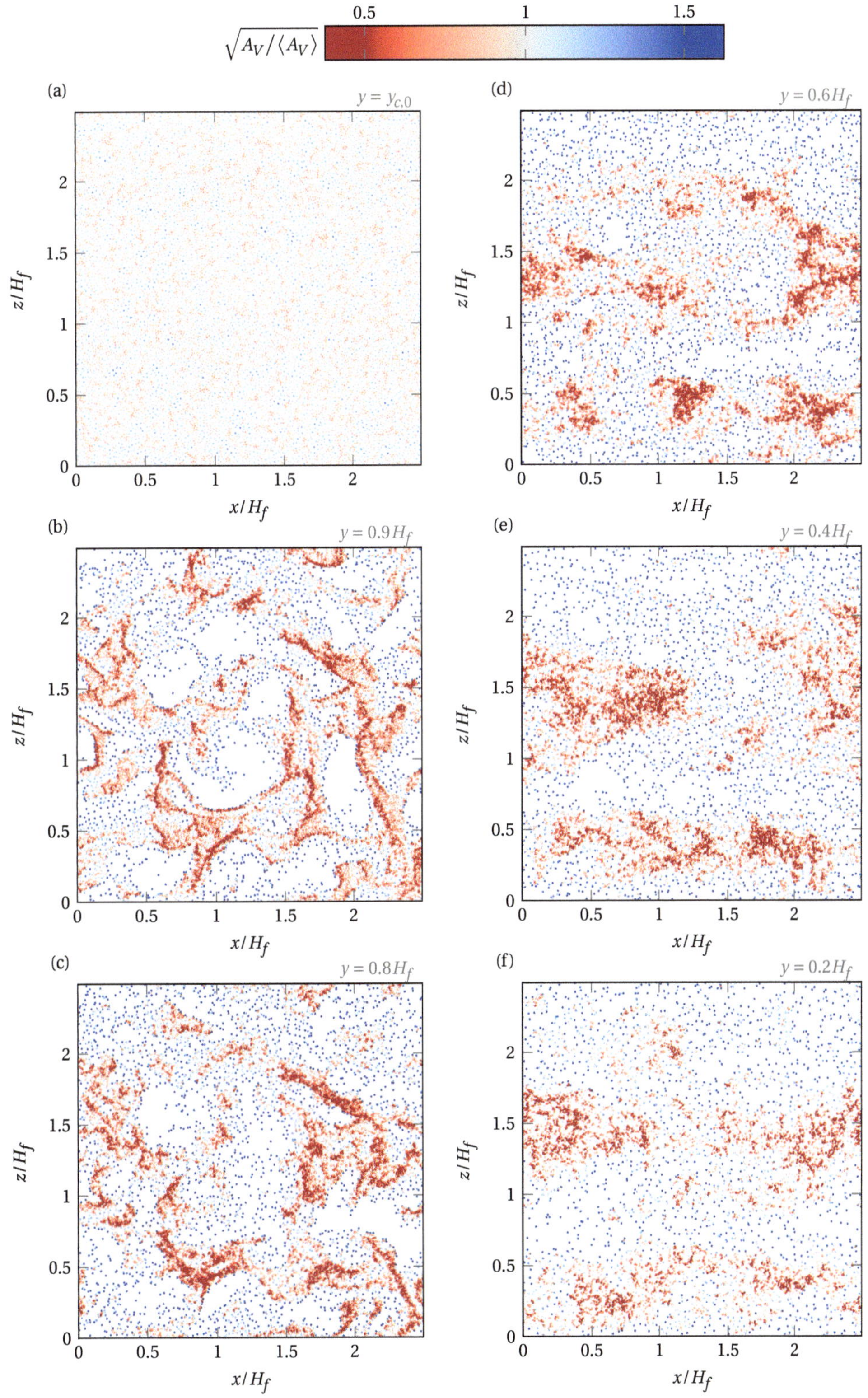

Figure 6.11: Visualisation of the time-compensated crossing locations $(\tilde{x}^{(i)}_{c,\mathbb{X}}, z^{(i)}_{c,\mathbb{X}})$ for Ga10-D10-HR at various wall-normal positions. Each dot represents a particle crossing where the diameter of the circle reflects the particle diameter. The colouring represents the local density of crossings to facilitate the visual detection of regions of preferential concentration. (a) The initial condition, (b-f) crossings at $y = \{0.9, 0.8, 0.6, 0.4, 0.2\}\,H_f$, respectively.

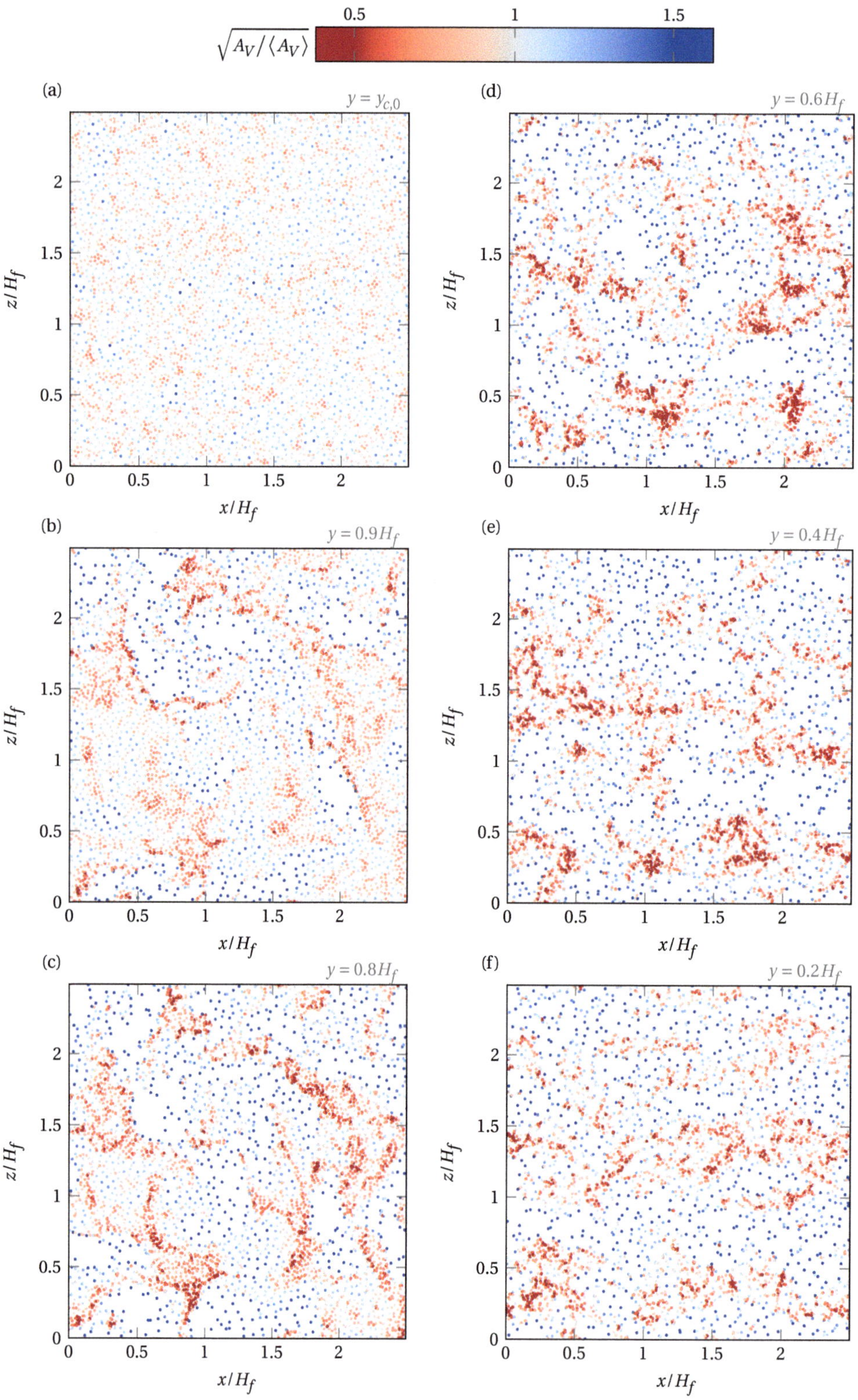

Figure 6.12: Same as fig. 6.11, but for Ga30-D15-HR. Note that both cases are initialised from the exact same flow field.

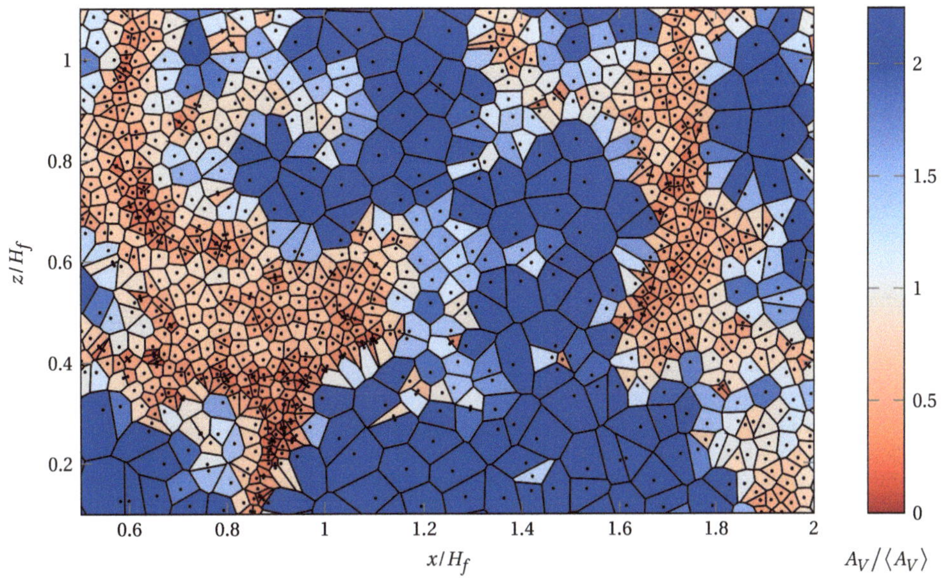

Figure 6.13: Examplary visualisation of the cells obtained from the Voronoï tessellation and the area contained by them. The seeds of the diagram which are depicted as black dots correspond to the crossing locations, i.e. the position of the particle centroid at the time of crossing in the moving frame of reference. The black lines surrounding the seeds constitute the boundaries of the corresponding Voronoï cell. The cells are coloured by their enclosed area. The figure shows a subset of the data displayed in fig. 6.12(c).

the effect of individual turbulent structures and the difference in response between the particle parameter points becomes obvious. The visualisations indicate that both parameter points respond to structures of roughly the same characteristic size, and suggest that the differences in preferential concentration is mainly due to inertial effects.

6.3.2 Voronoï analysis

In both fig. 6.11 and 6.12 the local concentration of crossings has been indicated by the colour of the dots. This colouring is based on the area A_V contained by the Voronoï cells attributed to each crossing point. The two-dimensional Voronoï tessellation partitions the wall-parallel plane into $N_{c,\mathbb{X}}$ cells each of which is attributed to a single crossing. These cells are defined as the subset of the plane in which the distance to the corresponding seed, i.e. the location of the crossing to which the cell is attributed, is shorter than to any other seed. This tessellation has already been employed tacitly in fig. 6.8 and 6.9 where the cells were drawn to emphasise the regular arrangement, and is displayed exemplary for the turbulent cases in fig. 6.13 to demonstrate how the area contained by the cell is representative for the local accumulations and voids.

Indeed, the probability density function of the Voronoï areas comprises information on the level of organisation, and thus may be used to distinguish random fluctuations from more ordered or intermittent configurations. In the following, the second moment of this distribution will be used as a characteristic quantity of lower dimensionality to quantify preferential concentration. The mean area of the Voronoï cells is dictated by the dimensions of the plane to be tessellated and the total number of seeds, and is given by

$$\langle A_V \rangle = \frac{L_x L_z}{N_{c,\mathbb{X}}} \tag{6.5}$$

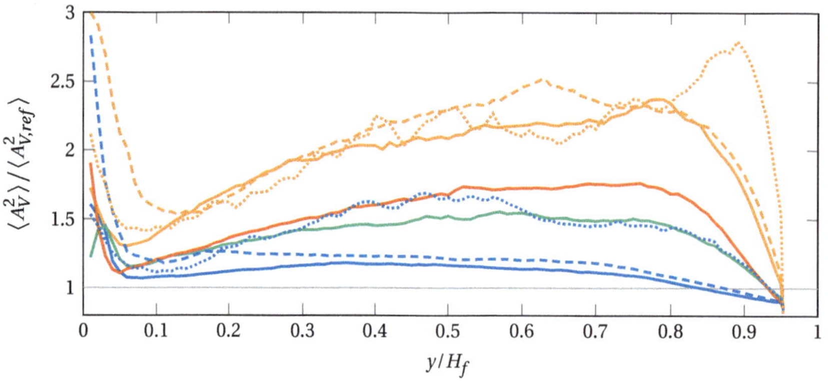

Figure 6.14: Second moment of the Voronoï cell area distribution $\langle A_V^2 \rangle$ where the contaminant crossing locations in the xz-plane at various wall-normal distances constitute the seeds of the Voronoï diagram. The curves are normalised by the corresponding value obtained for sets of points randomly distributed within the plane with uniform probability. Legend: Ga30-D10-LR (—), Ga10-D6-LR (—), Ga15-D10-LR (—), Ga15-D6-LR (—), Ga30-D10-RP (- - -), Ga10-D6-RP (- - -), Ga30-D15-HR (······) and Ga10-D10-HR (······).

for the current analysis. In order to detect deviations from random behaviour, a reference value for the second moment of the area distribution is required. Since for this analysis the directions of interest are homogeneous, it is sufficient to generate random distributions with uniform probabilities. For a uniformly distributed set of random points, the mean square of the Voronoï areas is given by

$$\langle A_{V,ref}^2 \rangle = 1.28031 \langle A_V \rangle, \tag{6.6}$$

where the value of the coefficient has been determined numerically by Tanemura (2003). Note that this value has been obtained for points which possess no length, width or thickness, whereas in this thesis particles with a finite size are considered which are not allowed to overlap, and hence, maintain a minimum distance between their centre locations at all times. Regarding the analysis of crossing locations, this distinction is most relevant for planes close to the plane of initialisation, as there particles will cross at similar times, and consequently their instantaneous distance in y may be small. For planes well below the plane of initial release, finite-size effects are expected to be negligible due to growing dispersion in three dimensions and the resulting spread in crossing time. A detailed discussion of the effect of non-overlapping and periodic domains on Voronoï statistics is provided by Uhlmann (2020). The deviation from a random arrangement can be assessed by comparing the mean square of Voronoï areas obtained for particle crossings to that obtained for the random distributions of points, whereby $\langle A_V^2 \rangle > \langle A_{V,ref}^2 \rangle$ indicates an organisation into clusters and voids. For simulations with multiple realisations, $\langle A_V^2 \rangle$ is evaluated for each realisation individually, and subsequently ensemble-averaged over all realisations.

When inspecting the ratio $\langle A_V^2 \rangle / \langle A_{V,ref}^2 \rangle$ as a function of vertical position (fig. 6.14), it can be observed that the initial arrangement is slightly more ordered than random points due to the initial distribution of all contaminants at exactly the same vertical position, the finite particle size and the impossibility of overlap. However, this order is lost already slightly below the plane of initialisation for most parameter points, and the rapidness of rearrangement is similar to what the cluster analysis suggested from a temporal point of view, cf. fig. 6.3. Furthermore, the organisation into streamwise-aligned stripes at $y \lessapprox 0.5H_f$ is reflected by the maintained relatively high values of $\langle A_V^2 \rangle$ at least for the HR simulations, although the values generally decrease with increasing distance from $y_{c,0}$. While it might be tempting to interpret fig. 6.14 in terms of the response of particles to the particular turbulent structures found in the various regions and layers of turbulence, such an interpretation is likely misleading. Apart from a potential non-locality of the preferential concentration, this is due to the fact that the further the plane of interest is located from the plane of initialisation, the higher the dispersion in cross-through times of

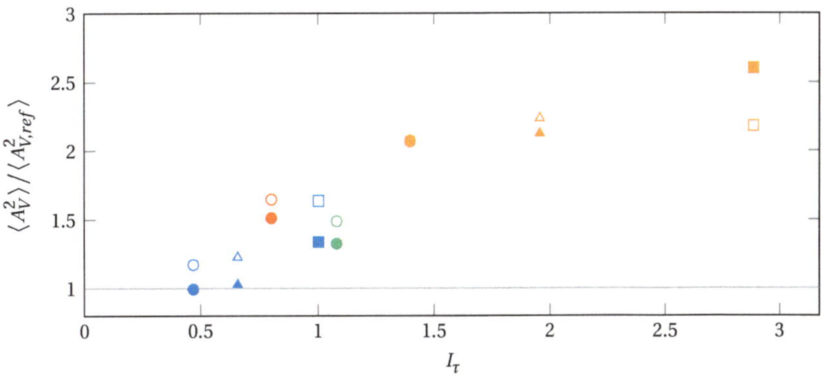

Figure 6.15: Normalised second moment of the Voronoï cell area distribution averaged over the intervals $y \in [0.4, 0.5]\,H_f$ (○,▲,□) and $y \in [0.8, 0.9]\,H_f$ (●,▲,■) as a function of relative turbulence intensity. Legend: G30-D10-LR (○,●), G10-D6-LR (○,●), G15-D10-LR (○,●), G15-D6-LR (○,●), G30-D10-RP (△,▲), G10-D6-RP (△,▲), G30-D15-HR (□,■), and G10-D10-HR (□,■).

the particles becomes, c.f. fig 4.13. This effectively acts as a filter for eddies with a lifetime significantly shorter than the range of cross-through times: if at a given time a particle interacts with a short-lived eddy and another particle passes through the same location at a later time after this eddy has vanished, its footprint will be blurred. This effect alone likely leads to a reduction of $\langle A_V^2\rangle$ with decreasing vertical position, and thus, the values obtained at different vertical positions should only be directly compared with caution.

Generally, all cases investigated exhibit signs of preferential concentration, albeit at different intensities. In an attempt to further synthesise the results, fig. 6.15 shows the intensity of preferential concentration averaged over the two intervals $y \in [0.8, 0.9]$ corresponding to the initial clustering near the surface, and $y \in [0.4, 0.5]$ corresponding to the phase of streamwise-alignment. A reasonable collapse may be achieved when the data is scaled as a function of relative turbulence intensity, although some scattering may be observed. The data has also been analysed with respect to a Stokes number dependence, however, St^+ fails to capture the trend of $\langle A_V^2\rangle$, and in particular, fails to collapse the curves for the LR, RP and HR cases (figure omitted).

6.3.3 Initial organisation

The rapid formation of accumulations and voids after the start of the settling process is now examined on the example of G10-D10-HR—the parameter point for which this phenomenon is the most pronounced. Figure 6.16 displays an aggregation of the flow conditions and contaminant trajectories at the onset of the settling process. The fluid flow at $t = 0$ in the plane of initialisation shows intermittent regions of upward and downward fluid flow which are related to the splats and antisplats on the free surface, cf. §2.2.3. The splats which impinge on the impermeable wall induce a diverging flow in the horizontal plane which is perceptible in the superimposed streamlines. The instantaneous flow field is characterised by two streamwise-aligned regions which differ regarding the intensity of fluid velocity fluctuations and are likely indicators for the presence of large-scale motions in the outer layer. When the contaminant trajectories in the horizontal plane are examined, see fig. 6.16(b), it becomes obvious that the movement of the particles matches the streamlines of the fluid flow well within the short time interval under investigation—the majority of the contaminants cross the plane at $y = 0.94 H_f$ at which the trajectories are truncated within $2t_b$. As the streamlines predominantly diverge from regions of upward fluid motion, the contaminants are transported away from these regions and cross the subjacent plane

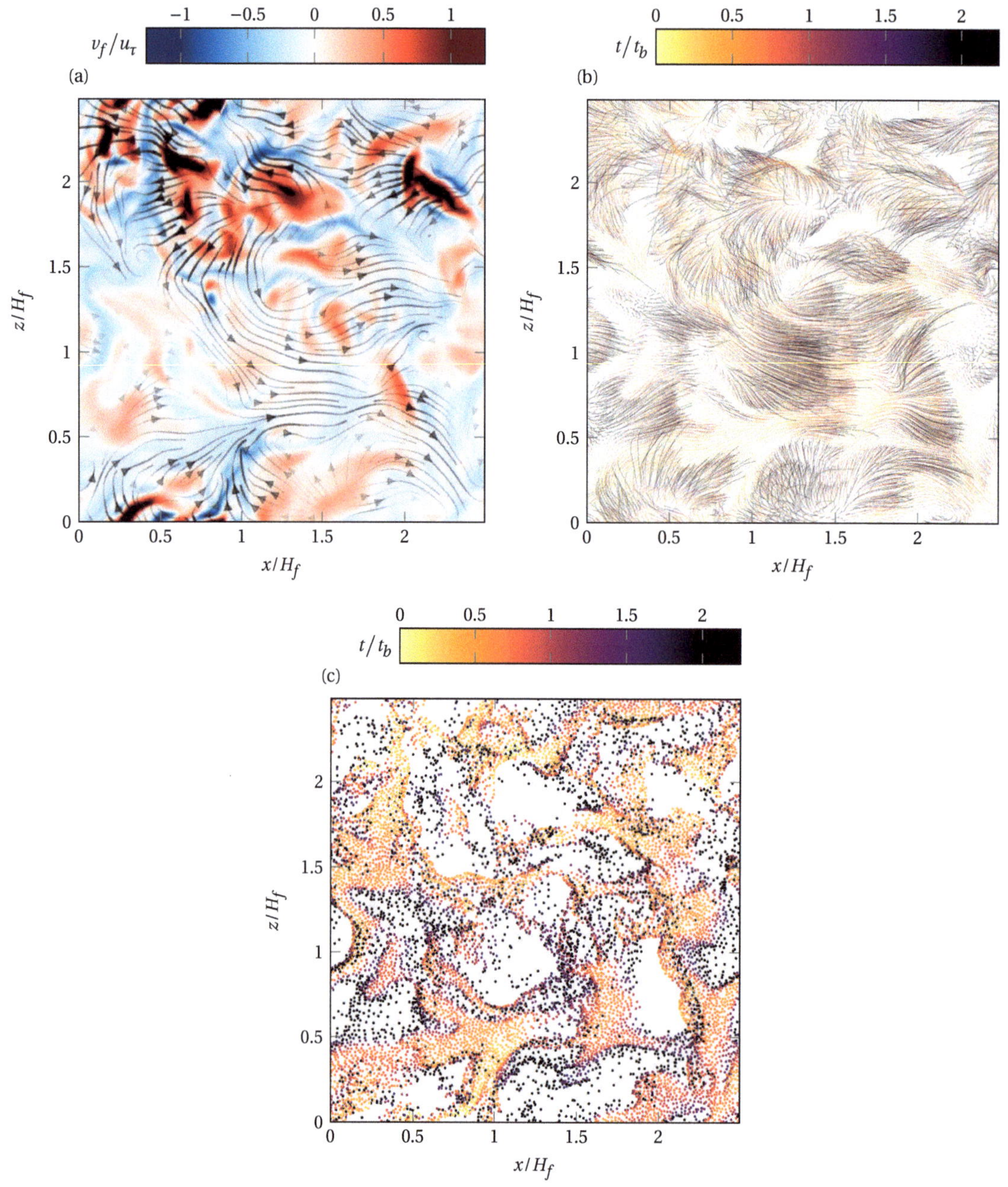

Figure 6.16: Flow conditions and particle trajectories at the very beginning of the transient settling phase. (a) Fluid flow at $t = 0$ in the plane of initialisation located at $y_{c,0}$. The coloured regions indicate the vertical component of the fluid velocity. The streamlines indicate the horizontal velocity components in the moving frame of reference with streamwise velocity $\langle u_f \rangle_{xz} \approx \langle u_c \rangle_{\mathrm{T}}$. The thickness and opacity of the streamlines indicate the magnitude of the horizontal fluid velocity $\sqrt{u_f'^2 + w_f^2}$. (b) Contaminant trajectories projected onto the wall-parallel plane starting from the initial position at $y \approx 0.95H_f$ and truncated at the location where the plane located at $y = 0.94H_f$ is crossed. The trajectories are shown in the same moving frame of reference as the fluid data. The progressive change in colour along a trajectory indicates the temporal evolution. (c) Horizontal locations of crossings through $y = 0.94H_f$ coloured by the cross-through time. The dots correspond to the endpoints of the trajectories previously shown and are likewise displayed in the moving frame of reference. The parameter point on display is G10-D10-HR.

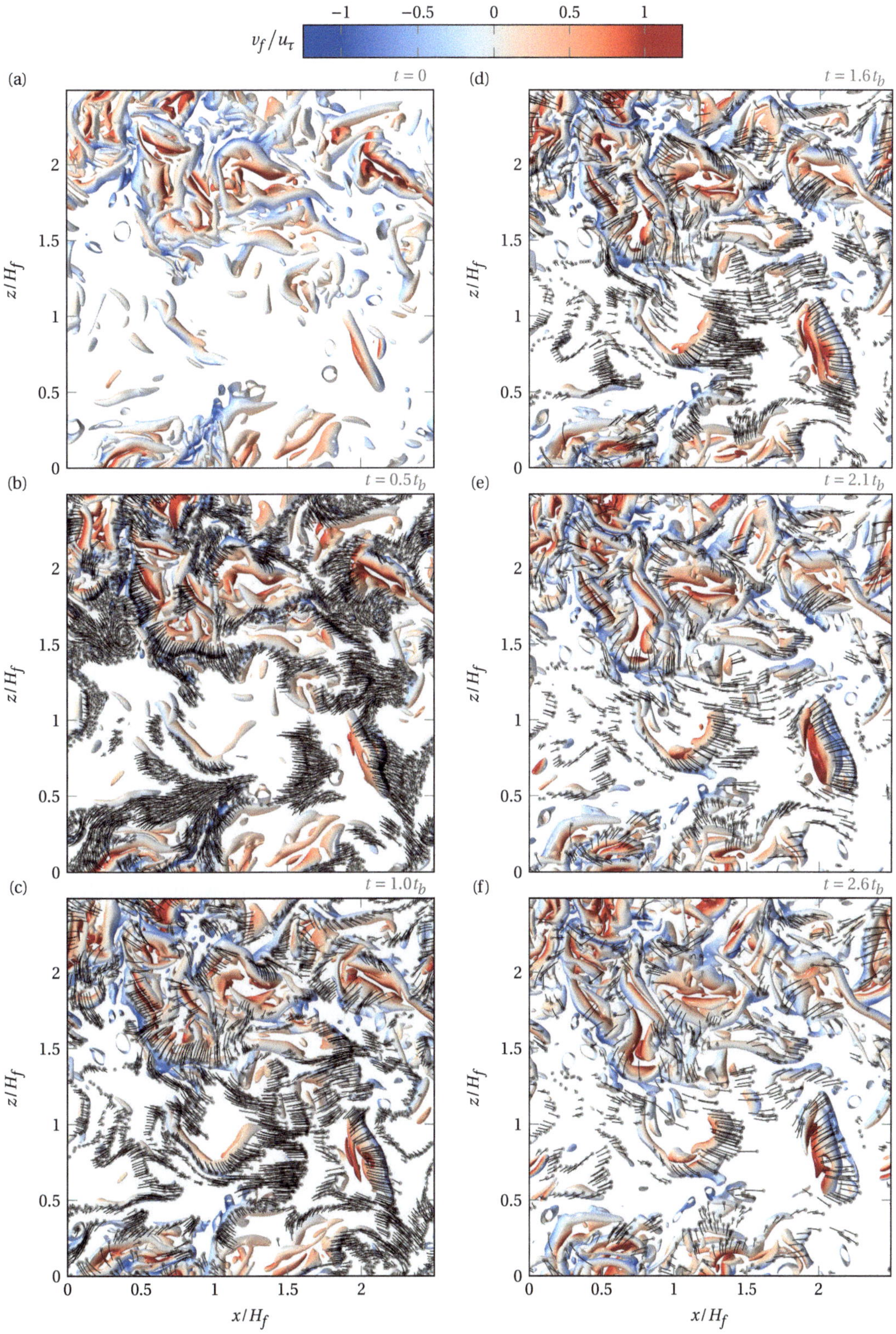

Figure 6.17: (a) Vortical structures near the upper boundary of the open channel visualised by the isosurfaces $Q^+ = 5 \cdot 10^{-5}$ just before contaminants are seeded into the flow. Only the region where $y > 0.875H_f$ is shown. The colouring indicates the direction of rotation of the vortices by showing the local vertical fluid velocity. (b-f) Vortical structures as before, but in the presence of settling contaminants at various times. Crossing locations through $y = 0.94H_f$ within $0.25t_b$ before and after the indicated time are displayed by black dots. The trajectories of the crossing particles for $1t_b$ before the crossing occurs are shown by black lines. The data is presented in a frame of reference moving with $\langle u_f \rangle_{xzt}(y_{c,0})$.

at locations where the vertical fluid velocity is small or negative. The result is the occurence of preferential concentration in downdrafts, and clustering in the form of three-dimensional accumulations—the particles which initially travel across the wall-parallel plane become vertically stacked behind those which spawned in a downdraft—and voids. The cross-through positions displayed in fig. 6.16(c) suggest that particles which cross at later times due to the rearrangement process predominantly do so near the edges between positive and negative v_f which manifests itself in well-defined edges in the distribution of cross-through locations. Similar sharp concentration fronts have been observed experimentally e.g. by Fessler et al. (1994).

As discussed in §2.2.3, splats and antisplats are understood to be manifestations of surface-parallel vortices which are located below the upper boundary of the open channel. Figure 6.17(a) portrays the vortical structures by means of isosurfaces of the Q-criterion introduced in eq. (2.51) at the time when contaminants are seeded into the flow. The vortices are preferentially located in the regions where intense fluid up- and downdrafts occur and the majority of them are aligned horizontally with the exception of few surface-attached vertical vortices. The trajectories of the contaminant particles are visualised in fig. 6.17(b-f) in batches for different time intervals alongside the evolving fluid flow. At all times it is evident that the particle motion is driven by the preferential sweeping mechanism. Virtually all contaminants are transported towards the downward facing side of nearby surface-parallel vortices and cross the subjacent wall-normal plane in downdraft regions, which explains the sampling bias in the fluid velocity sampled by the contaminants shown in fig. 5.8, and consequently, the rush enhancement in settling velocity. Moreover, since the particles are focused in the interstitial regions between the vortices, the origin of clustering and preferential concentration observed becomes evident, and indeed, these phenomena are driven by the interactions of the contaminants with turbulent structures.

6.3.4 Migration into large-scale streaks

While the interaction of the contaminants with vortical structures at early times was relatively straightforward to visualise due to the brevity of the time interval of interest, things become more complicated when the interaction with large-scale structures are concerned. In the following, the evolution of the scales of preferential concentration is analysed in a manner similar to the Fourier analysis presented in §5.5 for the fluid flow. For this purpose, the discrete crossing locations through a plane at height y are converted to an Eulerian field by introducing the probability density $\mathcal{P}_{\mathbb{X}}(x,z,t,y)$ of the crossing locations in the frame of reference moving with $\langle u_c\rangle_{\mathbb{X}}(y)$ as introduced in §6.3.1. The crossing probability density fulfils the normalisation property

$$\int_0^{L_x}\int_0^{L_z}\int_0^{t_{obs}} \mathcal{P}_{\mathbb{X}}\,\mathrm{d}t\,\mathrm{d}z\,\mathrm{d}x = 1, \tag{6.7}$$

and it should be acknowledged that the wall-normal position y serves a different purpose from the other two spatial dimensions in the sense that it defines the position of the plane under investigation rather than contributing to the definition of a three-dimensional probability of presence. For the analysis in this section, the temporal dependence is not of importance, and thus, as time-collapsed variant of $\mathcal{P}_{\mathbb{X}}$ is introduced as

$$\mathcal{P}_{\mathbb{X},t}(x,z,y) := \int_0^{t_{obs}} \mathcal{P}_{\mathbb{X}}\,\mathrm{d}t. \tag{6.8}$$

In its essence, $\mathcal{P}_{\mathbb{X},t}$ comprises the information visualised in fig. 6.11, but in terms of a Eulerian concentration field which is in practice constructed using a two-dimensional binning method. In order to

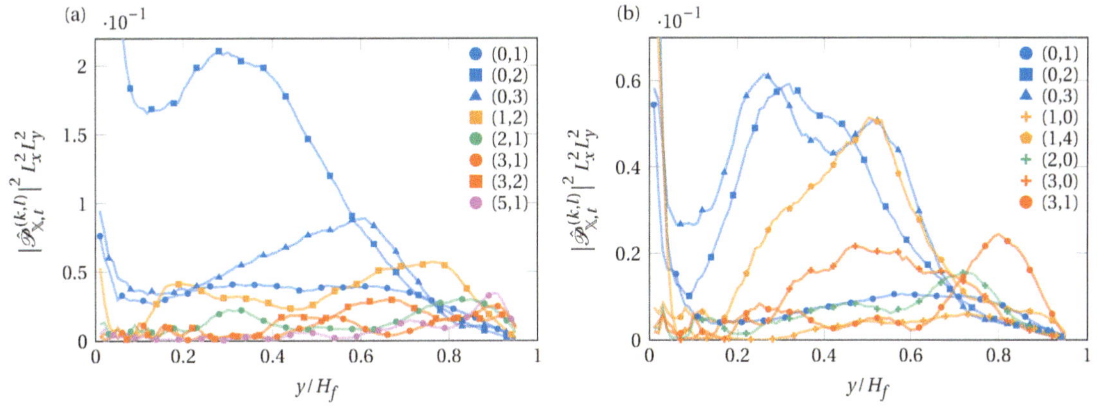

Figure 6.18: Evolution of the amplitudes of selected Fourier coefficients of the time-collapsed probability density of crossing locations as a function of the vertical position of the wall-normal plane. The number pairs (k, l) in the legend indicate the corresponding harmonics of the periodic system. (a) G10-D10-HR and (b) G30-D15-HR.

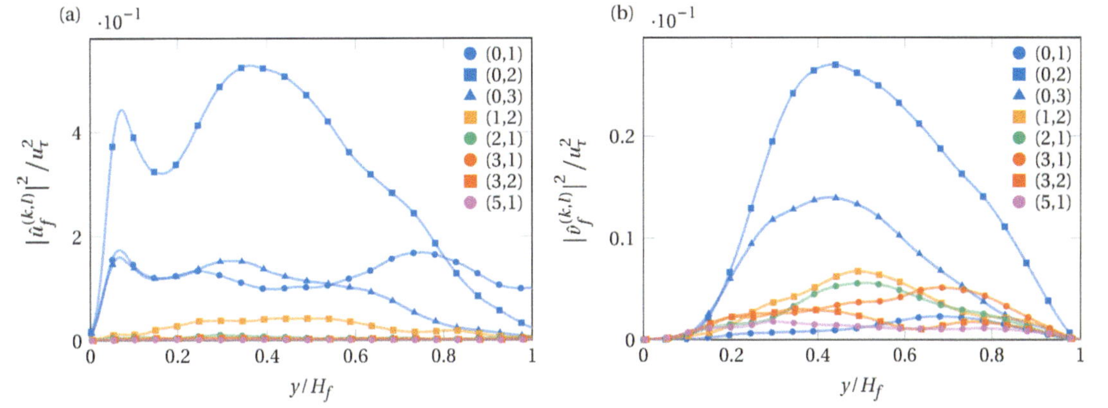

Figure 6.19: Spatial evolution of the amplitudes of selected Fourier coefficients of the (a) streamwise fluid velocity and (b) vertical fluid velocity. The number pairs (k, l) in the legend indicate the corresponding harmonics of the periodic system.

extract the spatial scales of preferential concentration, $\mathcal{P}_{\mathbb{X},t}$ is expanded by a Fourier series analogously to eq. (5.23) such that the backtransform reads

$$\mathcal{P}_{\mathbb{X},t}(x,z,y) = \sum_{k=-\infty}^{\infty} \sum_{l=-\infty}^{\infty} \hat{\mathcal{P}}_{\mathbb{X},t}^{(k,l)} \exp\left(-\mathrm{i}\left(\kappa_x^{(k)} x + \kappa_z^{(l)} z\right)\right). \tag{6.9}$$

Here, $\hat{\mathcal{P}}_{\mathbb{X},t}^{(k,l)}(y)$ is the discrete Fourier coefficient, and $\kappa_x^{(k)}, \kappa_z^{(l)}$ are the wavenumbers introduced in eq. (5.24) for l-th/k-th harmonics of the periodic system.

Figure 6.18 present the amplitudes of selected Fourier coefficients of $\mathcal{P}_{\mathbb{X},t}$ as a function of vertical position whereby the selection aims at representing the modes which most significantly contribute to $\langle \mathcal{P}_{\mathbb{X},t}^2 \rangle_{xz}$ throughout the various positions. The tendency of the accumulations to grow into larger patterns is well observed for the both cases presented. For positions close to the plane of initialisations, harmonics with wavelengths of a small integer fraction of the domain size contribute the most to the variations observed in the preferential concentration field. These modes subsequently lose importance in favour of the long streamwise modes which in the domain sizes investigated in this thesis span the entire streamwise length of the domain ($k = 0$). While for G10-D10-HR the $(0,2)$-mode clearly dominates below $y \lessapprox 0.6H_f$ (infinitely long, two spanwise undulations), the result is less conclusive for G30-D15-HR, but exhibits a similar tendency. These observations align well with the visual information provided by fig. 6.11 and 6.12.

An interesting question to be answered is whether the progressive migration into large-scale streaks is a result of particle inertia or rather driven by the inhomogeneity of turbulence in the wall-normal direction. In order to answer this, fig. 6.19 shows the same selection of modes as for fig. 6.18(a), but for the squared amplitudes of the Fourier coefficients of u_f and v_f. The similarity in the spatial evolution between the (0,2)-mode of u_f and $\mathcal{P}_{\mathbb{X},t}$ is striking, and suggests, that the progressive pattern formation is mainly driven by the inhomogeneous structure of wall-bounded turbulence. Yet, it may be observed that the progression of $\mathcal{P}_{\mathbb{X},t}$ lacks behind that for the large-scale streak which may imply non-locality in the mechanism responsible for preferential concentration. It is to be assumed that the driving mechanism for the migration of contaminants into large-scale streaks is the same mechanism which generates these structures in the first place. As it now stands, the interaction of particles with vortex clusters (del Álamo et al. 2006) or with packets of hairpin vortices (Kim and Adrian 1999) appear to be the prime candidates for the cause of this phenomenon.

6.4 Large-scale organisation and velocity enhancement

It was demonstrated quantitatively in §6.3.4 that contaminants arrange themselves mainly in structures of infinite streamwise length in the computational domains investigated. In the discussion on secondary flow patterns and spanwise-resolved crossing locations in §5.1 it was concluded that the streamwise-temporal averaged flow field is insufficient to explain the magnitude of particle velocity enhancement, especially in the direction of gravity. The introduction of the probability density of crossing locations enables a more detailed investigation using time-resolved flow fields which are solely collapsed in the streamwise direction in order to assess whether the discrepancy is caused a disregardance of time scales or length scales or both. To achieve this, a streamwise-collapsed variant of $\mathcal{P}_{\mathbb{X}}$ is defined as

$$\mathcal{P}_{\mathbb{X},x}(z,t,y) := \int_0^{L_x} \mathcal{P}_{\mathbb{X}}\, \mathrm{d}x, \tag{6.10}$$

and an example of the temporal evolution of this quantity is shown in fig. 6.20 which clearly reflects the migration of contaminants into the preferred spanwise locations during the settling process. A comparison with the deviations of $\langle u_f \rangle_x$ from the full spatio-temporal average shown in the same figure elucidates that this preferred location coincides with high-momentum pathways as expected. Essentially, fig. 6.20 depicts the time-resolved combination of fig. 5.1(a) and 5.3(a).

A simple estimation of the fluid velocity experienced by the contaminants under the assumption that only length scales with infinite extent in x are relevant is given by

$$\langle u_{f,\alpha} \rangle_{\mathbb{S}\mathbb{X}}^{est} := \int_0^{L_z} \int_0^{t_{obs}} \mathcal{P}_{\mathbb{X},x} \langle u_{f,\alpha} \rangle_x \, \mathrm{d}t\, \mathrm{d}z. \tag{6.11}$$

Figure 6.21 depicts a juxtaposition of the value obtained by this estimation and the actual value sampled by the contaminants, cf. §5.2.3. While for the streamwise velocity component, the disregardance of streamwise variations only results in a discrepancy by a factor of two, the vertical velocity is severely underpredicted by a factor of $\mathcal{O}(10)$. This again suggests that the sampling of fluid downdrafts of length scales smaller than H_f is essential in order to explain the enhancement in settling velocity, whereas the enhancement in streamwise velocity is predominantly a large-scale phenomenon.

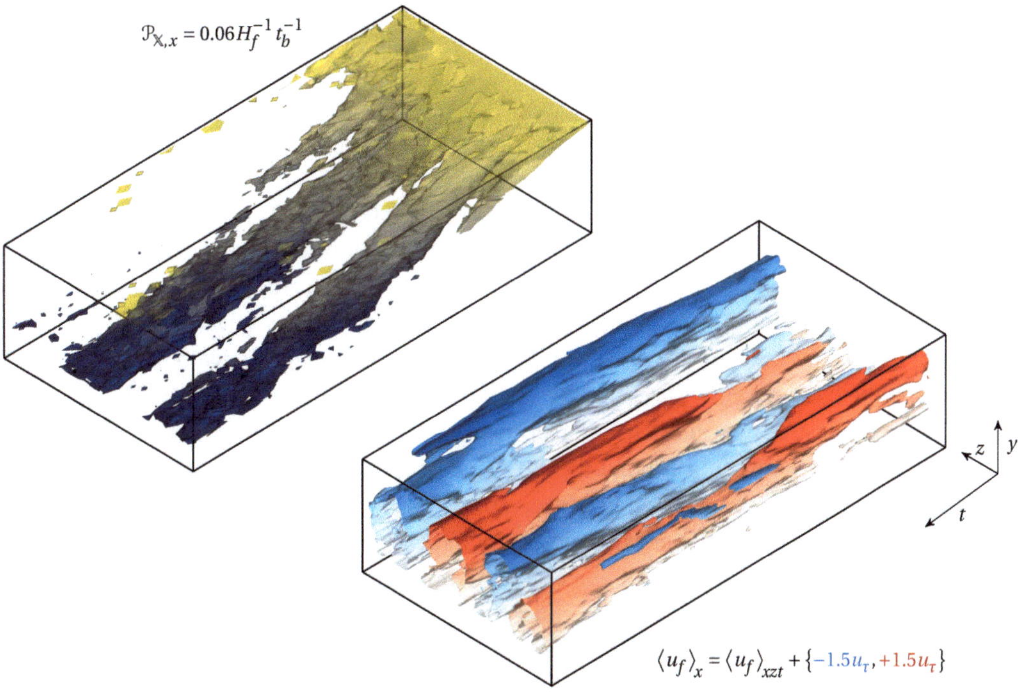

Figure 6.20: Visualisation of the streamwise-collapsed probability density of crossing locations (upper left) and the streamwise-averaged fluid velocity $\langle u_f \rangle_x$ (lower right). For $\mathcal{P}_{\mathbb{X},x}$ the colour scheme indicates the vertical position, while for $\langle u_f \rangle_x$ the sign of the deviation from $\langle u_f \rangle_{xzt}$ is shown. The parameter point on display is G10-D10-HR.

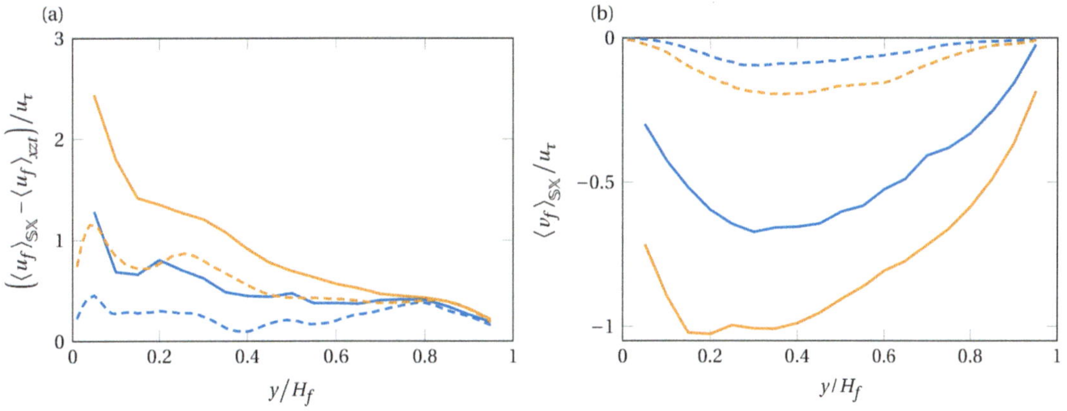

Figure 6.21: Comparison between the fluid velocity sampled by the contaminants and the prediction under the assumption that only long streamwise modes contribute. (a) Streamwise fluid velocity and (b) vertical fluid velocity. The line colours represent the cases G30-D15-HR (—) and G10-D10-HR (—), whereas the line styles indicate the actual value obtained from the DNS data using spherical-averaging $\langle u_{f,\alpha} \rangle_{\mathbb{S}\mathbb{X}\downarrow}$ (—) and the estimate from the streamwise-collapsed consideration $\langle u_{f,\alpha} \rangle_{\mathbb{S}\mathbb{X}}^{est}$ (- - -).

6.5 Interactions with the sediment bed

The focus is now placed on the long-term interactions between the contaminant particles and the pre-established sediment bed, especially with regards to the influence of bedforms. It was already observed in §4.6 and §6.1.2 that the transport characteristics and clustering behaviour depends considerably on the presence or absence of bedforms, and an evaluation of the cause of this discrepancy will be subject of this section.

Figure 6.22 portrays the positions of the contaminants relative to the featureless bed at various times for G10-D6-LR. During the settling phase and in the early interaction phase shown in fig. 6.22(a-c), the contaminants tend so align in streamwise stripes which concide with the location of sediment ridges. In fact, when the footprint of the large-scale roller structures is examined, i.e. the fluid velocities $\langle v_f \rangle_x$ and $\langle w_f \rangle_x$ in the zy-plane, it stands to reason that the contaminants are pushed beneath the large-scale low-speed streaks on the bed where they accumulate. An interesting observation to be made is that for long observation times this pattern becomes less perceptible, see fig. 6.22(d), although the nearest neighbour analysis of fig. 6.3(b) does not indicate that the clusters disappear. This peculiarity may be tied to the trend of the contaminant ensemble to slowly decrease its mean vertical position, cf. fig 4.19, as it appears that contaminants get buried beneath a layer of sediment particles, although this process not very clear in the visualisations. If this is indeed the case, contaminants which arranged themselves beneath a low-speed streak may eventually become trapped by other particles above them, which reduces their mobility and maintains their packed and clustered state. As can be deduced from fig. 6.22(a-d), the footprint of the large-scale rollers which presumably agitate the bed particles varies substantially within the time scales of interest in this analysis. Thus, the contaminants which have not been captured by the bed will tend to accumulate in different regions at different times, and as a result, the particle distribution may appear to become random in the projected view, albeit some clusters may be preserved due to immobilisation.

The deposition behaviour in the presence of ripples is vastly different from what has just been discussed. A visualisation of the contaminant-sediment interaction for the same parameter point as before, but in the presence of ripples (G10-D6-RP) is provided in fig. 6.23. While the arrangement of contaminants is rather homogeneous in the lateral direction, a marked inhomogeneity in the streamwise direction is developed. In fact, from the time series it becomes obvious that the contaminants tend to be deposited in the trough of the ripple bedform, likely due to the persistent zone of recirculation which is known to develop behind the crest of such bedforms, cf. Kidanemariam (2016) or Kidanemariam et al. (2022) for a detailed discussion on the flow patterns and hydrodynamic forces on particles for this exact configuration. After the deposition took place, a pronounced trapping phenomenon occurs: as the ripple propagates in the streamwise direction, the newly deposited contaminants are buried underneath a layer of particles which is much thicker compared to what has been observed for macroscopically flat beds due to the markedly larger variance in bed height which arises for the ripple bedform. Therefore, the captured contaminants are essentially immobilised until the overhead layer of particles has propagated far enough such that the trailing trough approaches them, which increases the probability of erosion. This observation satisfactorily explains the oscillating trend of the ensemble-averaged vertical position of the contaminants in time (fig. 4.19). Furthermore, the capturing effect appears to be very efficient: when the first contaminants have already been trapped, the remaining contaminants which are transported in the bedload or suspended load eventually also end up in the trough (fig. 6.23(b-d)). As a result, nearly all contaminants are immobilised at $t \approx 100 t_b$ for this parameter point (fig. 6.23(d)). This recognition provides a sound explanation of the strong temporary decrease in the streamwise contaminant flux observed in fig. 4.20. Since not all contaminants are trapped at the same time, the time of

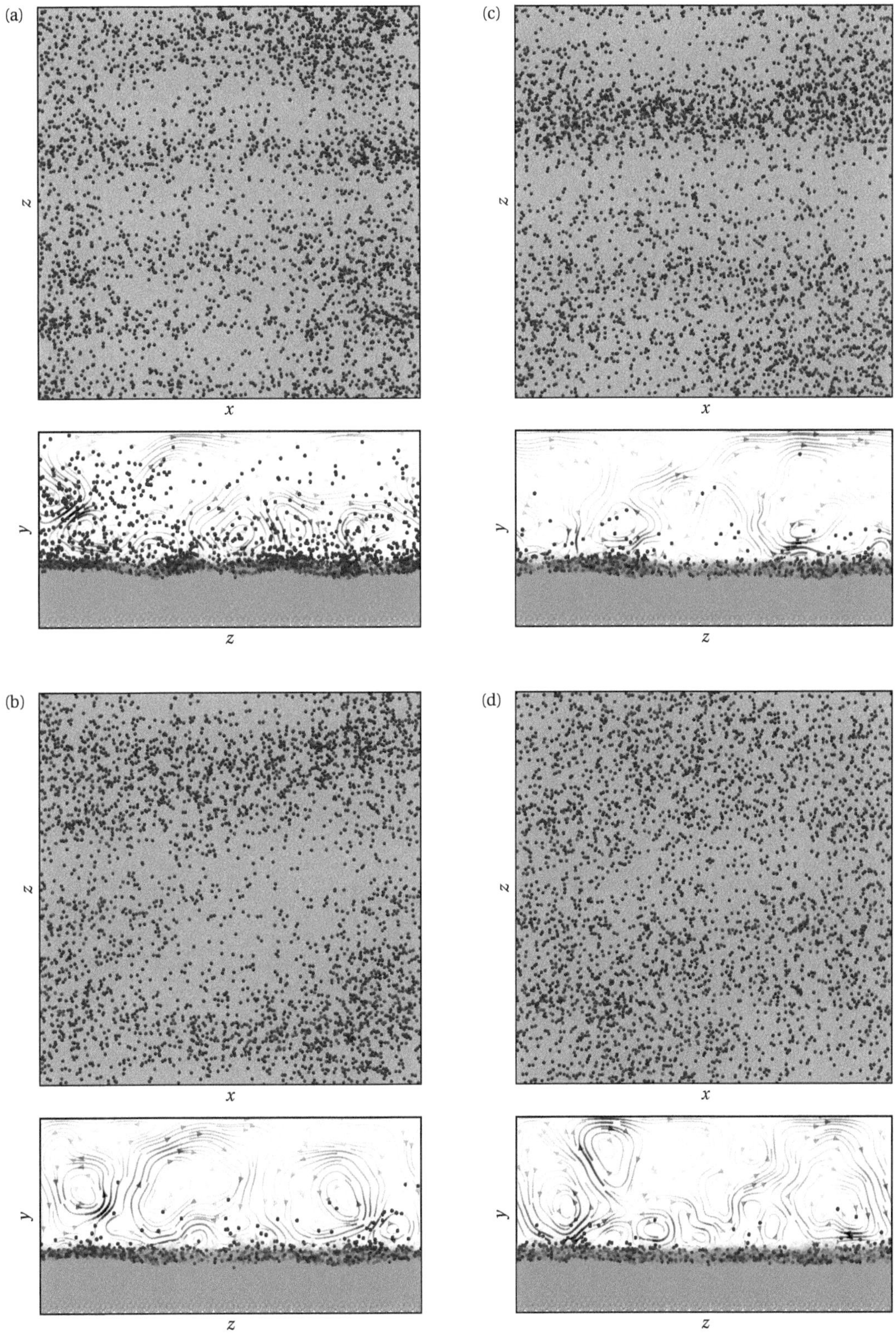

Figure 6.22: Visualisation of the interaction between contaminant particles (dark) and sediment particles (light, translucent) in the absence of bedforms. The upper part of each subfigure shows a top view, whereas the lower part depicts the same data from a frontal perspective. In the frontal view, the streamwise-averaged flow in the cross-section is shown in terms of streamlines whose thickness and opacity is scaled by the corresponding velocity magnitude. The simulation on display is the long realisation of G10-D6-LR at the times (a) $t = 25t_b$, (b) $t = 100t_b$, (c) $t = 200t_b$, and (d) $t = 400t_b$.

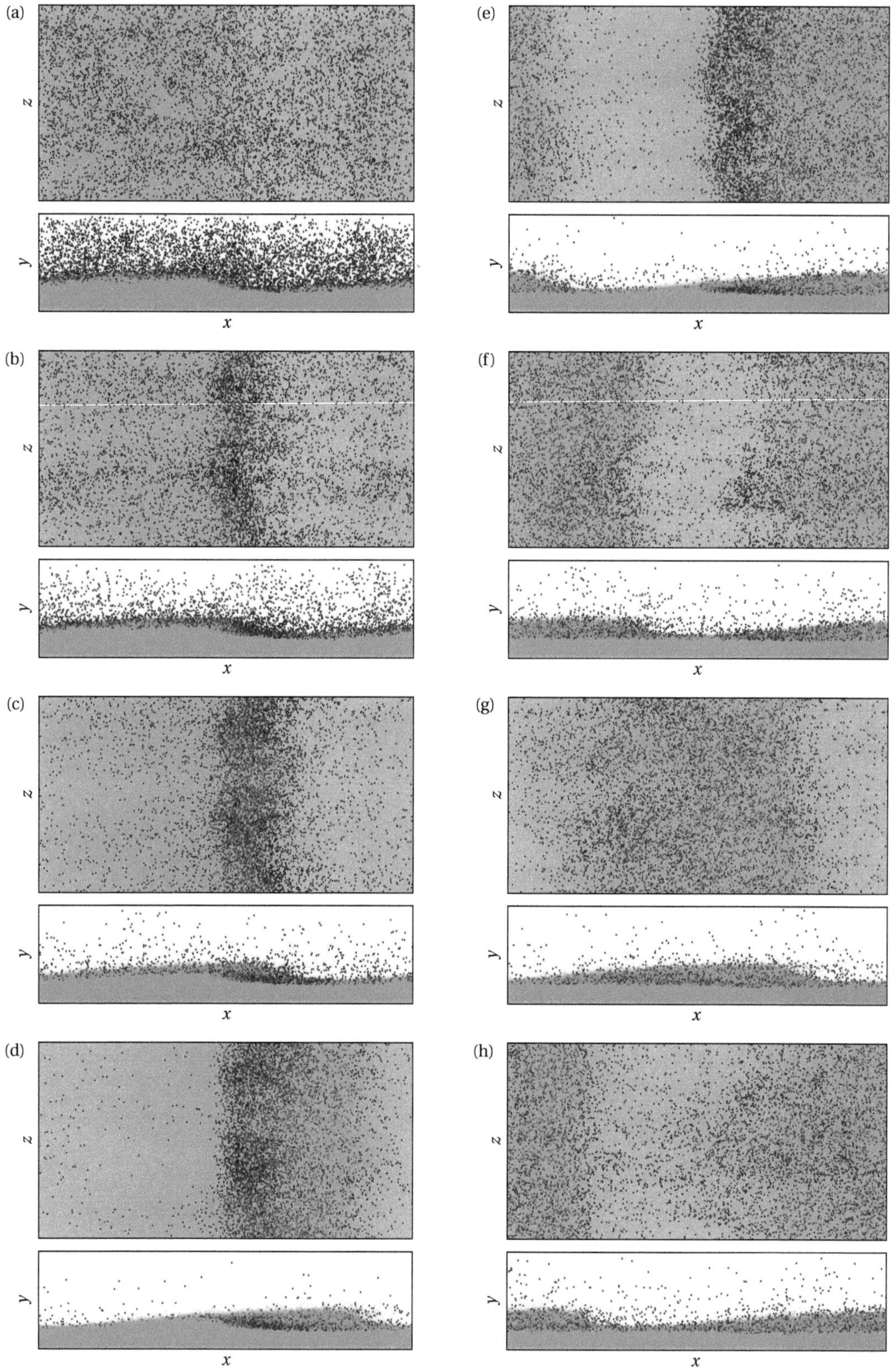

Figure 6.23: Visualisation of the interaction between contaminant particles (dark) and sediment particles (light, translucent) in the presence of ripple features. The upper part of each subfigure shows a top view, whereas the lower part depicts the same data from a lateral perspective. The simulation on display is the long realisation of G10-D6-RP at the times (a) $t = 10t_b$, (b) $t = 25t_b$, (c) $t = 50t_b$, (d) $t = 100t_b$, (e) $t = 150t_b$, (f) $t = 200t_b$, (g) $t = 300t_b$, and (h) $t = 400t_b$.

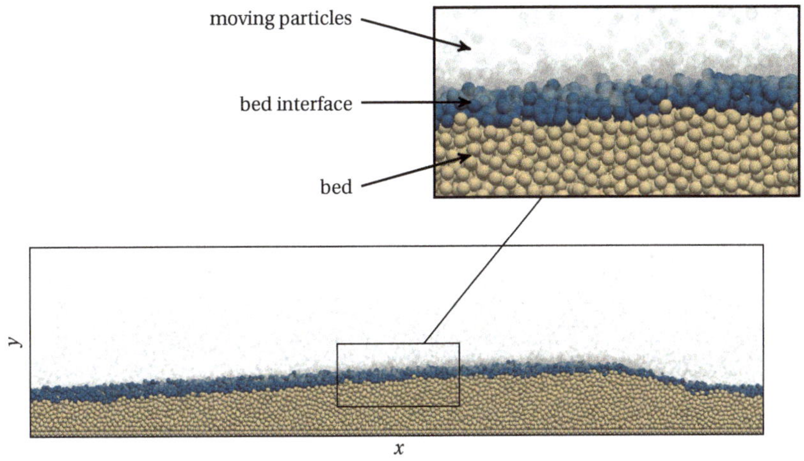

Figure 6.24: Visualisation of the classification of particles concerning their instantaneous positioning with respect to the bed. All contaminant and sediment particles which are not disregarded by the criteria stated in §3.4 are classified as bed particles (yellow & blue), while the dismissed particles constitute the bedload and suspended load (white, translucent). From the set of bed particles, those which constitute the fluid-bed interface are additionally classified as interface particles (blue).

resuspension also varies substantially, and for this reason, the cycle of deposition and release becomes blurred in time. However, at $t = 400\,t_b$ the impact of this mechanism is still perceptible.

It was demonstrated in fig. 4.19 that ripples are more effective at entraining contaminants beneath the mean position of the sediment-fluid interface, which is certainly at least partially a result of the varying height of the interface itself in conjunction with the capturing mechanism just described. In order to quantify when and where contaminants are deposited with respect to the instantaneous interface, a classification of the particle position is introduced based on the algorithm used to extract the interface. As described in §3.4, the set of particles which form the bed is defined by disregarding all particles which belong to the bedload or suspended load using criteria on the collision force and instantaneous particle velocity, and subsequently, the set of particles which form the interface is identified using geometric considerations. As this method is completely based on the positions of discrete particles, a classification can be introduced which attributes to each individual particle whether or not it belongs to the bed and/or the interface at a given time. A visual explanation of this classification is provided in fig. 6.24.

From this classification, the number of particles which belong to the bed, $N_{p,bed}(t)$, and to the interface, $N_{p,if}(t)$, are obtained. The classification takes into account both sediment and contaminant particles, and hence, these numbers can be further partitioned into contributions from each type of particles, viz.

$$N_{p,bed} = N_{c,bed} + N_{s,bed} \quad \text{and} \quad N_{p,if} = N_{c,if} + N_{s,if}. \tag{6.12}$$

The temporal evolution of the fraction of contaminants which are attributed to the bed and/or interface is shown in fig. 6.25(a) for the macroscopically flat bed. While shortly after the settling phase all deposited contaminants belong to the interface, a plateau is established quickly for $N_{c,if}$ at distinct values for each parameter point, whereas the number of bed particles increases. Generally, a high number of contaminants of around 90% is categorised as belonging to the bed at long observation times, and this number does not vary significantly between the two cases. What does vary, however, is the fraction of contaminants which belong to the interface, which is significantly higher for G30-D10-LR than for G10-D6-LR. This implies that the smaller and lighter particles are entrained deeper into the bed. Moreover, for this parameter point, $N_{c,bed}$ maintains a slowly increasing trend, whereas $N_{c,if}$ simultaneously

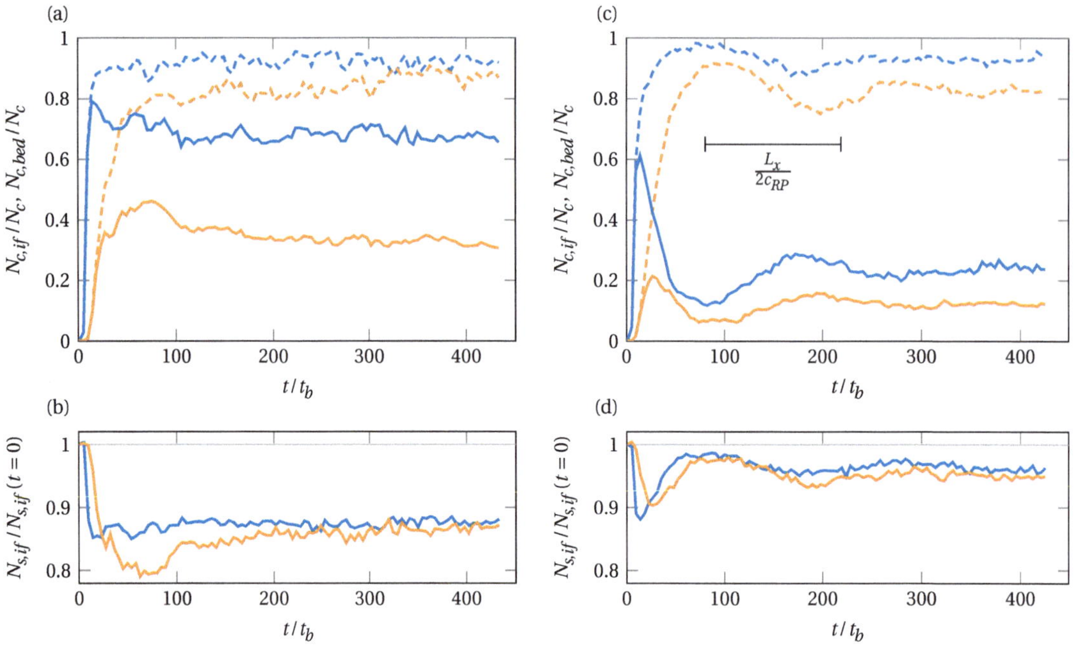

Figure 6.25: Temporal evolution of the number of particles which are classified as belonging to the bed and/or interface. (a,c) The share of contaminant particles belonging to the interface (—) or bed (- - ·). (b,d) Number of sediment particles constituting the interface relative to the respective number before contaminants are introduced to the flow. The cases shown in (a,b) are G30-D10-LR (—) and G10-D6-LR (—) which both feature a macroscopically flat bed. The data for ripples is given in (c,d) for G30-D10-RP (—) and G10-D6-RP (—). The interval shown in (c) indicates the duration it takes for the ripple to propagate over half its wavelength.

decreases. This may suggest that the contaminants are covered by sediment particles, which corroborates the hypothesis regarding the homogenisation of the arrangement in the wall-parallel plane stated earlier. Indeed, fig. 6.25(b) indicates that sediment particles restore their role as interface particles for G10-D6-LR.

Regarding the ripple bedform, the share of contaminants which become incorporated into the bed is of similar magnitude as before, yet, much fewer of these particles are part of the interface. This observation is clearly linked to the burying mechanism of the propagating ripple which is evidently much more effective at immobilising contaminants. The initial capturing after the settling phase is markedly perceptible, and the duration of the undulation between maximum deposition and maximum resuspension is slightly shorter than the time it takes the ripple to propagate over half its wavelength. Interestingly, although the amount of captured contaminants is substantially higher in the presence of ripples, it should be kept in mind that the streamwise flux is still comparable to or even higher than the flux in absence of bedforms, cf. fig 4.20. This is presumably a joint effect of the deposition and resuspension cycle and the increased likelihood of contaminants to be ejected into the outer flow due to the ramp-like shape of the ripple.

6.6 Settling length

A quantity of practical interest regarding the settling problem is the streamwise distance travelled by the contaminants before they are first deposited—often referred to as the *settling length*. In order to examine

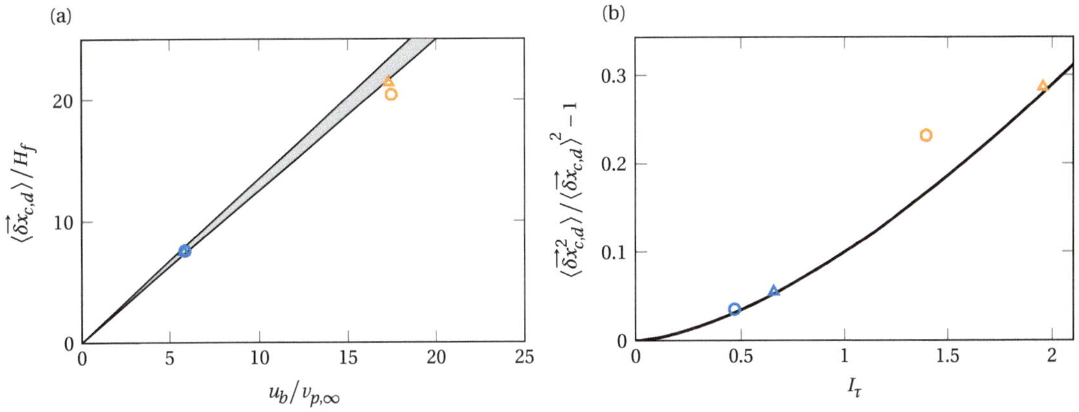

Figure 6.26: (a) Emsemble-average settling length of contaminant particles as a function of the ratio between bulk fluid velocity and contaminant terminal velocity in a quiescent environment. The two solid lines (—) display $1.25 u_b / v_{p,\infty}$ and $1.35 u_b / v_{p,\infty}$, respectively, where the slopes are chosen to represent the range of $\langle t_{c,d} \rangle / t_{p,s}$ observed across the parameter points presented. The shaded region indicates this range. (b) Normalised mean-squared displacement in streamwise direction until contaminants are deposited for the first time presented in terms of the relative turbulence intensity. The solid line (—) shows the fit of Li and Shen (1975) obtained by regression of the experimental data presented in Bayazit (1972), which is given by $0.1 I_\tau^{1.532}$ in the current representation. Legend: G30-D10-LR (○), G10-D6-LR (○), G30-D10-RP (△), and G10-D6-RP (△).

the settling length from the simulation data, the particle trajectories need to be expanded across the periodic boundary, viz.

$$\overrightarrow{x}_c^{(i)}(t) := x_c^{(i)} + N_{L_x}^{(i)} L_x, \tag{6.13}$$

where $\overrightarrow{x}_c^{(i)}(t) \in (-\infty, \infty)$ is the expanded streamwise position of the i-th particle, and $N_{L_x}^{(i)}(t) \in \mathbb{Z}$ the difference between the number of forward and backward crossings through the periodic boundary. The streamwise displacement of the i-th particle from its initial position $x_{c,0}^{(i)}$ is then given by

$$\overrightarrow{\delta x}_c^{(i)}(t) := \overrightarrow{x}_c^{(i)} - x_{c,0}^{(i)}. \tag{6.14}$$

In order to evaluate the settling length from the displacement, a deposition time $t_{c,d}$ needs to be defined which reflects the time at which a contaminant particle is first deposited into the bed. For this purpose, the classification procedure of §6.5 is employed, and the deposition time is then obtained from

$$t_{c,d}^{(i)} := \min\left\{ t : \chi_{c,bed}^{(i)}(t) > 0 \right\}, \tag{6.15}$$

where $\chi_{c,bed}^{(i)}(t)$ is an indicator function whose value is zero when a contaminant particle is not part of the bed and unity if it is. In principle, $t_{c,d}^{(i)}$ may be undefined if a contaminant never gets deposited within the observation time, however, for the four extended simulation runs discussed in this thesis this is never the case. Finally, the settling length $\overrightarrow{\delta x}_{c,d}$ is given by the streamwise displacement at the time of the initial deposition, viz.

$$\overrightarrow{\delta x}_{c,d}^{(i)} := \overrightarrow{\delta x}_c^{(i)}\left(t_{c,d}^{(i)}\right). \tag{6.16}$$

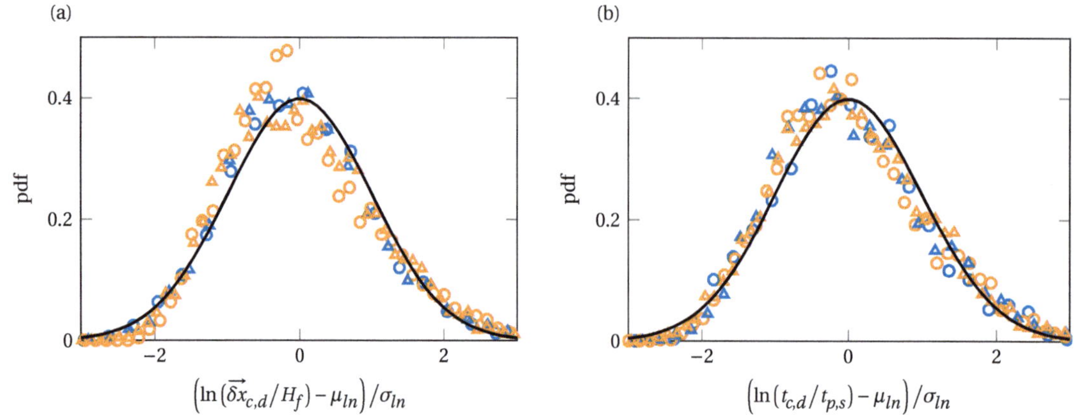

Figure 6.27: Comparison of normalised empirical data for (a) the settling length and (b) the deposition time to the normalised log-normal distribution. The corresponding parameters μ_{ln} and σ_{ln} have been obtained by calculating the sample mean and standard deviation of the logarithm of the sample data for each parameter point. Legend: G30-D10-LR (○), G10-D6-LR (○), G30-D10-LR (△), and G10-D6-LR (△).

As suggested by Li and Shen (1975) for smooth-wall open channel flows, the ensemble-averaged settling length

$$\langle \overrightarrow{\delta x}_{c,d} \rangle := \frac{1}{N_{c\Sigma}} \sum_{i=1}^{N_{c\Sigma}} \overrightarrow{\delta x}_{c,d}^{(i)} \tag{6.17}$$

is described well by the ratio between the bulk velocity in the channel and the ambient terminal velocity of the contaminants, see fig. 6.26(a), albeit only little variation in $u_b/v_{p,\infty}$ is at disposal from the present datasets. The available data may be approximated by a linear relationship with zero intercept and a slope given by the ratio of mean deposition time to characteristic settling time, which is found to not exhibit a clear scaling behaviour and to fluctuate in the relatively confined range of $\langle t_{c,d} \rangle / t_{p,s} = 1.25$–$1.35$ for the four parameter points. In fact, the inverse of this ratio ought to be related to the group velocity near the bed interface, which can be deduced from fig. 4.17 to be in a similar value range. This suggests that the average time until deposition is indeed longer than the naïve estimate of $H_f/v_{p,\infty}$, despite particles settling considerably faster individually and as an ensemble in the bulk of the channel. It can moreover be recognised that the presence of ripples does not substantially alter $\langle \overrightarrow{\delta x}_{c,d} \rangle$ despite the distinct difference in the deposition mechanism compared to macroscopically flat beds.

Regarding the spread around the mean, the mean-squared displacement is found to adhere to the empirical relationship

$$\frac{\langle \overrightarrow{\delta x}_{c,d}^{2} \rangle}{\langle \overrightarrow{\delta x}_{c,d} \rangle^{2}} \approx 0.1 I_\tau^{1.532} + 1 \tag{6.18}$$

established by Li and Shen (1975) based on the regression of experimental data from Bayazit (1972) for the settling length in turbulent smooth-wall open channel flows, see fig. 6.26(b). Interestingly, the aforementioned authors were able to reproduce this dependency using a stochastic model which assumes a random walk for the settling particles and a Gaussian distribution of turbulent velocity fluctuations, which suggests that the statistics of the settling length are rather robust with respect to the details of the settling problem.

To advance the comparison with this simplified model even further, Li and Shen (1975) discovered that the probability density distribution of the settling length follows a log-normal shape, which is given by

$$\mathrm{pdf}(X) = \frac{1}{X\sigma_{ln}\sqrt{2\pi}} \exp\left(-\frac{\left(\ln(X) - \mu_{ln}\right)^2}{2\sigma_{ln}^2}\right), \tag{6.19}$$

where X is a placeholder for any random variable. This distribution describes a Gaussian distribution in logarithmic scaling, and consequently its parameters are given by the mean and standard deviation of the logarithm of X, viz.

$$\mu_{ln}(X) := \mathrm{E}\left(\ln(X)\right), \tag{6.20a}$$

$$\sigma_{ln}(X) := \sqrt{\mathrm{E}\left(\ln(X^2)\right) - \mathrm{E}\left(\ln(X)\right)^2}, \tag{6.20b}$$

where $\mathrm{E}(\bullet)$ is the expected value operator.

Figure 6.27(a) depicts the empirical probability density of $\ln\left(\overrightarrow{\delta x}_{c,d}/H_f\right)$ scaled by the corresponding logarithmic mean and standard deviation obtained from the samples, i.e.

$$\mathrm{E}\left(\ln(\overrightarrow{\delta x}_{c,d}/H_f)\right) = \langle \ln\left(\overrightarrow{\delta x}_{c,d}/H_f\right)\rangle, \tag{6.21a}$$

$$\mathrm{E}\left(\ln(\overrightarrow{\delta x}_{c,d}/H_f)^2\right) = \langle \ln\left(\overrightarrow{\delta x}_{c,d}/H_f\right)^2\rangle. \tag{6.21b}$$

The simulation data follows the normalised log-normal distribution to a reasonable degree corrobating the recognition of Li and Shen (1975) and the results from their simple statistical model. Moreover, the distribution of deposition times exhibits an equivalent behaviour as can be seen in fig. 6.27(b), and indeed, the scaling applied to the mean in fig. 6.26(a) can be used to approximately obtain the settling length distribution from that of the deposition time, i.e.

$$\mathrm{pdf}\left(\overrightarrow{\delta x}_{c,d}/H_f\right) \approx \mathrm{pdf}\left(t_{c,d}u_b/H_f\right). \tag{6.22}$$

It should be noted that all parameter points depicted in fig. 6.27 appear to exhibit a slight skewness towards smaller values, whereas the skewness of the transformed Gaussian vanishes. Moreover, it stands to reason that the probability density of $t_{c,d}$ ought to be related to that of the cross-through times during the settling phase discussed in §4.4. For the latter a generalised Gamma distribution was employed to describe the data, which indeed yielded a more accurate representation of the data than a log-normal distribution, presumably due to its increased number of degrees of freedom. In fact, it can be demonstrated that the log-normal distribution is a specialisation of this distribution (Prentice 1974) raising the question whether the higher complexity of the generalised Gamma distribution is justified, or one should prefer the simpler log-normal distribution following the principle of Occam's razor (of Ockham 1495).[5]

[5] Apart from pure aesthetic considerations, the advantages of choosing a model of lower complexity lies in the better ability to potentially falsify its assertions (Popper 1935), and in its increased "informativeness" (Sober 1975) in the sense that the ratio of the required ad hoc information to its explanatory capabilities is minimised. The difficulty of choosing an appropriate representation of the data is a manifestation of the fundamental problem of underdetermination (Mill 1856) in the philosophy of science.

The expressiveness of the log-normal distribution conveys itself by recognising that its parameterisation can be deduced solely from the knowledge of the mean and mean square of the distribution in the non-transformed sample space, viz.

$$\mu_{ln}(X) = \ln\left(\mathrm{E}(X)\right) - \tfrac{1}{2}\ln\left(\frac{\mathrm{E}(X^2)}{\mathrm{E}(X)^2}\right), \tag{6.23a}$$

$$\sigma_{ln}(X) = \sqrt{\ln\left(\frac{\mathrm{E}(X^2)}{\mathrm{E}(X)^2}\right)}, \tag{6.23b}$$

albeit it should be acknowledged that an estimation using these relations differs from that obtained by eq. (6.20) for data which does not perfectly follow the distribution. Thus, the information required to reconstruct the complete probability distribution of deposition times (and settling lengths) is contained in fig. 6.26, which in conjuction with the evidently robust scaling laws provide a large predictive power.

6.7 Lateral displacement and long-term dispersion

In accordance with its definition, the discussion on the settling length in §6.6 was limited to the streamwise displacement of contaminants in the time interval between their release and their first deposition. In the context of river pollution events other characteristics such as the lateral displacement and the long-term transport beyond the first deposition are also of interest, and their characterisation is the subject of this section.

To begin with, the lateral displacement of the i-th contaminant particle from its initial spanwise position $z_{c,0}^{(i)}$ is defined analogously to eq. (6.14) as

$$\overrightarrow{\delta z}_c^{(i)}(t) := \overrightarrow{z}_c^{(i)} - z_{c,0}^{(i)}, \tag{6.24}$$

where $\overrightarrow{z}_c^{(i)}(t) \in (-\infty, \infty)$ is the instantaneous spanwise position expanded across the periodic boundary, cf. eq. (6.13) for the streamwise analogue. The lateral displacement at deposition is then given by

$$\overrightarrow{\delta z}_{c,d}^{(i)} := \overrightarrow{\delta z}_c^{(i)}\left(t_{c,d}^{(i)}\right). \tag{6.25}$$

The lateral dispersion differs from that in the streamwise direction in the sense that no advection in the direction of interest occurs, and theoretical considerations suggest a normal distribution around a zero mean (Fischer 1973) after an initial transient phase, i.e. the probability density distribution is given by

$$\mathrm{pdf}(X) = \frac{1}{\sigma_{nd}\sqrt{2\pi}} \exp\left(-\frac{X^2}{2\sigma_{nd}^2}\right), \tag{6.26}$$

$$\sigma_{nd}(X) := \sqrt{\mathrm{E}\left(X^2\right) - \mathrm{E}\left(X\right)^2}. \tag{6.27}$$

A comparison of the normalised empirical distribution of $\overrightarrow{\delta z}_{c,d}$ obtained from the numerical simulations with a normal distribution confirms this supposition for the displacement at deposition, see fig. 6.28(a), as well as the displacement at crossings of the wall-parallel planes during the preceding transient settling phase as shown in fig. 6.28(b) for a variety of vertical positions and parameter points.

The lateral dispersion during the settling phase is displayed in fig. 6.28(c) in terms of the spatial evolution of the standard deviation of the spanwise particle locations at cross-through. In fact, as the data is

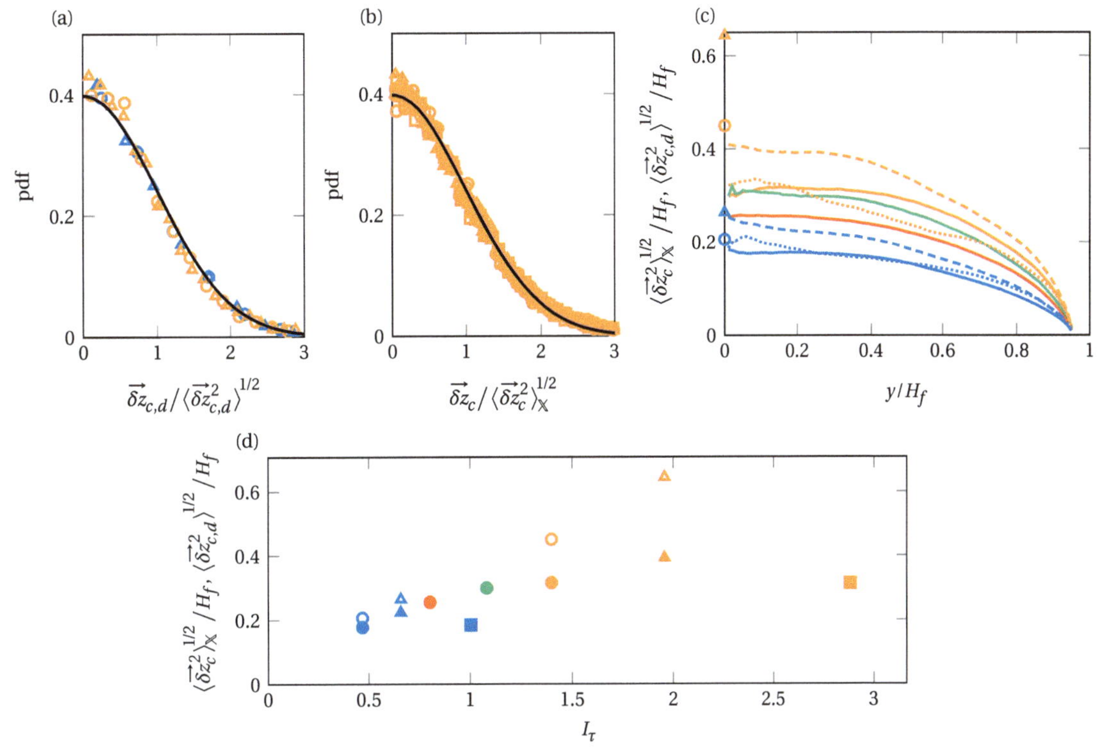

Figure 6.28: (a) Lateral contaminant displacement at deposition normalised by the sample standard deviation and assuming symmetry around a zero mean. (b) Same as before, but for crossings of wall-parallel planes during the settling. All cases with $Ga \approx 10$ (LR, RP, HR) are displayed for $y = \{0.3, 0.5, 0.7\}\, H_f$. (c) Sample standard deviation $\langle\overrightarrow{\delta z}_c^2\rangle_{\mathbb{X}}^{1/2}$ as a function of vertical position during the initial settling process before deposition (——,- - -,······). Also shown is the standard deviation of the displacement at deposition $\langle\overrightarrow{\delta z}_{c,d}^2\rangle^{1/2}$ for the runs where an extended observation time is available (○,△). (d) Same data as before at $y = 0.2H_f$ for the spatially-averaged data as a function of relative turbulence intensity (●,▲,■). Also shown is the standard deviation at deposition (○,△). Legend: G30-D10-LR (——,○,●), G10-D6-LR (——,○,●), G15-D10-LR (——,●), G15-D6-LR (——,●), G30-D10-RP (- - -,△,▲), G10-D6-RP (- - -,△,▲), G30-D15-HR (······,■), and G10-D10-HR (······,■).

normally distributed and no mean drift occurs, $\langle\overrightarrow{\delta z}_c^2\rangle_{\mathbb{X}}^{1/2}$ is the sole parameter characterising the lateral spread. Interestingly, virtually the entire lateral dispersion during the transient occurs in the upper two-thirds of the channel, with little to no increase in variance close to the sediment bed. This is presumably a result of the migration into large-scale fluid streaks described in §6.3.4 which is supported by the spatial trend observed in fig. 6.18. Moreover, since the lateral wavelength of these structures is of $\mathcal{O}(H_f)$, the magnitude of the standard deviation—which stays consistently below $H_f/2$—may indicate that contaminants seldom travel across these structures, but rather converge towards a streak in their initial periphery. Thus, turbulent large-scale motions can be hypothesised to act as lateral transport barriers during the settling. Nevertheless, there exists evidence that this effect is not absolute: when the lateral distribution at deposition is investigated, it is found that for $Ga \approx 10$ the contaminants may spread significantly further in lateral direction before deposition than what is suggested by the settling data, whereas for $Ga \approx 30$ the values approximately coincide. A potential explanation for this discrepancy may be related to the tendency of the particles with higher relative turbulence intensity to be transported as bedload before deposition. In this case, it may be speculated that the large-scale roller structures whose footprint is depicted in fig. 6.22 push the contaminants into the neighbouring low-speed streak before they are deposited, resulting in additional dispersion in the near-bed region. Figure 6.28(d) depicts the standard deviation of the lateral displacement for crossings at $y = 0.2H_f$ and at the time of deposition as a function of the relative turbulent intensity. For the crossings during the settling where the stagnation in dispersion has been observed, the increasing trend with I_τ appears to level off for $I_\tau \gg 1$. In case of the

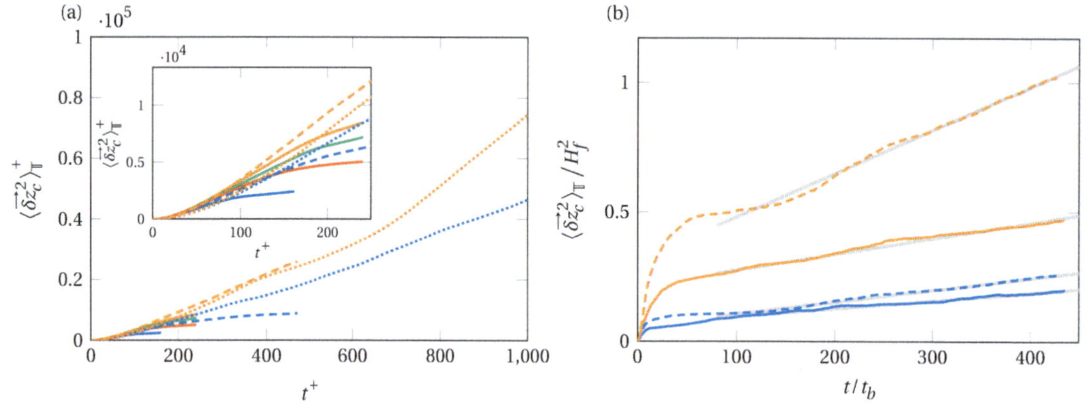

Figure 6.29: Temporal evolution of the mean-squared displacement in lateral direction. (a) During the transient settling phase and scaled in inner units. (b) Long-term evolution for the extended simulation runs in outer units. The grey lines (—) indicate the result of the linear regression performed to obtain the lateral particle diffusion coefficient. Legend: G30-D10-LR (—,○), G10-D6-LR (—,○), G15-D10-LR (—), G15-D6-LR (—), G30-D10-RP (- - ·,△), G10-D6-RP (- - ·,△), G30-D15-HR (······), and G10-D10-HR (······).

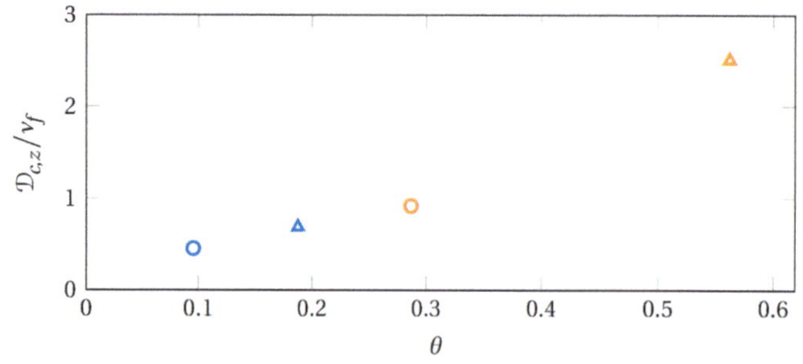

Figure 6.30: Particle diffusion coefficient in lateral direction in the bed-interaction phase as a function of the Shields parameter of the contaminants. The value has been obtained by linear regression of the mean-squared displacement in the time interval $t \in [80t_b, t_{obs}]$. Legend: G30-D10-LR (○), G10-D6-LR (○), G30-D10-LR (△), and G10-D6-LR (△).

displacement at deposition, a different trend is observed in accordance with the previous recognitions, and an approximately linear increase is found. This discrepancy may suggest a change in the dispersion behaviour in the near-bed region in the sense that the lateral mobility is restricted during the initial settling, but not for the bedload transport.

From a temporal perspective, a marked change in dispersion behaviour is perceptible between the settling phase and the sediment-interaction phase. The mean-squared lateral displacement of the contaminants during the settling phase is depicted in fig. 6.29(a). In the viscous scaling, the various parameter points exhibit a similar trend, albeit no complete collapse is achieved. The primary distinction between the cases is the time at which the increase in displacement commences to slow down. This transition in behaviour corresponds to the onset of contaminant deposition on the sediment bed, and thus, the time of transition is approximately given by $t^+_{p,s}$. The long-term displacement is shown in fig. 6.29(b) for the extended simulation runs, and a marked difference in the rate of dispersion compared to the initial phase is evident for all cases.

Given the Gaussian distribution of the samples and the absence of advection in lateral direction, a diffusion coefficient may be defined which characterises the spreading rate. Assuming a constant diffusivity $\mathcal{D}_{c,z}$, the diffusion equation solved for a Gaussian yields (Crank 1975)

$$\sigma_{nd}^2 = 2\mathcal{D}_{c,z} t, \qquad (6.28)$$

and hence, the lateral diffusion coefficient of the particles may readily be obtained from linear regression of the data presented in fig. 6.29(b). As the diffusivity is clearly not constant within the entire observation time, the time interval for regression is constraint to $t \in [80t_b, t_{obs}]$ in order to determine its value exclusively for the bed-interaction phase. Figure 6.30 displays $\mathcal{D}_{c,z}$ scaled by the fluid viscosity as a function of the Shields parameter. In this scaling, the data collapses satisfactorily across the cases, and a dependency on Shields parameter is not unreasonable due to its role of describing the tendency of particles to stay in suspension. However, as the availability of data is scarce, further input is required to judge the conclusiveness of the detected scalings, especially with regards to the characterisation of the sediment particles in the polydisperse case.

6.8 Summary and concluding remarks

The overarching topic of this chapter was the investigation of particle organisation, collective effects, and the formation of patterns within the dispersed phase. An investigation of cluster formation has been performed by means of a nearest neighbour analysis, and a quantification of clustering could be established by comparing the statistics obtained from the simulation data to a random reference distribution which takes into account the inherent spatial inhomogeneity of the ensemble of contaminants. It was found that intermittent accumulations of contaminant particles are formed at all parameter points under investigation, and that the temporal evolution of this organisation correlates well with the increase in mean settling velocity reported in §4. A causal relation between the two phenomena was investigated, but it was concluded that it is more likely that both phenomena are caused by the preferential sampling phenomenon described in §5. By examining the microstructure of particle accumulations, it was noted that for the cases with the lowest level of relative turbulence intensity the horizontally aligned arrangement imposed as the initial condition may persist over significant time and settling distance. Using auxiliary settling simulations at similar parameter points, but in absence of turbulence, it was deduced that such arrangements likely cause a hindrance effect similar to the effect described by Richardson and Zaki (1954). The auxiliary simulations also suggested that the significance of this hindrance is tied to tendency of the contaminants to form clusters, i.e. that the level of retardation considerably reduces when the horizontal alignment breaks apart. The formation of clusters can therefore be regarded as essential for the contaminants to obtain the observed enhancement in vertical velocity, but due to a mechanism which avoids retardation rather than one which actively promotes an enhancement.

The effect of preferential sampling on the arrangement of contaminants was studied by assessing the preferential concentration of crossing positions through wall-parallel planes. A visual examination as well as a quantification based on two-dimensional Voronoï tessellation indicated that the spatial distribution of crossing locations in a suitable frame of reference exhibits marked patterns. The clarity of these patterns depend on the contaminant parameters, and it was found to be most intense for high relative turbulence intensities. Close to the plane of initialisation, the accumulations and voids are of intermediate scales compared to H_f forming sharp fronts in the horizontal directions with little indication of preferential alignment. A comparison with the vortical structures present near the free surface yielded that the preferential concentration at this position is a result of the preferential sweeping mechanism, and in particular, the interaction with surface-aligned vortices. As a consequence, contaminants are found to settle predominantly in fluid downdrafts already at an early stage of the transient. In the lower half of the channel, the dominant accumulation patterns are streamwise-aligned stripes which coincide with the large-scale streaks. The progressive reorganisation into this arrangement was recognised to be a result of the wall-normal inhomogeneity of the turbulent flow rather than caused by inertial effects, although a lag in adaption of the contaminants to changes in the experienced flow conditions could be perceived. It was suggested that the cause for the accumulation in high-speed regions is the same as the cause for the existence of the latter, and hence, the interaction of particles with vortex clusters (del Álamo et al. 2006) or with packets of hairpin vortices (Kim and Adrian 1999) appear to be the prime candidates to explain this phenomenon. Moreover, it was ascertained once again that the enhancement in settling velocity ought to be caused by interactions which exhibit length scales significantly smaller than H_f.

Regarding the deposition of contaminants at the end of the settling phase, distinct phenomena are observed depending on the bedform. For a macroscopically flat bed, the contaminants tend to be deposited below the large-scale high-speed streaks of the outer layer, and subsequently tunnel the large-scale rollers to accumulate near the sediment ridges which are associated to the large-scale low-speed streaks. For high relative turbulence intensities, evidence exists that this mechanism may also act while the contaminants are transported in the bedload layer without prior deposition. While shortly after the settling phase distinct streamwise-aligned stripes of contaminants are perceivable on the bed, the distribution becomes more random in the projected view on the horizontal plane, although the degree of clustering is not found to be reduced. It was conjectured that some clusters of contaminants are immobilised by being buried by sediment particles, whereas the remaining mobile particles tend to accumulate in varying positions in the long term due to the lifecycle of the large-scale motions. When a ripple feature is present in the sediment bed, the contaminants are efficiently captured on its leeward side to be deposited into the trough. As the ripple propagates, the contaminants are immobilised by a thick layer of sediment particles to be finally released by erosion on the windward side when the pattern has been displaced sufficiently far downstream. Not all contaminants are captured at the same time. However, as the ripple continuously removes them from the bedload and stores them for a time roughly equivalent to half the ratio between its wavelength and its propagation speed, it may occur that a large share of contaminants is temporarily removed from the flow, resulting in a burst in suspended contaminant concentration after they are released again. This storage phenomenon serves as an explanation for the oscillating behaviour of the streamwise contaminant flux observed in chapter §4. While the ripple-induced cycle of deposition and erosion appears to continue indefinitely, the flux tends to stabilise as the time at which the particles get captured becomes more diffuse.

The displacement of contaminants from their initial position until they are first deposited into the bed was quantified in terms of the probability distribution of the settling length. A log-normal distribution is detected in accordance with the proposition by Li and Shen (1975), and its mean and standard deviation is found to be well predicted by their scaling arguments and the empirical fit obtained from experimental data in a smooth-wall open channel flow. In the lateral direction, the displacement follows a normal distribution around a zero mean whose variance increases in the upper parts of the channel during the settling, but stagnates in the lower third. This stagnation is conjectured to be related to the convergence into high-speed streaks which might act as a transport barrier during the settling phase. The rate of dispersion is significantly lower in the bed-interaction phase than in the settling phase. From the evolution of the mean-squared displacement, a lateral particle diffusion coefficient could be determined for the former which enables an estimation of the long-term spreading. The obtained values of this diffusion coefficient are of the same order as the kinematic viscosity. The data might indicate a scaling proportional to $St^+ I_T^2$, however, more evidence is required to make a judgement on its conclusiveness.

7 Scales of particle-turbulence interaction

It was speculated at several points throughout this thesis that a disparity exists in the characteristic length scales of the turbulent structures which cause streamwise or vertical enhancement of the instantaneous particle velocity. The purpose of this chapter is to gain more insights on this interaction, and to quantify the flow scales which contribute the most to the preferential sampling phenomenon. To achieve this, a method of scale decomposition is employed which is largely based on the work of Lozano-Durán et al. (2016), Motoori and Goto (2021) and Motoori et al. (2022).

7.1 Scale decomposition

For the purpose of isolating ranges of scales of velocity fluctuations, the three-dimensional instantaneous flow field is filtered using a Gaussian filter kernel which is given by

$$K_\sigma\left(\mathbf{x}',\mathbf{x},\sigma,\boldsymbol{\ell}\right) := \exp\left(-\frac{1}{2}\left(\frac{(x'-x)^2}{(\ell_x\sigma)^2}+\frac{(y'-y)^2}{(\ell_y\sigma)^2}+\frac{(z'-z)^2}{(\ell_z\sigma)^2}\right)\right), \tag{7.1}$$

where $\sigma > 0$ is the standard deviation of the kernel which acts as the free parameter determining the cut-off scales of the filtering procedure, and

$$\boldsymbol{\ell} = \left(\ell_x, \ell_y, \ell_z\right)^T \tag{7.2}$$

with $\ell_x, \ell_y, \ell_z > 0$ is the aspect ratio of the effective standard deviation in three dimensions which provides the possibility of having an anisotropic kernel. A low-pass filtered fluctuating fluid velocity field is obtained by convoluting the kernel with $u'_{f,\alpha}(\mathbf{x},t)$, viz.

$$\mathcal{L}\{u'_{f,\alpha},\sigma,\boldsymbol{\ell}\}\left(\mathbf{x},t\right) := \frac{1}{C_\sigma}\int_{-\infty}^{\infty}\int_{-\infty}^{\infty}\int_{-\infty}^{\infty} u'_{f,\alpha}(\mathbf{x}',t)K_\sigma \,\mathrm{d}z'\,\mathrm{d}y'\,\mathrm{d}x', \tag{7.3}$$

where the normalisation prefactor is determined from the condition that the kernel integrates to unity, viz.

$$C_\sigma\left(\sigma,\boldsymbol{\ell}\right) := \int_{-\infty}^{\infty}\int_{-\infty}^{\infty}\int_{-\infty}^{\infty} K_\sigma \,\mathrm{d}z'\,\mathrm{d}y'\,\mathrm{d}x' = (2\pi)^{\frac{3}{2}}\ell_x\ell_y\ell_z\sigma^3. \tag{7.4}$$

While in principle K_σ has an infinite support, it is in practice truncated at a distance of three standard deviations while upholding its scaling properties. The convolution is evaluated for the entire simulation domain Ω, and in particular does not distinguish between the solid or fluid phase, i.e. the filter is employed on the full field as obtained by the immersed boundary method described in §3.2.1. In the wall-bounded direction, the velocity field is extended by reflecting the velocity field about the wall,

Table 7.1: The properties of the scale decomposition using high-/band-/low-pass filters with a Gaussian kernel which is employed in this chapter.

m	$\sigma^{(m)}/H_f$	filter type	λ_σ/H_f	$\ell_x\lambda_\sigma/H_f$	$\ell_z\lambda_\sigma/H_f$	λ_σ/d_c (G30-D15-HR)	λ_σ/d_c (G10-D10-HR)
1	0.008	high-pass	0.03	0.10	0.05	1.5	2.2
2	0.008	band-pass	0.05	0.16	0.08	2.5	3.6
3	0.016	band-pass	0.11	0.32	0.16	4.9	7.2
4	0.032	band-pass	0.21	0.63	0.32	9.8	14.4
5	0.064	band-pass	0.42	1.26	0.63	19.7	28.8
6	0.129	band-pass	0.84	2.52	1.26	39.3	57.6
7	0.257	low-pass	1.68	5.04	2.52	78.6	115.2

which requires the sign of the vertical component v_f to be inverted to fulfil incompressibility (Lozano-Durán et al. 2016). This extension is natural for the free-slip wall at the upper boundary, as its boundary condition can be understood as an imposed mirror symmetry about the centreline of a closed channel, cf. §2.2.1. In case of the lower no-slip boundary, the treatment of the boundary is of little significance due to the presence of the sediment bed which essentially behaves like an extension of the fluid domain with vanishing $u'_{f,\alpha}$ with the difference that momentum is conserved throughout the domain.[1]

The low-pass filtered field is complemented by a residuum which comprises all length scales smaller than the cut-off. In fact, the residuum may be understood as a high-pass filtered velocity field given by

$$\mathcal{H}\{u'_{f,\alpha},\sigma,\boldsymbol{\ell}\}(\mathbf{x},t) := u'_{f,\alpha} - \mathcal{L}\{u'_{f,\alpha},\sigma,\boldsymbol{\ell}\}. \tag{7.5}$$

As a result, $u'_{f,\alpha}$ has essentially been decomposed into two fields which contain either large-scale ($\mathcal{L}\{u'_{f,\alpha}\}$) or small-scale ($\mathcal{H}\{u'_{f,\alpha}\}$) fluid fluctuations. The decomposition procedure can be repeated to isolate intermediate scales which yields the band-pass filtered field

$$\mathcal{B}\{u'_{f,\alpha},\sigma^{(m)},\sigma^{(n)},\boldsymbol{\ell}\}(\mathbf{x},t) := \mathcal{L}\{u'_{f,\alpha},\sigma^{(m)},\boldsymbol{\ell}\} - \mathcal{L}\{u'_{f,\alpha},\sigma^{(n)},\boldsymbol{\ell}\}, \tag{7.6}$$

where $\sigma^{(m)} < \sigma^{(n)}$ are the values of the filter parameter which restrict the range of the band. The generalised decomposition containing N_σ subdivisions is given by

$$\mathcal{H}\{u'_{f,\alpha},\sigma^{(1)},\boldsymbol{\ell}\} + \sum_{m=1}^{N_\sigma-1} \mathcal{B}\{u'_{f,\alpha},\sigma^{(m)},\sigma^{(m+1)},\boldsymbol{\ell}\} + \mathcal{L}\{u'_{f,\alpha},\sigma^{(N_\sigma)},\boldsymbol{\ell}\} = u'_{f,\alpha} \tag{7.7}$$

and requires $N_\sigma - 1$ values of σ to be specified.

In order to enable an informed choice for the values of σ, the kernel defined in eq. (7.1) is analysed for an idealised setup using the Fourier transformation in order to obtain its wavelength spectrum, see §C in the appendix for details. Since the Fourier transform of a Gaussian distribution in physical space also yields a Gaussian distribution in spectral space, the wavelengths comprised by each subdivision generally overlap. This is indeed desired behaviour, as one of the motivations behind the choice of a

[1] The Gaussian filter behaves similar to a diffusion problem in the sense that with increasing σ gradients are progressively equalised. In fact, this analogy has been exploited in §6.7 to define a diffusion coefficient for Gaussian-like dispersion. The mirror-reflection boundary condition is thereby equivalent to a zero-flux boundary condition in the diffusion analogue, and thus, the quantity to be filtered is conserved and merely redistributed. If the domain is extended using a vanishing field, a sink is introduced to the diffusion problem in the form of a Dirichlet-type boundary condition. However, as the sediment bed is part of the simulation domain, the conservation property is maintained. Moreover, this distinction is of little significance in the current context as the velocity field is filtered in terms of its fluctuations around an approximately zero mean.

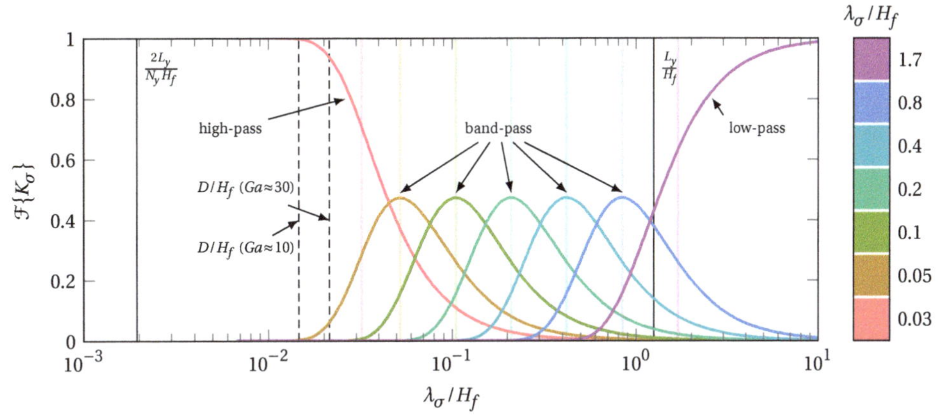

Figure 7.1: Support of a one-dimensional Fourier-tranformed Gaussian filter kernel $\mathcal{F}\{K_\sigma\}$ in spectral space under idealised conditions for the decomposition employed in this chapter. Each band contains a weighted distribution of wavelengths with overlapping content. The characteristic wavelengths for each subdivision are indicated by vertical lines in the corresponding colours. The solid black vertical lines indicate the fundamental wavelength and the smallest wavelength captured by the numerical discretisation of the HR simulation runs. The black dashed lines state the relative size of the contaminant particles for the parameter points G30-D15-HR and G10-D10-HR.

Gaussian kernel was that it provides the most balanced trade-off between localisation in physical space and spectral space, which makes it suitable for the investigation of local flow conditions (in the vicinity of particles) with regards to the scales involved (due to coherent motions which are not strictly localised). Nonetheless, one has to acknowledge that length scales derived from the following analyses are diffuse in their nature.

Figure 7.1 depicts the spectral properties of the decomposition described in table 7.1 which is employed in the following exclusively for the two HR parameter points. For this decomposition, a doubling of the filter parameter between each subdivision was employed, i.e. $\sigma^{(m+1)} = 2\sigma^{(m)}$, as this ensures a uniform distribution of information among the bands. The value of $\sigma^{(1)}$ was chosen such that the high-pass filtered field contains most of the fluid scales which are smaller than the particle sizes under investigation. Consequently, the number of subdivisions which can be accomodated is bounded by the fundamental wavelength of the simulation domain, and for the current choice of parameters, $N_\sigma = 7$ results in a low-pass field which predominantly contains global deviations from the spatio-temporal mean. In addition, it should be acknowledged that an anisotropic kernel with $\boldsymbol{\ell} = (3, 1, 1.5)^T$ is utilised, which results in a variation in the cut-off frequency depending on the direction. This choice was motivated by the typical aspect ratio reported for the bounding box of vortex clusters (del Álamo et al. 2006), as these structures have been conjectured to play a dominant role in contaminant organisation in high-speed streaks.

Regarding the characterisation of scales which are comprised by each band, the characteristic wavelength in case of the band-pass filter is defined as the corresponding position of the peak portrayed in fig. 7.1 which is given by

$$\lambda_\sigma = \pi\sigma\sqrt{\frac{6}{\ln(4)}} \approx 6.54\sigma. \tag{7.8}$$

In case of the low-pass and high-pass filters, the wavelength at which the power spectrum is attenuated by half is selected, which reads

$$\lambda_\sigma = \pi\sigma\sqrt{\frac{2}{\ln(\sqrt{2})}} \approx 7.55\sigma \tag{7.9}$$

for the low-pass filter and

$$\lambda_\sigma = \pi\sigma\sqrt{-\frac{2}{\ln(1-\sqrt{2}/2)}} \approx 4.01\sigma \tag{7.10}$$

for the high-pass filter. Using these definitions, the characteristic wavelengths are approximately evenly distributed in the logarithmic scaling as can be seen in fig. 7.1.

7.2 Description of the decomposed velocity field

As this analysis aims at explaining the enhancement of the streamwise and vertical contaminant velocity, the focus is laid on the two corresponding fluid velocity components. The filtered variants of these fields are denoted as $\check{u}_f(\mathbf{x}, t)$ and $\check{v}_f(\mathbf{x}, t)$, and both are to be understood as fluctuating fields about their respective spatio-temporal mean since they are derived from $u'_{f,\alpha}$.

Figure 7.2 provides visualisations of the streamwise velocity fluctuations for HR just before contaminants are seeded into the flow. The unfiltered field clearly depicts the long streamwise-aligned streaks which the contaminants were found to preferrentially sample for $u'_f > 0$. When the decomposition is applied, these structures are only observable for the filter levels with high characteristic wavelengths as expected. Their footprint begins to be perceivable at around $\lambda_\sigma \boldsymbol{\ell} \approx \left(0.6, 0.2, 0.3\right)^T H_f$ which is in reasonable agreement with the results of the Fourier analysis presented in fig. 5.23. As suggested by the same reference, the fluctuations of the vertical velocity at the same filter levels exhibit much less anisotropy as can be deduced from fig. 7.3.

For the high-pass filtered field and the lowest bands, the fluctuations are considerably less aligned with the flow direction, and regularly emerge in long thin tubes which presumably indicates that they are the result of the small-scale vorticity referred to as "worms", cf. §2.1.5. The visibly increased randomness and isotropy of the small scales is in good agreement with the expectations from classical turbulence theory, cf. §2.1.3, which increases faith in the plausibility and utility of the decomposition strategy.

In the visualisations, the values of the isosurfaces were selected based on the variance of the fluctuations for each filter level. Indeed, the intensity of the fluctuations may differ considerably in absolute terms among the various scales, and thus, ought to be quantified in order to correctly judge the relevance of the observed motion. In fig. 7.4 the root-mean-square of $\check{u}_f$ and $\check{v}_f$ at each level is provided in terms of the friction velocity. Generally, large scales are found to be of high relative significance in the upper parts of the channel, whereas small scales dominate near the sediment bed as one would expect from Townsend's attached eddy hypothesis, cf. §2.2.3. A substantial difference between $\check{u}_f$ and $\check{v}_f$ is given by the fact that for the streamwise component the root-mean-square generally increases with increasing length scale in the bulk, whereas for the vertical component a peak can be observed at around $\lambda_\sigma \approx 0.2H_f$ which again is in accordance with the Fourier spectra presented earlier in fig. 5.24. The low-pass filtered streamwise velocity exhibits a peculiar trend which demands further investigation: as suggested by fig. 7.2(h), this level represents the turbulent motion at the largest scales which is not accurately captured by the short streamwise length of the computational domain. As discussed in §5.5, the unresolved portion of the energy spectrum which usually resides in the very-large scale motions is contained by the modes of infinite streamwise length for the current cases, which is predominantly contained by the low-pass portion of the decomposition. At the same time, the domain is sufficiently wide to capture the characteristic pattern of alternating streaks in which this energy is arranged, which

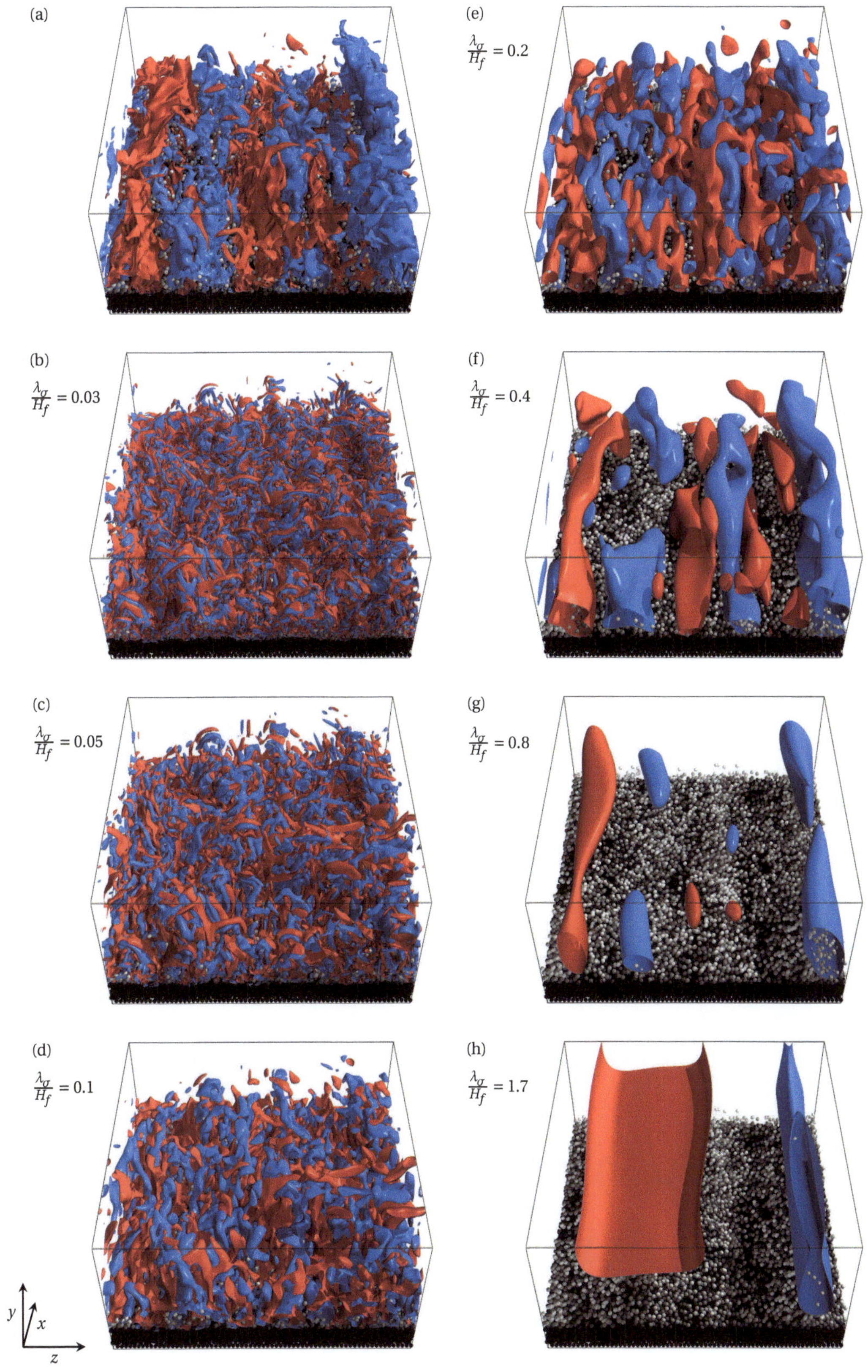

Figure 7.2: Visualisation of the decomposition of the streamwise fluid velocity at $t = 0$ for the HR simulations. (a) Unfiltered field. (b) High-pass filter with $\lambda_\sigma = 0.03H_f$. (c-g) Band-pass filter with $\lambda_\sigma = \{0.05, 0.1, 0.2, 0.4, 0.8\}\, H_f$. (h) Low-pass filter with $\lambda_\sigma = 1.7H_f$. The unfiltered field shown in (a) corresponds to the sum of (b-h). Blue isosurfaces depict negative fluctuations and red positive ones. Isosurfaces with $\check{u}_f / \langle \check{u}_f \rangle_{xyzt} = \{-1.25, 1.25\}$ are displayed.

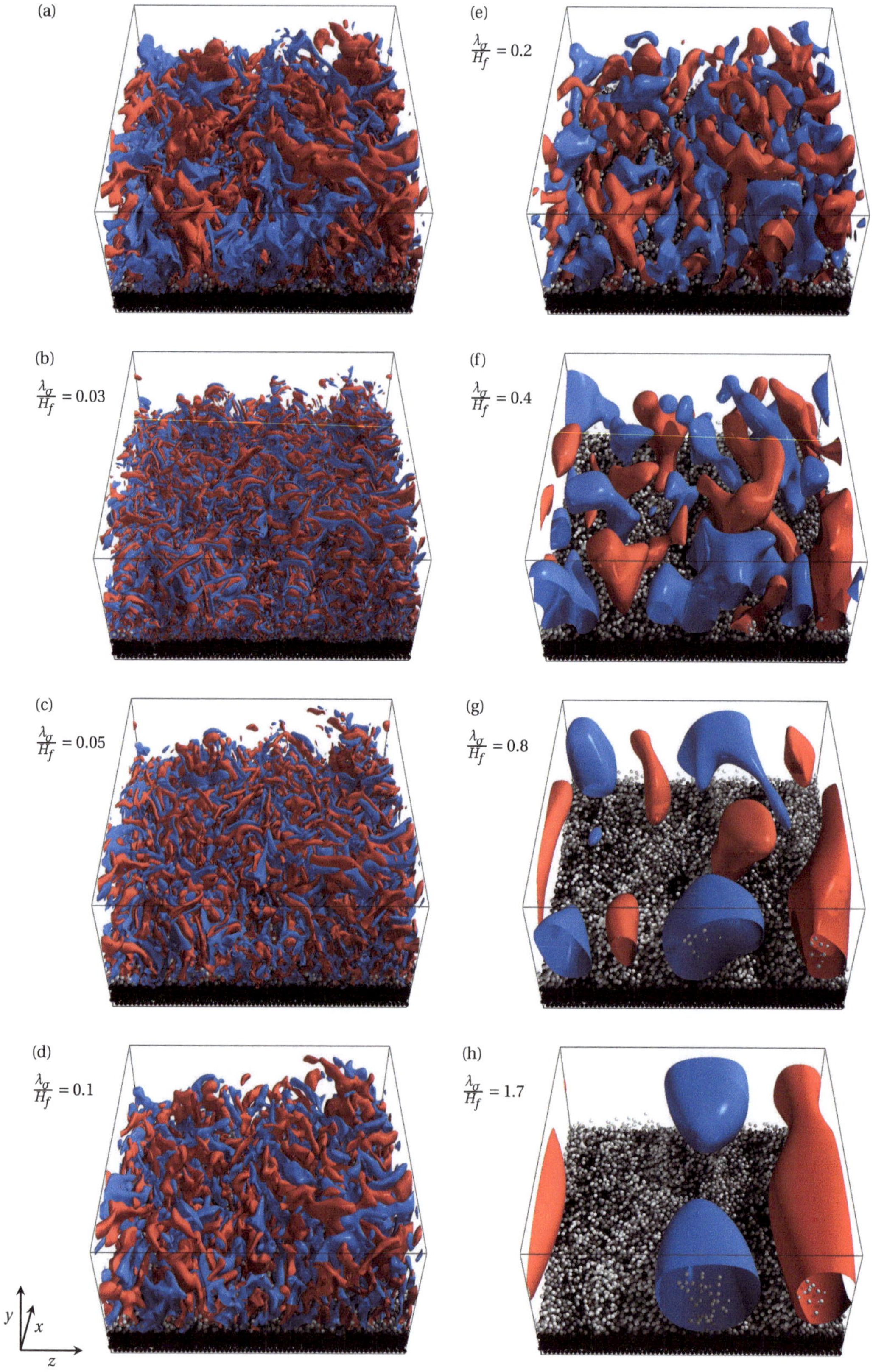

Figure 7.3: Visualisation of the decomposition of the vertical fluid velocity at $t = 0$ for the HR simulations. (a) Unfiltered field. (b) High-pass filter with $\lambda_\sigma = 0.03H_f$. (c-g) Band-pass filter with $\lambda_\sigma = \{0.05, 0.1, 0.2, 0.4, 0.8\}\, H_f$. (h) Low-pass filter with $\lambda_\sigma = 1.7H_f$. The unfiltered field shown in (a) corresponds to the sum of (b-h). Blue isosurfaces depict negative fluctuations and red positive ones. Isosurfaces with $\breve{v}_f / \langle \breve{v}_f \rangle_{xyzt} = \{-1.25, 1.25\}$ are displayed.

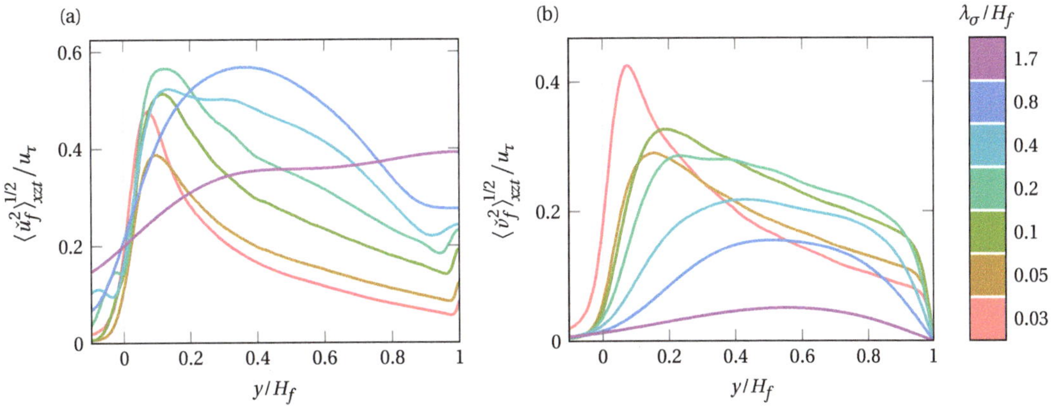

Figure 7.4: Root-mean-square of the filtered velocity field at different scales. (a) Streamwise component $\check{u}_f$. (b) Wall-normal component $\check{v}_f$. The characteristic wavelength of each scale can be inferred from the colourbar.

results in a spanwise variance of $\check{u}_f$ at $\lambda_\sigma \approx 1.7H_f$. Moreover, the input to the decomposition is the fluctuating field with respect to the spatio-temporal average, such that temporal fluctuations of the spatial mean also contribute to the low-pass portion of the decomposition. The presence of the velocity structure shown in fig. 7.2(h) can be identified from the two-dimensional visualisation of the same flow field previously presented in fig. 6.16. The distinct spatial evolution of $\langle \check{v}_f^2 \rangle^{1/2}_{xzt}$ might then be a result of the distinct location of LSMs and VLSMs at higher positions in the channel.

7.3 Quantification of the scales of enhancement

The main concern of the current chapter is the identification of the scales of fluid motion which contribute the most to the enhancement of u_c and v_c. In order to achieve this, the concept of the sphere-averaged fluid velocity in the vicinity of particles introduced in §5.2.3 is applied to the decomposed fields resulting in a characteristic velocity experienced by each particle at each individual scale. The explanatory power of this procedure originates partly from the additivity of the decomposition and the linearity of the sphere-averaging operator: the property that the sum of contributions from all decomposed fields adds up to the value obtained for the unfiltered field is maintained for $\langle \check{u}_{f,\alpha} \rangle^{(i)}_{\mathbb{S}}$. For this reason, the ratio of $\langle \check{u}_{f,\alpha} \rangle^{(i)}_{\mathbb{S}} / \langle u_{f,\alpha} \rangle^{(i)}_{\mathbb{S}}$ immediately reflects the contribution of the corresponding filter scale to the overall value.

In fig. 7.5(a,b) this ratio is shown ensemble-averaged over all downward crossings through the $y = 0.3H_f$ wall-parallel plane as a function of λ_σ for the two HR parameter points. The hypothesis that the enhancement is induced at distinct scales for streamwise and vertical components is corroborated: while the peak contribution to $\langle \check{u}_f \rangle_{\mathbb{S}\mathbb{X}\downarrow}$ occurs around $\lambda_\sigma = 0.4$–$0.8H_f$, the maximum of $\langle \check{v}_f \rangle_{\mathbb{S}\mathbb{X}\downarrow}$ is found at $\lambda_\sigma = 0.1$–$0.2H_f$.[2] Consequently, the scales of enhancement differ from each other by roughly a factor of four. One arrives at a similar conclusion when the correlation between the particle velocity and the nearby decomposed fluid velocity is examined, see fig. 7.5(c,d). While the previously shown contributions to the overall sampled value are certainly also affected by the intensity of velocity fluctuations available at this particular position, the correlation coefficient disregards this effect due to its normalisation properties, and hence, presumably more accurately reflects the actual response of contaminants

[2] Due to the anisotropy of the filter kernel, these numbers have to be multiplied by ℓ to obtain the three-dimensional characteristic wavelengths.

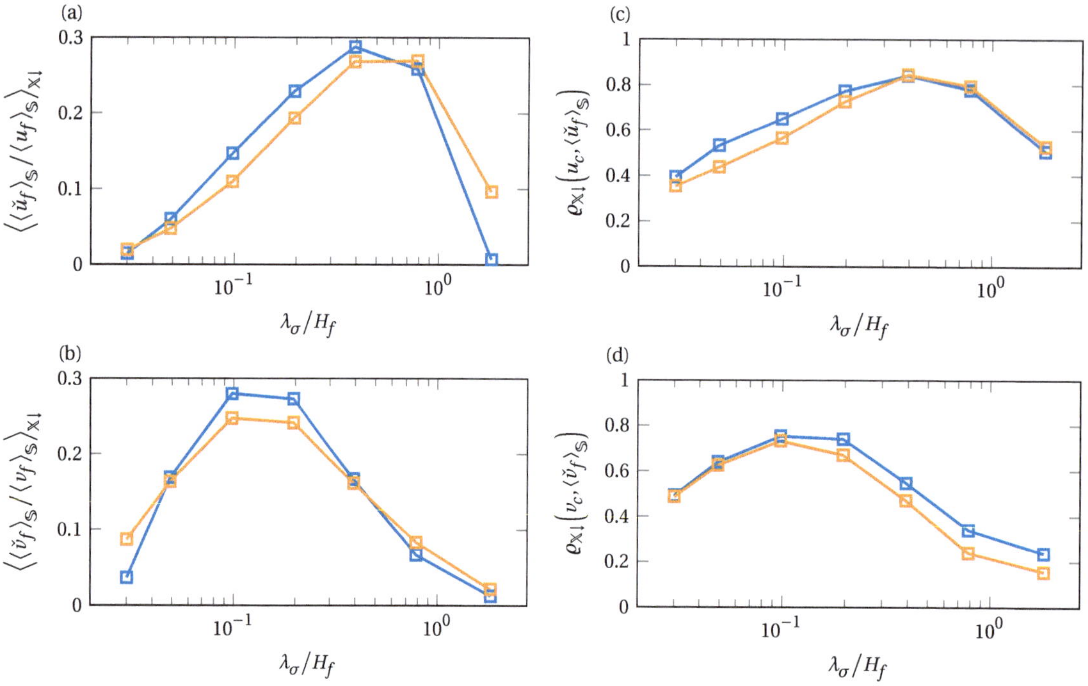

Figure 7.5: Evaluation of the contribution of different flow scales to the enhancement of streamwise and vertical momentum of settling contaminant particles. All statistics are obtained from samples gathered from all downward crossings of the wall-parallel plane at $y = 0.3H_f$. (a) Ratio of the sphere-averaged streamwise velocity evaluated on a filtered scale to that obtained for the unfiltered field. As the decomposition is additive and the sphere-averaging operation linear, this ratio immediately expresses the contribution of a given scale to the overall value. (b) Same as before, but for the vertical velocity. (c) Linear correlation coefficient between the streamwise particle velocity and the sphere-averaged streamwise fluid velocity at a given scale. (d) Same as before, but for the vertical velocity. Legend: G30-D15-HR (—) and G10-D10-HR (—).

to fluid motions at a given scale. Interestingly, the correlation suggests that the two particle parameter points behave very similar except for a slightly decreased overlap of the relevant streamwise and vertical scales for G10-D10-HR.

The spatial evolution of the sphere-averaged velocity is depicted in fig. 7.6. Concerning the streamwise component, the general trend and distribution of momentum across the various scales resembles that observed in the fluid statistics shown in fig. 7.4(a). Significant deviations can be observed close to the bed where effects other than preferential sampling gain importance, and close to the plane of initialisation where the sphere-average velocity is governed by the initial condition and the initial organisation. In particular, the only scale which does not vanish at initialisation is that obtained from the low-pass filter which also contains the temporal fluctuations of the xz-averaged momentum. Therefore, it is likely that the initial field happens to contain higher streamwise momentum than the temporal average in the plane of initialisation, which also explains the offset previously observed in fig. 5.9(a). This peculiarity demonstrates the benefit of ensemble averaging the result over multiple realisations at different instances in time if small domains are investigated as has been done for the simulations at lower Reynolds numbers. Regarding the vertical component displayed in fig. 7.6(b), it can again be noted that the amplitude of enhancement for a given scale appears to be proportional to the corresponding root-mean-square of $\check{v}_f$ at the same height which suggests that the scale of enhancement is predominantly determined by the distribution of the fluid momentum rather than by properties of the particles themselves—at least for the parameter points currently under investigation.

In contrast to the sampled velocity, the correlation coefficient between the particle velocity and the filtered fluid velocity is relatively constant at all vertical positions, see fig. 7.7. This indicates that the scales

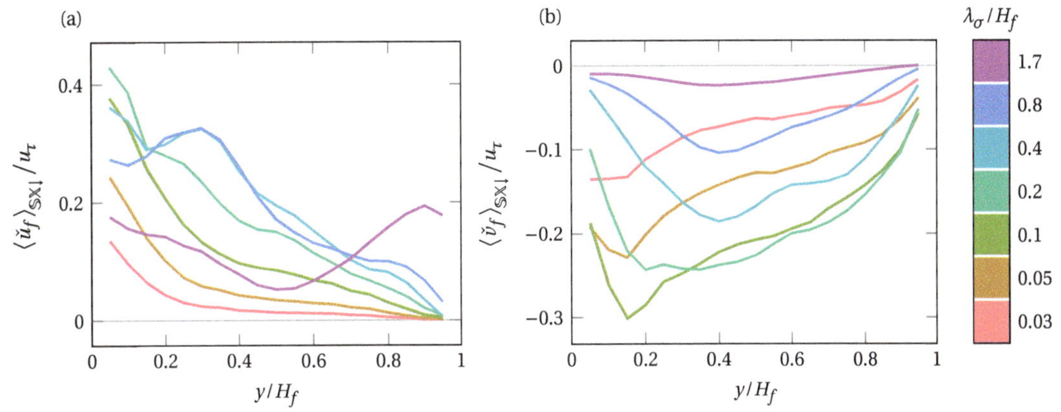

Figure 7.6: Evolution of the sphere-averaged fluid velocity as a function of vertical position for each filter scale. (a) Streamwise component. (b) Wall-normal component. The parameter point on display is G10-D10-HR.

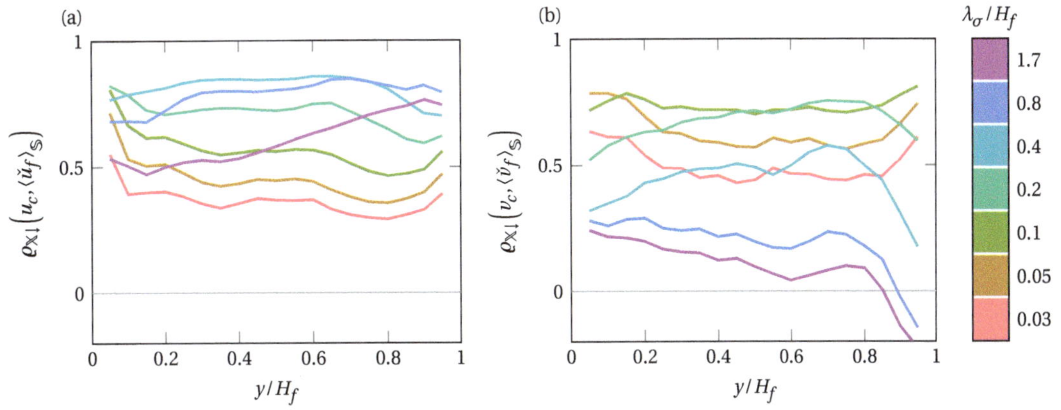

Figure 7.7: Evolution of the crossing-conditioned correlation coefficient between the instantaneous particle velocity and the corresponding sphere-averaged fluid velocity. The data is shown as a function of vertical position for each filter scale. (a) Streamwise component. (b) Wall-normal component. The parameter point on display is G10-D10-HR.

at which particle-turbulence interactions occur are not solely dependent on the share of fluid momentum carried by that scale (as this proportion varies with y), but are rather linked to the specific particle parameter point, which ought to be represented by a quantity which quantifies the relative response time of a particle to a structure such as the scale-local Stokes number proposed by Motoori et al. (2022). However, the difference in particle diameter, density and Galileo number appears to be too insignificant to assess this hypothesis from the current dataset.

7.4 Summary and concluding remarks

The aim of this chapter was to quantify the characteristic length scales of the turbulent velocity structures which are preferentially sampled by the contaminants and lead to the enhancement in streamwise and wall-normal particle velocity. In order to do so, a decomposition method was introduced based on Gaussian filters which subdivides the fluid velocity field into ranges of scales which can be examined individually, but possesses the property that the complete flow field can be reconstructed from the sum of its subdivisions. Using this method, the velocity experienced by the particles characterised by the sphere-averaging procedure introduced in chapter §5 is decomposed into the contributions from various flow scales. The result corroborates the hypothesis that the enhancement in streamwise and vertical particle velocity occurs on different length scales. Regarding the streamwise velocity, structures with spatial dimensions of $(1.2, 0.4, 0.6)^T H_f$ to $(2.4, 0.8, 1.2)^T H_f$ in $(x, y, z)^T$ were found to be most significant, whereas for the vertical velocity the cardinal size range lies between $(0.3, 0.1, 0.15)^T H_f$ and $(0.6, 0.2, 0.3)^T H_f$. These ranges coincide approximately with the length scales which are the most active for the respective fluid velocity component in terms of their root-mean-square as obtained from Fourier analysis. Nonetheless, it was shown that the particle velocity exhibits an approximately constant correlation to each decomposed flow scale throughout the entire channel, which might suggest that there exist structures of preferred scales with which the contaminants interact.

8 Investigation of settling using an exact coherent state

Throughout this thesis, coherent vortical motions have been identified to be the root of phenomena such as clustering, preferential concentration and settling velocity enhancement ocurring in the transient settling problem through mechanisms such as preferential sweeping. Hereby, the interactions take place on a multitude of scales: while particles were demonstrated to interact directly with individual vortices (§6.3.3), the self-organised vorticity at larger scales such as vortex clusters or hairpin packets was hypothesised to play a key role in the explanation of organisation of contaminants into large-scale velocity streaks (§6.3.4) which leads to an anisotropy in the scales which cause the streamwise and vertical velocity enhancement observed for the contaminants (§7). The interactions which lead to the large-scale organisation and enhancement are difficult to describe as they are likely the collective result of many smaller-scale interactions with vortices and eddies which evolve over time, and in the case of transient settling, meaningful observations of these interactions are difficult to obtain as the number of particles which interact with a structure at a given time is limited and the effects may be non-local. However, a thorough description of this process might also not be required for most purposes, as the net effect on the ensemble of particles appears to be well described by the macroscopic emergent features of the organised vorticity.

In a pursuit of simplicity, the question arises whether or not the essential features of transient settling in a turbulent channel can be reproduced—qualitatively and perhaps even quantitatively—in a simpler surrogate system which facilitates a more systematic study of the scaling of the various phenomena identified with respect to the particle parameters. Such "ruthless simplifications"[1] were employed already in the discussion on the effect of shear on the settling velocity (§5.3.3) and regarding the collective hindrance effect at the beginning of the settling process (§6.2), where they have contributed considerably to the understanding of the full system. It is needless to say that the laminar flows investigated in these contexts are incapable of reproducing the defining features of turbulent settling. Nonetheless, the dynamical systems approach to turbulence has provided access to non-trivial flow states which contain some essentials of turbulent flows, while at the same time being substantially easier to analyse due to their defined behaviour in space and time—the exact coherent states described in §2.2.4. So far, the usage of these exact solutions as a simplification of turbulence in particle-laden flows is limited to the pioneering study of Pestana et al. (2020), who were able to reproduce the near-wall migration of particles into low-speed streaks utilising the solution of Nagata (1990) in Couette flow. The objective of the current chapter is to assess whether or not the transient settling in an open channel can be examined in the same manner, and if new knowledge can be gained from this approach.

[1] The writer's usage of this term has been coined by Steven Strogatz who in an interview stated that his preferred method to tackle a problem is to "keep simplifying the problem—actually fairly ruthlessly—until it's minimal. Until it's the simplest problem that contains the nugget of the difficulty—the essential difficulty."

8.1 Description of a hairpin-like exact coherent state

Regarding channel flows, various exact coherent states (ECS) have been identified and extracted in literature in the form of travelling waves (TWs) and relative periodic orbits (POs), see e.g. Waleffe (2001), Toh and Itano (2003), Park and Graham (2015), Wall and Nagata (2016), Shekar and Graham (2018) and Yang et al. (2019). Thereby, not every structure is equally relevant to turbulent flows, or more specifically in the current context, does not reflect the properties of developed turbulence equally well. One exact coherent state which is especially appealing for the use in this thesis is the travelling wave first published by Shekar and Graham (2018) as its topology resembles that of a hairpin vortex—or more precisely, a pair of staggered hairpins—which in the theory of Adrian (2007) plays an important role in the explanation of LSMs and VLSMs. Indeed, hairpin vortices are found to occur in the turbulent simulations with fig. 8.1 showing two examples of such structures in the initial fields of the HR run obtained by applying the Q-criterion to the decomposed fields presented in §7. Apart from its topological relevance, this ECS is favourable because it emerges at a relatively high value of the friction Reynolds number and its domain-averaged statistics reflect that of a turbulent flow rather well.

For the use in this thesis, the hairpin TW has been obtained independently from Shekar and Graham (2018) by conducting constrained DNSs in a closed channel from randomly perturbed initial fields using the publicly available pseudo-spectral solver CHANNELFLOW (Gibson et al. 2008, Gibson 2014). In particular, the constraints consisted of imposing instantaneous symmetries which are equal to the symmetries which are exhibited by the target solution: a mirror symmetry about the xy-plane at the periodic boundary, a mirror symmetry about the xz-plane at the centreline, and a shift symmetry with respect to a shift of $[L_x/2, L_z/2]$. Here, the mirror symmetry about the centre plane plays a crucial role in the current context as it ensures that the TW is indeed also a solution to open channel flow. The simulations were conducted in the same domain size for which the original solution was identified, namely $[L_x \times L_z]/H = [\pi \times \pi/2]$, and at a bulk Reynolds number of $Re_b = 1600$. After a suitable recurrent state has been identified, the invariant state was obtained using the Newton-Krylov-hookstep algorithm of Viswanath (2007) which was readily available in the DNS code. In fact, the state which was first identified by this procedure is a previously unreported relative periodic orbit which is connected to the family of travelling waves by a Hopf-type bifurcation which occurs at $Re_b \approx 1304$. This orbit shares most of the defining features of the TW and its dynamics are limited to a weakly pulsating behaviour.

For technical reasons, the extracted TW has been transferred to a domain size of $[L_x \times L_z]/H = [3 \times 1.5]$ by means of homotopy as this aspect ratio is more favourable for the particle-laden flow solver. Moreover, the bulk Reynolds number was reduced to its corresponding value at the inflection point ($Re_b = 960.8$) which results in a friction Reynolds number of $Re_\tau = 91.1$, see fig. 8.2 for a visualisation of the structure and fig. 8.3 for the bifurcation diagram of the ECS in the new domain. Although the TW was found to be linearly stable in some regions of the parameter space if the aforementioned symmetries are imposed, it is generally unstable in the important case of a sole symmetry about the centre plane (unconstrained open channel flow). However, the number of unstable directions remains low, such that the TW is maintained for $\mathcal{O}(100\, t_b)$ for small random perturbations before relaminarisation occurs. The propagation velocity of the TW in streamwise direction reads

$$c_{TW} = 1.116 u_b, \tag{8.1}$$

and consequently, the solution is time-invariant in a frame of reference which moves at this velocity.

Figure 8.4 shows a comparison of the velocity statistics obtained for the TW to those obtained for a turbulent smooth-wall open channel flow at $Re_\tau = 200$. While the mean streamwise velocity scaled in inner

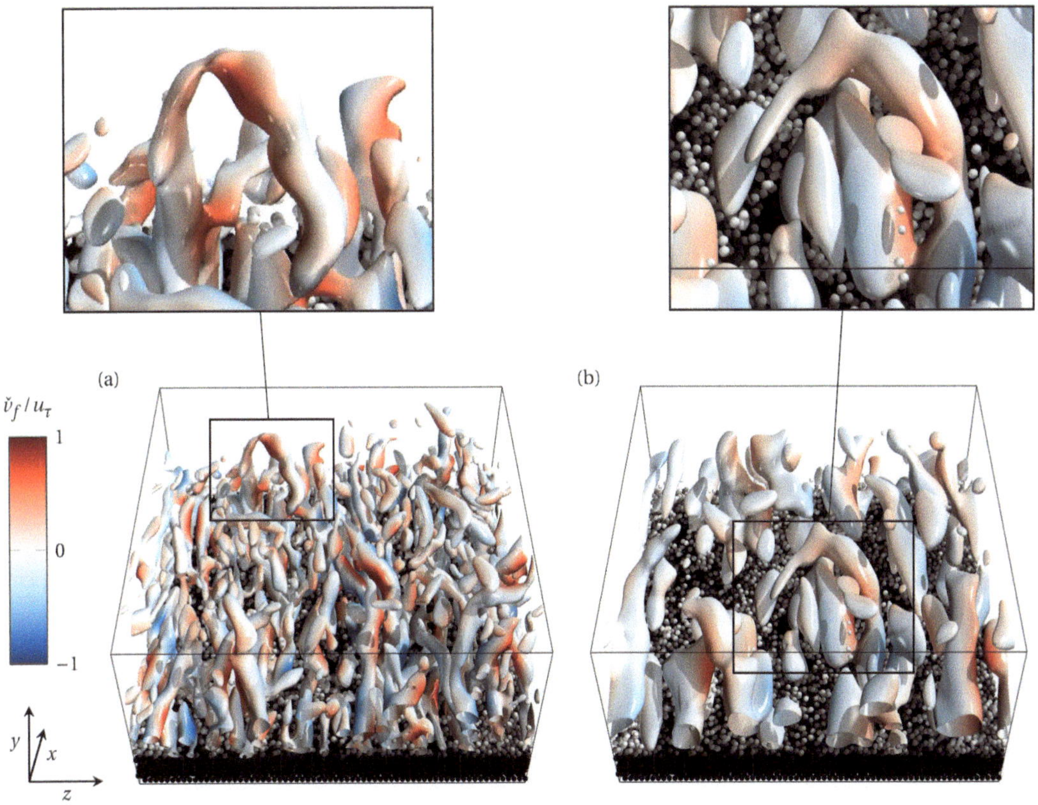

Figure 8.1: Examples of hairpin vortices found in the initial field of the HR simulation series. The vortices are visualised by applying the Q-criterion on the band-pass filtered fluctuating velocity fields. Isosurfaces at $\check{Q}'^{+} = 1.1 \cdot 10^{-5}$ are depicted and the colouring indicates the local value of the band-pass filtered vertical fluid velocity. (a) Band-pass filtered field with $\lambda_\sigma = 0.2H_f$. (b) Band-pass filtered field with $\lambda_\sigma = 0.4H_f$.

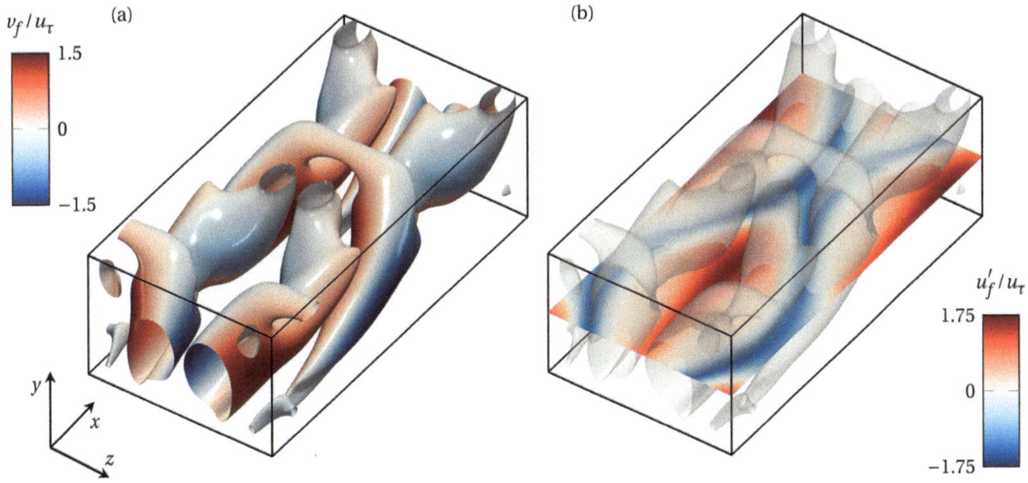

Figure 8.2: Visualisation of the travelling wave solution employed in this chapter. (a) Regions of prevailing vorticity as indicated by the Q-criterion with a threshold of $Q^{+} = 1 \cdot 10^{-3}$. The colouring corresponds to the local value of the vertical fluid velocity. (b) A slice at $y = 0.5H$ showing the fluctuations of the streamwise fluid velocity. The isosurfaces of the Q-criterion are overlayed translucently to indicate the influence of the vortices.

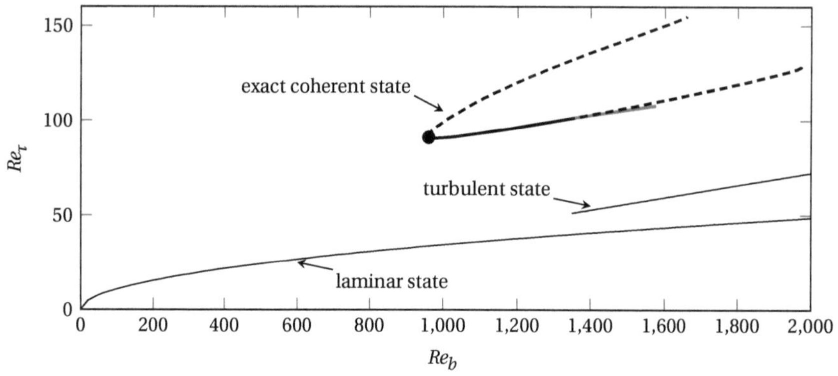

Figure 8.3: Bifurcation diagram of the hairpin travelling wave in the domain $[L_x \times L_z]/H = [3 \times 1.5]$ in terms of the bifurcation parameter Re_b and the observable Re_τ. The solid line (—) indicates the region where the TW is linearly stable in the symmetric subspace, whereas the dashed lines (- - ·) show unstable regions. A stable periodic orbit bifurcates from the lower branch at $Re_b = 1349.6$ as indicated by the grey line (—). All states in this family of solutions are unstable if any of the imposed symmetries is revoked. As a reference, the curve $Re_\tau = \frac{2}{3}\sqrt{8Re_b/3}$ for laminar Poiseuille flow is shown, as well as $Re_\tau = 0.09Re_b^{0.88}$ for turbulent flows according to the empirical correlation of Dean (1978).

units deviates quantitatively from the standard log-law and the turbulent data in the outer region, the scaling in outer units indicates similar behaviour for both types of flows. A similar observation can be made for the Reynolds stress. In fact, all second order statistics suggest that the near-wall behaviour is similar for the ECS and the turbulent flow if the data is examined as a function of y^+, while the trend in the outer region is reflected more appropriately in a y/H scaling. This prompts the conclusion that many of the discrepancies which may be observed in either scaling are a result of a reduced separation of inner and outer units due to the inherently small value of Re_τ for the TW. Aside from this lack of scale separation, the root-mean-square of the velocity fluctuations in streamwise and vertical direction exhibit a good qualitative, and at times even quantitative, agreement. This is especially true for $\langle v_f^2 \rangle^{1/2}$ whose deviation seems to be almost entirely explained by the lower value of the friction Reynolds number, and which plays a vital role for the settling velocity enhancement phenomenon, cf. §4.3.1. Regarding the spanwise fluctuations, the statistics deviate rather significantly apart from close to the centre plane, however, so far the role of $\langle w_f^2 \rangle^{1/2}$ is unclear concerning the particle settling phenomena to be discussed.

8.2 Sampling method and particle parameters

Now that the basic ECS to be employed in the settling simulations has been characterised, the procedure of seeding particles into the flow will be delineated. To begin with, the spectral flow field obtained by the `CHANNELFLOW` solver is interpolated onto a uniform staggered mesh as required by the immersed boundary simulation code, cf. §3.2.1. Here, the target mesh only comprises half of the closed channel since open channel boundary conditions are to be imposed in the particle-laden simulations which are satisfied due to the centre line mirror symmetry of the TW solution. No further symmetries are enforced in the presence of particles such that the flow remains unconstrained.

Similar to the turbulent simulations, the settling particles are seeded near the top boundary of the domain at $y_{p,0} = 0.95H$. However, in contrast to them, the wall-parallel plane is sampled systematically

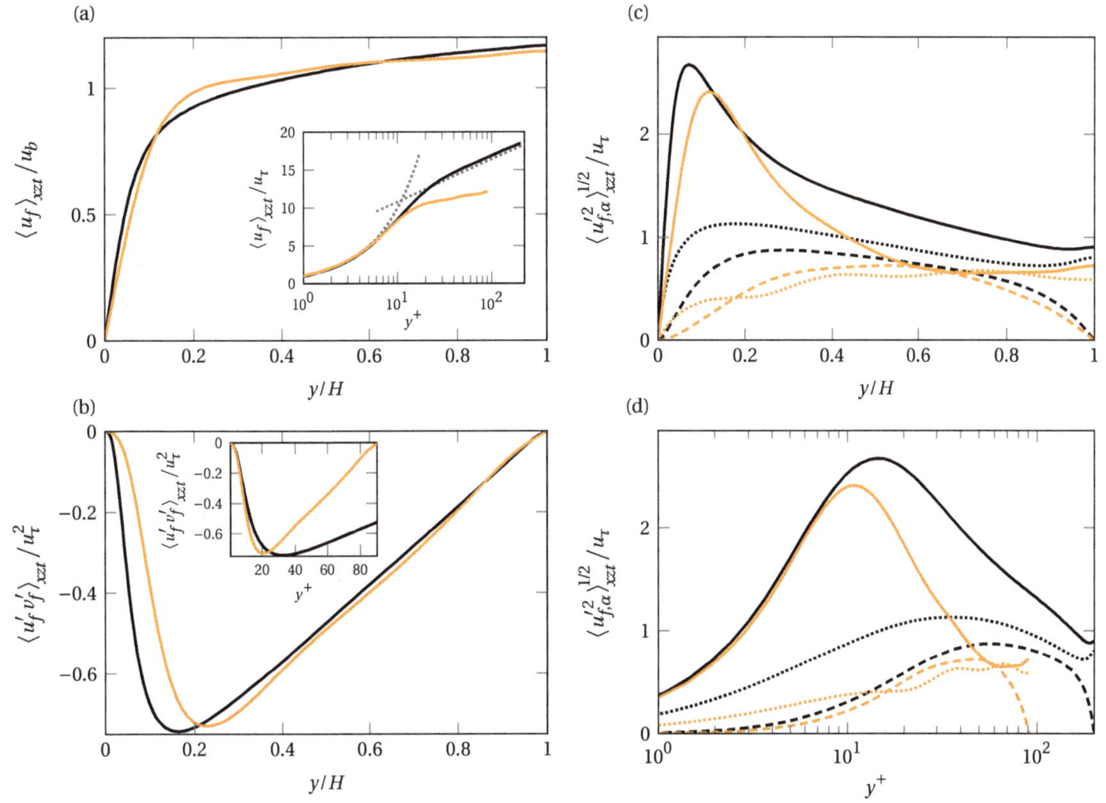

Figure 8.4: Comparison of the velocity statistics between the hairpin travelling wave in the domain used for the settling simulations at $Re_\tau = 91.1$ (—) and a turbulent open channel flow at $Re_\tau = 200$ (—) as computed by Bauer et al. (2022). For the TW, only the lower half of the full channel is displayed with $y = 0$ corresponding to the lower wall and $y = H$ to the centre plane (the free-slip boundary of the equivalent open channel). (a) Mean streamwise velocity in outer units. The inset shows the same data in inner units with dotted lines indicating the law of the wall and the log-law. (b) Reynolds stress $u_f' v_f'$ as a function of y/H. The inset shows the same data as a function of y^+. (c) Root-mean-square of the fluid velocity fluctuations as a function of y/H for u_f' (—), v_f' (- - -) and w_f' (⋯⋯). (d) Same as before, but as a function of y^+. As the TW is time-invariant in an appropriate frame of reference, its statistics correspond to the spatial average within the simulation domain.

due to the widely reduced sample space of the ECS compared to the full turbulent simulations.[2] To be precise, the plane of initialisation is discretised by a regular grid with a uniform spacing of $h_{p,0}$ whose nodes mark the initial horizontal positions $(x_{p,0}, z_{p,0})$ of the sample particles. Due to the spanwise mirror symmetry and the streamwise-spanwise shift symmetry of the initial flow condition, only a quarter of the xz-plane is unique, and hence, it is sufficient to only sample the domain $x \in [0, L_x/2)$, $z \in [0, L_z/2)$ in order to reconstruct any initial condition from the symmetries. Under these preconditions, the total number of sample particles is given by

$$N_{p\Sigma} = \frac{L_x L_z}{4 h_{p,0}^2}, \tag{8.2}$$

[2] The outcome of the experiment is only a function of the initial x and z position of the particles under the aforementioned preconditions and under the assumption that interparticle interactions are negligible. Thus, in contrast to turbulent settling, the outcome is not a function of the seeding time as the initial velocity field is time-invariant. Nonetheless, the particle trajectories are still anticipated to be chaotic in the sense that initially neighbouring particles may exhibit diverging trajectories. However, due to the reduced range of scales in the ECS it can be hypothesised that this divergence occurs at a slower rate, and hence, due to the limited observation time less samples might be required to form an adequate estimate of the sample space.

Table 8.1: Summary of the common physical and numerical parameters of all simulations conducted within this chapter.

Re_b	Re_τ	$[L_x \times L_z]/H$	$[N_x \times N_y \times N_z]$	H/d_p	d_p^+	$d_p/\Delta x$	N_p	$N_{p\Sigma}$
960.8	91.1	$[3.0 \times 1.5]$	$[576 \times 192 \times 288]$	16	5.7	12	16	288

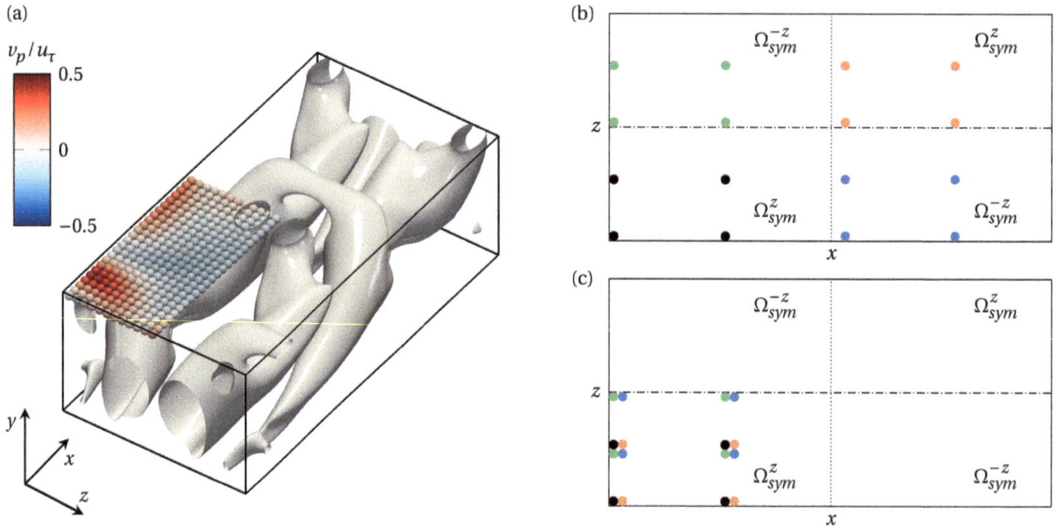

Figure 8.5: Initial distribution of sample particles. (a) Visualisation of all $N_{p\Sigma} = 288$ initial particle positions to be sampled on a regular grid with a uniform spacing of $h_{p,0} = d_p$. Due to the symmetric properties of the TW, only a quarter of the wall-parallel plane needs to be sampled in order to reconstruct all possible trajectories. The colouring of the particles indicates the vertical particle velocity at initialisation which corresponds to the value of the fluid velocity at the centroid of each particle. (b) An example for the actual arrangement of $N_p = 16$ particles for a single realisation of the numerical experiment. The initial positions are chosen such that the distance between the particles is approximately maximised in order to avoid collective effects and a destabilisation of the TW. The particles are spread within the entire domain, and hence, care has to be taken in order to avoid duplicate positions. The mirror symmetry about the xy-plane is indicated by the dashdotted line (–·–·), whereas the dotted line (······) indicates the streamwise-spanwise-shift symmetry. The minimal unique subdomain is marked by Ω_{sym}^{z}, and Ω_{sym}^{-z} denotes its spanwise-mirrored duplicate. (c) The same positions as before, but mapped into the region $x \in [0, L_x/2)$, $z \in [0, L_z/2)$. The colours of the dots are matched with those of (b) in order to reveal the unmapped positions.

where $h_{p,0}$ is set to be equal to the particle diameter in the following. The choice of d_p is an intricate task as one would generally like the ratio H/d_p to be sufficiently large to evade a strong influence of the initial particle acceleration on the result, but also avoid an all too small value of d_p^+ in order to not disregard finite size effects and to keep the computational costs minimal as for these simulations the resolution of the numerical mesh is dictated by the ratio $d_p/\Delta x$. For the current study, a diameter of $d_p := H/16$ was deemed to be an adequate compromise and yields a total number of sample particles of $N_{p\Sigma} = 288$ per parameter point. The distribution of the initial positions in the manner previously elaborated is visualised in fig. 8.5(a) and a summary of the relevant base parameters of the simulations is provided in table 8.1.

Not all particles are introduced to the domain within a single simulation, but rather multiple independent realisations are performed for two major reasons. First, the primary aim of this study is to assess the interaction between individual particles and the vortical structure of the ECS, and therefore collective effects resulting from interparticle interaction or particle-induced modifications of the flow field are generally undesired. Second, as previously mentioned, the TW solution is linearly unstable which necessitates that perturbations are kept to a minimum in order to avoid a significant deviation of the flow from its initial state during the observation time. For these reasons, only N_p particles are added to the flow at once and $N_{run} = N_{p\Sigma}/N_p$ realisations are required to obtain all samples. There are numerous ways to distribute these N_p particles in the plane of initialisation, though it seems natural to arrange

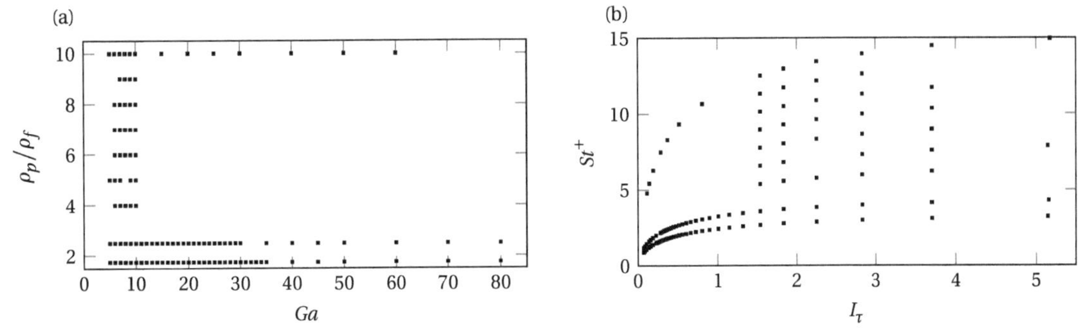

Figure 8.6: Summary of the particle parameter points which are investigated in this chapter. (a) Variation of the input parameters. (b) The resulting values of the relative turbulence intensity and the Stokes number.

them in a manner that approximately maximises the distance between them for the aforementioned reasons. Figure 8.5(b) displays an example of the distribution of $N_p = 16$ particles in a four-by-four arrangement which is employed henceforth. As can be seen, the entire cross-section of the simulation domain has been exploited, and thus, care needs to be taken in order to avoid duplicate samples due to the symmetric properties of the domain. In fig. 8.5(c), the initial positions are mapped back into the region $x \in [0, L_x/2)$, $z \in [0, L_z/2)$ which corresponds to the minimal unique subdomain Ω_{sym}. As suggested by the mapped positions, the remaining sample distributions are obtained by repeatedly shifting this distribution by $2h_{p,0}$ in streamwise and $h_{p,0}$ in spanwise direction.

The choice of $N_p = 16$ is based on a supplementary study outlined in §D in the appendix which compares the results obtained with this arrangement to those obtained from 288 realisations of isolated settling particles sampling the same initial conditions for a single particle parameter point. The results were contrasted with respect to the robustness of key quantities such as the mean settling velocity, the deviations of the TW from its initial state, and the divergence of particle trajectories for the same initial condition due to the presence of other particles. While the outcome was indeed still acceptable for 32 particles per simulation, the selected setup was deemed to provide the best compromise between the computational effort and the prevention of undesired secondary effects. Generally, the preservation of the ECS was not found to be problematic for any of the investigated parameter points.

Finally, the parameter points which are investigated in the following are presented. The major advantage of studying this simplified problem as opposed to fully turbulent flows is the severly reduced computational effort required for the simulations which enables an extensive parameter sweep. Consequently, if the outcome of this study is qualitatively comparable to the preceding observations made in turbulence, the extended parameter range may possibly lead to conjectures on scaling laws which can subsequently be tested with targeted fully turbulent simulations. A comprehensive summary of the simulated parameter points is provided in fig. 8.6. The input parameters are varied in the range $Ga \in [5, 80]$, $\rho_p/\rho_f \in [1.75, 10]$, which results in a range of $I_\tau \in [0.07, 5.15]$ for the relative turbulence intensity and $St^+ \in [0.9, 15.0]$ for the Stokes number.

8.3 Settling velocity

To begin with, the ensemble-averaged settling velocity discussed in §4.1 for the turbulent settling is evaluated for the current setup. For this purpose, the temporal evolution of this quantity for three of the LR parameter points is compared to that for the TW settling at matched densities and approximately

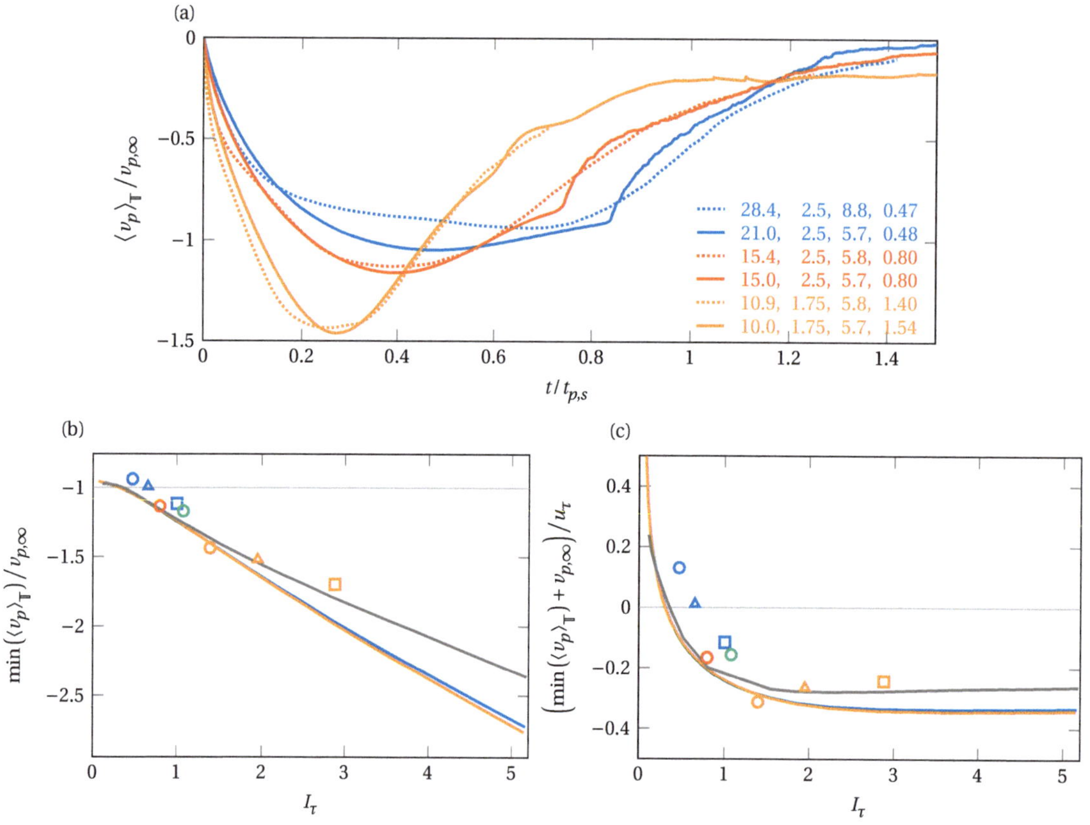

Figure 8.7: Ensemble-averaged settling velocity as defined in eq. (4.1). (a) Direct juxtaposition of the three turbulent parameter points G30-D10-LR (······), G15-D6-LR (······), and G10-D6-LR (······) with their respective counterparts for settling in the ECS (—, —, —). The parameter points are matched in their density and are approximately matched with respect to their relative turbulence intensity. The legend summarises the parameter quadruplet $\{Ga, \rho_p/\rho_f, d_p^+, I_\tau\}$. (b) Quantification of the peak enhancement as in fig. 4.2 showing all turbulent parameter points as well as the values obtained from the TW for the density ratios $\rho_p/\rho_f = 1.75$ (—), $\rho_p/\rho_f = 2.5$ (—), and $\rho_p/\rho_f = 10$ (—). The data is presented as a function of the relative turbulence intensity and scaled by the respective ambient terminal velocity. (c) Same as before, but presented as the excess in vertical velocity with respect to the terminal velocity and scaled by the friction velocity.

matched values of the relative turbulence intensity. As can be observed in fig. 8.7(a), the similarity between the settling in the ECS and the fully turbulent flow is extraordinary: not only does the simplified system reproduce the qualitative trend remarkably well, but in the cases of G15-D6-LR and G10-D6-LR, the evolution is moreover reproduced with outstanding quantitative accuracy. For G30-D10-LR, the agreement is markedly inferior, however, it was elaborated in §6.2 that this case is most prominently subject to the collective hindrance effect which is not present in the simplified system, and hence, there is reason to believe that this phenomenon might be the major source of the discrepancy.

Figure 8.7(b) provides a more extensive comparison between all available turbulent cases and the ECS by extracting the peak values of $\langle v_p \rangle_{\mathbb{T}}$ and presenting them as a function of I_τ. The general agreement is reasonable and the quantitative behaviour is reflected well, albeit the turbulent parameter point at the highest value of the relative turbulence intensity (G10-D10-HR) deviates rather significantly from the corresponding line. As the predictions of the simplified system turned out to be adequate, the more extensive and densely populated parameter range may be exploited to make an educated guess on how the peak value of $\langle v_p \rangle_{\mathbb{T}}$ evolves in regard to I_τ. Interestingly, for low values of I_τ the settling velocity is found to be reduced with respect to its ambient value even in absence of collective effects. The explanation for this observation is readily obtained by examining the trajectories of the particles for a respective parameter point as is portrayed in fig. 8.8. Due to the low relative turbulence intensity, the particles

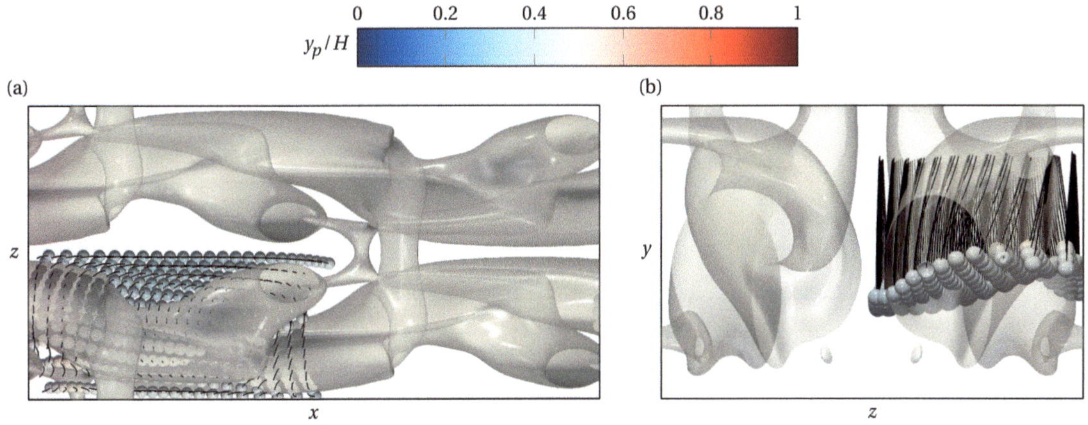

Figure 8.8: Visualisation of particle trajectories for a parameter point which exhibits a reduced mean settling velocity ($Ga = 35$, $\rho_p/\rho_f = 1.75$, $I_\tau \approx 0.23$). Particle positions are shown at an arbitrary point in time for which the transient has progressed enough to judge its characteristics. The colouring of the particles indicates their current vertical position, and the preceding trajectory of each particle over a limited time interval is indicated by a black line. A frame of reference is employed which moves with the propagation velocity of the TW, and the isosurfaces of the Q-criterion indicating the vortical structure corresponds to the undisturbed ECS which is steady in this perspective. Furthermore, all initial positions are aggregated from the individual realisations and mapped onto the minimal unique subdomain.

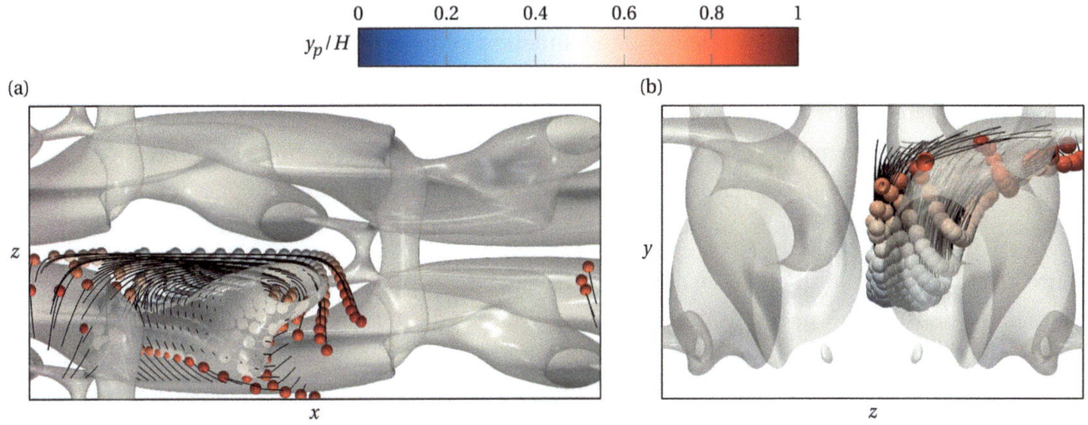

Figure 8.9: Visualisation of particle trajectories for a parameter point which exhibits an enhanced mean settling velocity ($Ga = 10$, $\rho_p/\rho_f = 1.75$, $I_\tau \approx 1.54$). The details of the depiction are given in the caption of fig. 8.8.

do not preferentially sweep the vortex, but rather cross it even in the presence of a counteracting flow which leads to the loitering effect described in §2.3.3. Concerning the trend towards high value of I_τ, it is more suitable to analyse the excess in settling velocity with respect to the terminal velocity scaled by u_τ, see fig. 8.7(c). As can be extracted from this figure, the excess does not increase indefinitely, but rather saturates at a value which appears to be a function of the particle density—at least in the simplified system. Consequently, the ratio between the mean settling velocity and the ambient terminal velocity is linearly proportional to I_τ at sufficiently high values thereof. In order to promote the understanding of this behaviour, the trajectories for a case with enhanced settling velocity are portrayed in fig. 8.9. As may be anticipated, the enhancement is a result of the preferential sweeping of the vortical structure, and hence, it might be hypothesised that the plateau in enhancement might be a result of particles sampling the optimal trajectory around the vortex for the current setup.

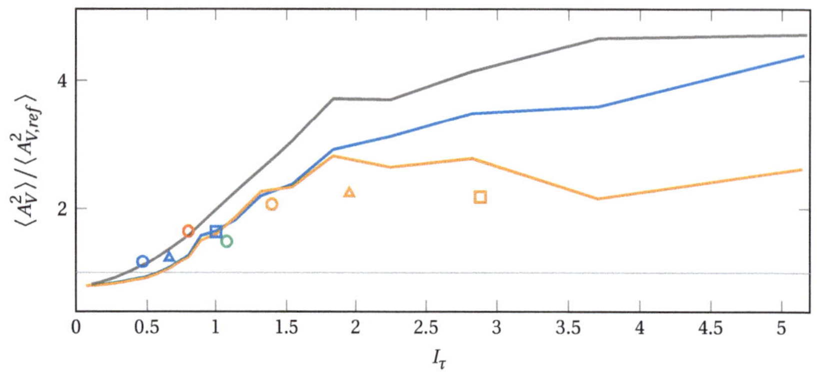

Figure 8.10: Normalised second moment of the Voronoï cell area distribution at $y = 0.5H$ as a function of relative turbulence intensity for $\rho_p/\rho_f = 1.75$ (—), $\rho_p/\rho_f = 2.5$ (—), and $\rho_p/\rho_f = 10$ (—). Also shown are the turbulent data points averaged over the interval $y \in [0.4, 0.5]\, H_f$, cf. fig 6.15, which are denoted by the following markers: G30-D10-LR (○), G10-D6-LR (○), G15-D10-LR (○), G15-D6-LR (○), G30-D10-RP (△), G10-D6-RP (△), G30-D15-HR (□), and G10-D10-HR (□).

8.4 Preferential concentration

The preferential concentration in wall-parallel plane is now revisited in order to examine the scaling of the mean square of the Voronoï cell areas. From the turbulent data previously presented in §6.3.2, it was conjectured that this quantity is a non-linear function of the relative turbulence intensity, albeit the scattering of the data points impeded any conclusions on other dependecies. In order to potentially improve the understanding, the same analysis is repeated for the settling in the ECS. For this purposes, the locations of crossings through a given plane have been determined according to §4.2 and are analysed in the frame of reference described in §6.3.1.[3] The Voronoï cells are subsequently obtained from the time-compensated crossing locations taking into account all duplicate positions due to symmetries.

Figure 8.10 shows the evolution of the normalised mean square of A_V with respect to I_τ at three distinct values of the density ratio, as well as the data points obtained for the turbulent simulations. The trend of the turbulent data is reflected reasonably well by the TW data, especially in consideration of the fact that the simplified system comprises much more structure since the initial placement of particles takes place on a regular grid and, due to the lack of temporal dynamics and existence of spatial symmetries, the horizontal displacement of this initial condition is significantly less complex. In fact, it can be observed that the order of the initial arrangement is maintained far beyond the plane of initialisation for $I_\tau \to 0$ due to the inability of the flow to efficiently disperse the particles within the wall-parallel plane, cf. fig. 8.8 for a visual example of this behaviour. Interestingly, both the turbulent and TW data suggest that the increase in $\langle A_V \rangle^2$ levels off for $I_\tau \gtrsim 1.5$. Moreover, the TW data prompts the conjecture that the ultimate degree of organisation depends quite considerably on the density ratio in a way that for the currently discussed parameter points an increase in particle intertia—which is also reflected by the respective value of St^+ as can be deduced from fig. 8.6—indeed leads to a more intermittent spatial distribution of the particles.

[3] In the original discussion it was mentioned that the propagation velocity of the frame of reference is motivated by the advection speed of coherent structures with which the particles interact. As this speed cannot be ultimately defined in turbulence, the reference velocity was set to the streamwise particle velocity averaged over all crossings. It should be noted, that for the TW the exact propagation velocity of the structure is known and given by c_{TW}. However, for the sake of comparability, the Voronoï analysis is performed using the original definition.

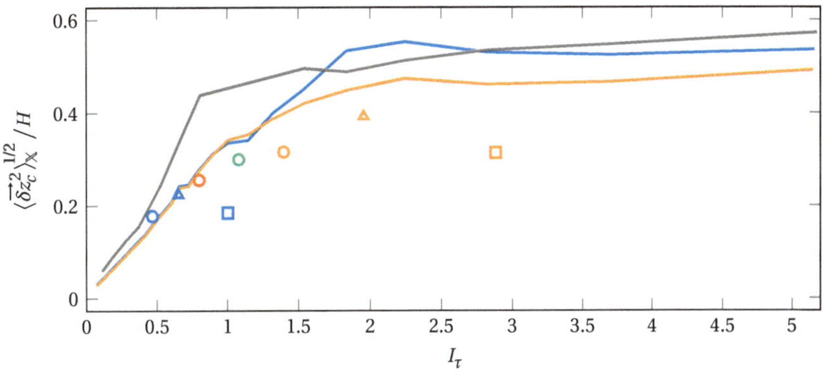

Figure 8.11: Mean-squared lateral displacement at the time of crossing the wall-parallel plane located at $y = 0.2H$ as a function of relative turbulence intensity for $\rho_p/\rho_f = 1.75$ (—), $\rho_p/\rho_f = 2.5$ (—), and $\rho_p/\rho_f = 10$ (—). Also shown are the corresponding turbulent data points, cf. fig 6.28(d), which are denoted by the following markers: G30-D10-LR (○), G10-D6-LR (○), G15-D10-LR (○), G15-D6-LR (○), G30-D10-RP (△), G10-D6-RP (△), G30-D15-HR (□), and G10-D10-HR (□).

As the enhancement in settling velocity is driven by the ability of particles to sample downdrafts, it stands to reason that the evolution of $\langle A_V\rangle^2$ is linked to that of the peak velocity enhancement, and indeed, when fig. 8.10 is compared to fig. 8.7 similarities can be noted in the convergence behaviour for high values of I_τ. This corroborates the hypothesis formulated in §8.3 that the ultimate level of enhancement is the outcome of particles sampling the most optimal preferential sweeping trajectories at a given value of ρ_p/ρ_f. Interestingly, the data suggests that a stronger focusing of trajectories (as observed for $\rho_p/\rho_f = 10$) does not necessarily lead to faster settling, or in other words, increased inertia may increase the likelihood of converging trajectories, but this convergence might not necessarily occur in the most intense downdrafts. Nonetheless, while the foregoing inferences appear to be conclusive for the settling in the ECS, their applicability to the turbulent problem remains to be demonstrated.

8.5 Lateral displacement

Regarding the mean-squared displacement in the lateral direction, it was hypothesised in §6.7 that the value of $\langle\overrightarrow{\delta z}_c^2\rangle_{\mathbb{X}}^{1/2}$ is bounded by the spacing between the large-scale structures which might act as lateral transport barriers for particles during the settling phase. The aim of this section is to assess this phenomenon in the TW as the legs of the hairpin-like vortices essentially act similar to large streamwise-aligned roller structures, cf. fig. 8.2, and therefore potentially induce a likewise effect.

The standard deviation of the lateral displacement is depicted in fig. 8.11 in comparison to the turbulent cases. The data for the TW is in agreement with the turbulent data, and in particular, with the LR and RP simulations. Moreover, the deviating values for HR suggest that the expected value significantly depends on the details of the flow which justifies the discrepancies. The trend observed for the settling in the ECS reinforces the hypothesis that particles are seldomly transported across multiple rollers even at high values of I_τ. Since the hairpin-like ECS does not feature laterally isolated velocity streaks as is the case for the turbulent cases in short domains, but rather exhibits interconnected wavy streaks due to the staggered nature of the hairpins, see fig. 8.2(b), this recognition aids the speculation that this barrier effect is still relevant in turbulent flows even when the streamwise extent of the LSMs and VLSMs is accurately captured.

8.6 Summary and concluding remarks

In this chapter, the feasibility of investigating the turbulent settling of particles in an open channel using a simplified proxy system based on the family of exact coherent states described in Shekar and Graham (2018) is examined. The selected ECS corresponds to a travelling wave whose topology resembles staggered hairpin vortices, i.e. the vortical structures which are abundant in wall-bounded turbulence and which were hypothesised by Adrian (2007) to play a substantial role in the explanation of LSMs and VLSMs—the large-scale turbulent features which were conjectured in this thesis to propel preferential concentration and velocity enhancement.

The TW solution was retrieved independently of Shekar and Graham (2018) and transferred by means of homotopy to a parameter point which is suitable for the investigation of particle dynamics using the immersed boundary method. In particular, the domain size was reduced to achieve a favourable aspect ratio and the value of the bulk Reynolds number was decreased to its value at the inflection point ($Re_b = 960.8$, $Re_\tau = 91.1$). The low-order fluid velocity statistics of the TW at this parameter point were compared to a turbulent open channel at $Re_\tau = 200$ (Bauer et al. 2022) and found to be in reasonable agreement, especially if the reduced separation between inner and outer scales is taken into account. This is especially true for the vertical velocity component which is known to be relevant in regard to the enhancement in particle settling velocity.

Similar to the settling in sediment-laden turbulent flow, particles were seeded near the free-slip boundary of the flow domain in order to let them settle transiently. However, in contrast to it, the sampling of initial positions was done systematically (rather than in random arrangements) since the set of all possible initial positions is greatly reduced due to symmetric properties and the lack of temporal dynamics in the TW solution. Moreover, particles were added in batches over multiple realisations in order to avert collective effects.

The temporal evolution of the ensemble-averaged settling velocity was compared between the simplified and the turbulent system for three parameter points, and an outstanding agreement was found when the relative turbulence intensity and the density ratio are matched. Due to the considerably smaller computational effort required for the simplified system, an extensive parameter sweep could be performed mainly in terms of the relative turbulence intensity. It was found that for $I_\tau \to 0$ a regime exists where loitering can be observed and the peak settling velocity is attenuated compared to the ambient terminal velocity. For $I_\tau \gg 1$, the enhancement was found to saturate when scaled in terms of u_τ, or correspondingly, to increase linearly with I_τ when scaled in terms of $v_{p,\infty}$.

The saturation of the enhancement was conjectured to be caused by the settling particles to preferentially sample the path of maximum enhancement, i.e. an increase of I_τ beyond its critical value does not lead to significantly different trajectories relative to the TW anymore. Evidence for this behaviour was provided by a Voronoï analysis of the crossing locations of a given wall-normal plane which yielded that the mean-squared area also levels off for sufficiently high values of I_τ. The ultimate value for both the settling velocity enhancement and the spatial intermittency of the crossing locations was found to be dependent on the density ratio.

Ultimately, the existence of lateral transport barriers during the settling phase was assessed for the simplified system. It was found that the mean-squared lateral displacement evaluated in the lower part of the channel does not increase indefinitely with I_τ which suggests that settling particles rarely cross through the large-scale roller structures within which they tend to focus. A similar trend was indicated by the turbulent datasets, although in this case the parameter range was insufficient to make a definite statement. It remains to be demonstrated whether the inferences made from the simplified system are transferable to high-Reynolds-number turbulent flows.

In conclusion, this chapter demonstrated that ECS may be employed as a valuable numerical laboratory to study particle-turbulence interactions to a degree that valuable quantitative predictions can be conjectured based on observations of the simplified system. These conjectures may assist in the choice of the parameter range to be investigated in more complex simulations or experiments in order to corroborate or falsify the inferences. Moreover, the use of ECS enables analyses which are difficult to realise, or even inaccessible, in the fully turbulent setup. In particular, since the complete velocity field is known a priori, a velocity experienced by the particle at any given position, and even at previous times, may be obtained unambiguously even for finite-size particles. In the future, this knowledge may be exploited to gain more insights on particle-vortex interactions, especially in regard to non-local effects.

9 Implications on bacteria transport modelling

While the preceding chapters predominantly focused on hydrodynamic details of the abstracted particle settling and dispersion problem, the purpose of this chapter is to discuss selected results in regard to their implication on bacteria transport, or the modelling thereof.

9.1 Evaluation of a one-dimensional transport model

The first model which explicitly distinguished between the suspended and particle-attached mode of bacteria transport is the stream model by Jamieson et al. (2005). In this one-dimensional streamwise-resolved model both modes of transport are described by a respective advection-diffusion-reaction equation. The equations remain uncoupled given the argument that in freshwater streams bacteria adsorption to particles is predominantly irreversible due to the low-ionic strength environment (Jamieson et al. 2005), and thus, may be solved individually. Therefore, only the particle-attached transport will be discussed in the following.

The ADR equation for the suspended sediments which carry the attached bacteria is given by

$$\frac{\partial C_{PB}}{\partial t} + u_b \frac{\partial C_{PB}}{\partial x} = \mathcal{D}_{PB} \frac{\partial^2 C_{PB}}{\partial x^2} + q_{PB}, \tag{9.1}$$

where $C_{PB}(x)$ is the number concentration of suspended particles per unit streamwise length,[1] u_b is the bulk velocity corresponding to the discharge of the stream divided by its cross-sectional area, $\mathcal{D}_{PB}$ the dispersion coefficient of the particles, and q_{PB} the net rate of particle removal due to deposition. Regarding the sink term, the authors follow the approach of Thomann and Mueller (1987) who suggest the first-order kinetic equation, viz.

$$q_{PB} = -\frac{v_p}{H_f} C_{PB} \tag{9.2}$$

where the rate constant is equivalent to the negative inverse of characteristic settling time $t_{p,s}$ which was introduced in eq. (4.3). The value of the dispersion coefficient is treated as a free parameter and is calibrated to best reflect available measurements. In the original study, a possibly time-varying Dirichlet boundary condition is imposed at the upstream boundary, whereas at the downstream boundary a zero-flux Neumann boundary condition is employed which allows for bacteria to be advected out of the domain, but prevents a diffusive flux. Generally the location of the downstream boundary was chosen

[1] In the original publication, C_{PB} is defined as a mass concentration. However, in the current context where the contaminants/suspended solids are monodisperse, a number concentration can be used equivalently and will be more convenient in the following.

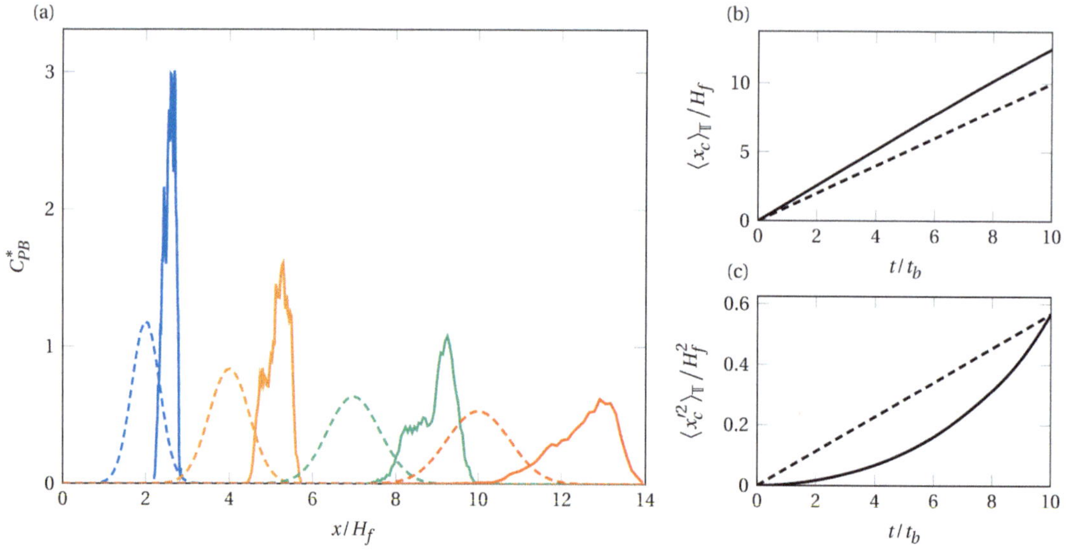

Figure 9.1: Comparison between the DNS data for G10-D10-HR and the predictions of the stream model by Jamieson et al. (2005) in the case of negligible deposition ($q_{PB} = 0$). The data is shown for the time interval $t \in [0, 10]\, t_b$ within the transient settling phase during which no significant deposition may be observed in the DNS. (a) Concentration profiles as a function of streamwise distance from the origin of the localised contamination. Solid lines depict the DNS data (—), while dashed lines show the Gaussian distributions (- - -) obtained from eq. (9.4). The line colours indicate the time at which the concentration is evaluated, namely $t = 2t_b$ (—), $t = 4t_b$ (—), $t = 7t_b$ (—), and $t = 10t_b$ (—). (b) Ensemble-mean of the streamwise contaminant position as a function of time. (c) Variance of the streamwise contaminant position. The dispersion coefficient of the stream model was calibrated such that the variance matches at $t = 10t_b$.

in their study such that it has little influence on the simulation results. As an initial condition, C_{PB} is set to zero everywhere within the domain.

The stream model assumes that suspended particles are removed from the flow at all times, i.e. q_{PB} is constant in space and time. However, in the case of combined sewer overflows where the contamination enters the stream near the free surface, deposition only occurs after the initial settling phase, and thus, q_{PB} is neither homogeneous in space nor time. Therefore, for the sake of comparison, the streamwise transport of particle-attached bacteria as predicted by the model in absence of q_{PB} is contrasted with the DNS data restricted to the settling phase in the following. In order to compare the transport, the domain of the stream model is set to be unbounded with a localised initial condition in space in time, viz.

$$C^*_{PB}(x, 0) = \delta(x), \tag{9.3}$$

where $C^*_{PB}(x,t) := C_{PB}(x,t) / \int_{-\infty}^{\infty} C_{PB}(x,0)\,\mathrm{d}x$ is the normalised concentration and $\delta(x)$ is the Dirac delta distribution. For the unbounded domain, the pure advection-diffusion problem exhibits a simple analytic solution (Crank 1975, Lehnigk 1976) of the form

$$C^*_{PB}(x,t) = \frac{1}{2\sqrt{\pi \mathfrak{D}_{PB}\, t}} \exp\left(-\frac{1}{2} \frac{(x - u_b t)^2}{2\mathfrak{D}_{PB}\, t} \right), \tag{9.4}$$

which corresponds to a normal distribution whose mean position propagates at a constant speed of u_b and whose variance increases linearly with time and is proportional to the dispersion coefficient, cf. eq. (6.27) and (6.28). In order to emulate a streamwise-localised initial condition for the DNS data, the data is henceforth analysed in terms of the displacement from the initial position, i.e. the trajectory unrolling described in §6.6 is employed, cf. eq. (6.14).

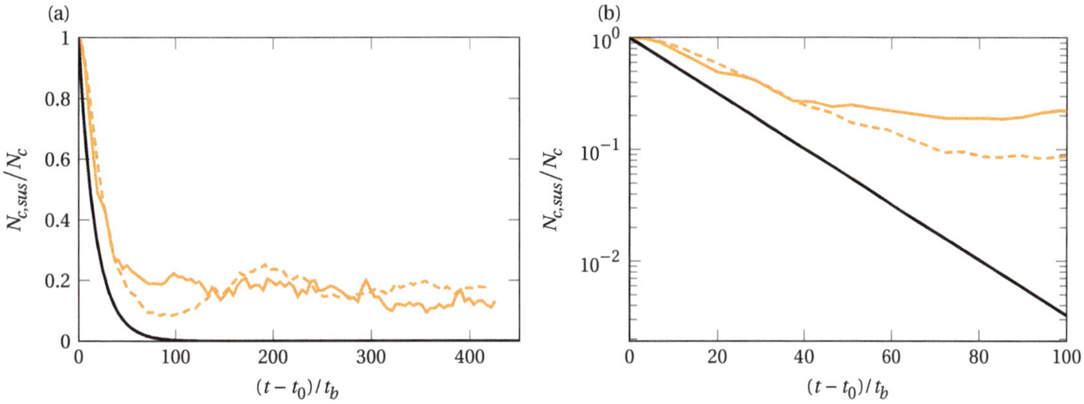

Figure 9.2: Evaluation of the global deposition behaviour as predicted by the stream model of Jamieson et al. (2005) (—) and evaluated for G10-D6-LR (—, macroscopically flat bed) and G10-D6-RP (- - -, ripple features). The DNS data has been shifted in time by t_0 which corresponds to the time at the onset of deposition. (a) Linear axes for the full time interval covered by the DNS. (b) Semi-logarithmic scale with emphasis on the initial deposition.

Figure 9.1(a) shows a comparison between the temporal evolution of the streamwise dispersion of particles obtained from the stream model and the DNS for the parameter point G10-D10-HR within a time interval during which no significant deposition occurs, i.e. where the disregardance of q_{PB} is justified. First, it can readily be acknowledged, that the streamwise distribution of particles is not Gaussian, but exhibits significant skewness. In fact, the concentration profile approaches a distribution which resembles that for a vertical line source of dissolved tracers at early times after injection as shown e.g. in Fischer (1973). Thus, the distribution is likely the result of the wall-normal inhomogeneity of the streamwise fluid velocity, and the distribution of cross-through times of the contaminants as presented in §4.4. Second, the propagation velocity of the distribution is significantly faster than u_b, see fig. 9.1(b). Again, this discrepancy may be partly explained by the vertical distribution of contaminants. However, aside from this effect, the tendency of contaminants to accumulate in high-speed velocity streaks also contributes to this effect, cf. §5.2.2, §5.2.3 and §6.3.4. Third, the variance in particle position does not increase linearly in time, and thus, does not follow the diffusion process described by eq. (9.1), see fig. 9.1(c). For the comparison, the dispersion coefficient has been calibrated to arrive at the same variance as the DNS data at the end of the time interval under investigation. The actual variance increases approximately quadratic in time, and hence, the rate of diffusion is overpredicted over the entire observation time. As a result, the localisation of the contamination is maintained for longer times than the stream model suggests.

In order to assess the accuracy of the deposition term q_{PB}, eq. (9.1) is reformulated into its conservative form. Recognising that there is no advective or diffusive flux across the boundaries for the unbounded domain yields

$$\frac{\partial N_{c,sus}}{\partial t} = -\frac{1}{t_{p,s}} N_{c,sus}, \tag{9.5}$$

where $N_{c,sus}(t) = \int_{-\infty}^{\infty} C_{PB}\,\mathrm{d}x$ is the total number of suspended contaminant particles within the domain. Using the classification procedure introduced in §6.5, this number is given by $N_{c,sus} = 1 - N_{c,bed}$ which is readily evaluated from the DNS. Given that all contaminants are initially suspended, the solution of eq. (9.5) describes an exponential decay of the form

$$\frac{N_{c,sus}}{N_c} = \exp\left(-\frac{t}{t_{p,s}}\right). \tag{9.6}$$

A juxtaposition of this decay model and the DNS data in absence and presence of ripples is displayed in fig. 9.2. Here, the empirical data has been shifted in time such that the origin of the abscissa corresponds to the onset of deposition to enable a fair comparison. When the entire observation interval is evaluated, the exponential decay qualitatively describes the rapid decrease in the number of suspended particles in an adequate manner. However, the nature of the sink term causes all particles to be removed from the suspended mode of transport in the long term, which is naturally not the case for the DNS data as particles still propagate streamwise due to bedload transport and resuspension, cf. §4.6 and §6.5. Consequently, the stream model is only capable of reproducing the initial transient and does not adequately capture the long-term transport of particle-attached bacteria in the vicinity of a river bed. Moreover, when the initial deposition of the contaminants is examined in detail (fig. 9.2(b)), it can be noticed that the exponential decay greatly overpredicts the rate of deposition at early times. In fact, the first-order kinetic demands that the removal rate of the highest at early times (when the concentration is still maximum), while it was demonstrated in §6.6 that the times of initial deposition actually follow a log-normal distribution, i.e. the peak deposition rate occurs significantly after the first particle arrived at the bed. This distinction is found to be rather significant as, for instance, the DNS data suggests a half-life which is twice as long as what is predicted by the model. In addition to the difference in the distribution of deposition times, the exponential decay rate is likely overpredicted when the ambient settling velocity is employed to compute the characteristic settling time, since it was reported in §4.5 and §6.6 that the net deposition time is reduced by roughly 20% across all cases as compared to the value obtained by $H_f/v_{p,\infty}$.

It is self-evident that a minimal model such as the stream model of Jamieson et al. (2005) cannot capture all subtleties of suspended particle transport in the same way a fully-resolved simulation does. Nonetheless, the comparison with the empirical data obtained within this thesis provides two major suggestions on how this particular model could be improved significantly in its ability to capture the streamwise spreading of particle-attached bacteria. First, it would be beneficial to consider a third mode of transport which depicts the bedload transport of settled contaminated particles which may be formulated in terms of a classical sediment transport formula (e.g. Meyer-Peter and Müller 1948).[2] Second, the deposition term, which would act as a sink in the suspended solid concentration and as a source for the bedload, should be modified to reflect the log-normal distribution of the settling length. In fact, §6.6 provides relatively robust empirical relationships which facilitate an adequate estimation of this distribution which only requires additional knowledge on the relative turbulence intensity as compared to the current requirements.

9.2 Solar irradiance

A physical process which may play a decisive role in the fate of bacteria is the exposure to sunlight, especially with regard to its ultraviolet components, cf. §1.1. For a given ambient intensity of ultraviolet irradiance $I_{UV,0}$ at the surface of a water body, the intensity to which particle-bound bacteria immersed inside a water body are exposed to is prevailingly a function of the distance covered by a ray inside the water body (absorption) and the turbidity (scattering). A widespread description of light attenuation

[2] Indeed, the inclusion of sediment transport in bacteria fate and transport models is not a new idea and has been implemented in various studies, see §2. However, it is rare that both settling particles and sediment transport are described separately within a single model.

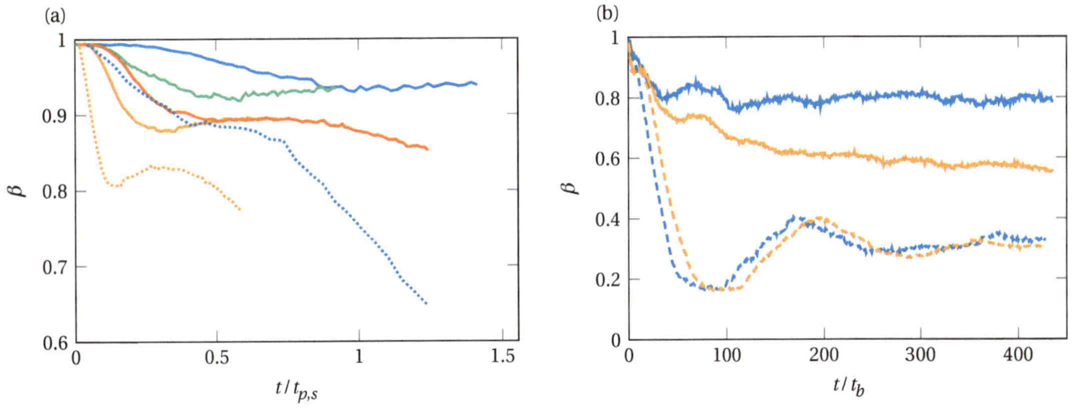

Figure 9.3: Proportion of the contaminant surface area which is not shaded by other particles, and hence, is directly exposed to solar radiation. (a) During the initial transient settling phase portraying the impact of clustering. (b) The long-term trend which displays the effect of sediment bedforms and contaminant entrainment into the sediment bed. Legend: G30-D10-LR (—), G10-D6-LR (—), G15-D10-LR (—), G15-D6-LR (—), G30-D10-RP (- -), G10-D6-RP (- -), G30-D15-HR (······), and G10-D10-HR (······).

inside a medium is given by Beer's law (Beer 1852) which reads under the assumption that the sunlight enters the water body from a strictly vertical direction

$$\frac{I_{UV}}{I_{UV,0}} = \exp\left(-k_{att}\left(H_f - y\right)\right), \tag{9.7}$$

where $I_{UV}(y)$ is the local UV intensity and k_{att} is the attenuation coefficient. As this law states that the UV intensity decays exponentially with the distance to the surface, the rate at which particles settle in the initial phase may be an important factor to consider under conditions where sunlight exposure is a dominant inactivation mechanism.[3] In fact, §4.5 suggests that for high relative turbulence intensities the contaminants generally exhibit an increased collective settling velocity until they are in the vicinity of the sediment bed.

Concerning the scattering of light by suspended solids in the water column, Qian et al. (2016) found that turbidity can be a decisive factor for the prediction of inactivation rates in the aftermath of a combined sewer overflow. Turbidity reduces the exposure to irradiance due to scattering or blockage of light by particles which are located in between the light source and the position of interest. Typically, this effect is parametrised by the concentration of total suspended solids present in the water column. This approach implicitly assumes that these solids are homogeneously distributed which was demonstrated not to be the case in this thesis. In order to highlight the effect of spatial intermittency on the exposure to sunlight, an exposure factor β is defined as the ratio between the cross-sectional surface area of contaminant particles which is directly exposed to radiation to the corresponding total area. Accordingly, $1-\beta$ is the share of the cross-sectional area of contaminant particles which is in the shade of other contaminant or sediment particles assuming that the ray direction is strictly vertical. Figure 9.3(a) depicts this exposure factor during the transient settling phase for various parameter points which all share the same concentration of suspended solids. While the exposed surface area is reduced by around 20% for the case with the highest relative turbulence intensity, it is found to decrease only slowly for more inert particles. A comparison of the evolution of β with the findings of §6.1.2 strongly suggests that the

[3] The discussion on UV inactivation of bacteria is not limited to the exposure to sunlight, but may also be of relevance for technical disinfection procedures using UV radiation which are at times applied to the storm sewage before it is released into rivers, cf. §1.1.

reduction in sunlight exposure is proportional to the tendency of contaminants to form clusters. A spatially inhomogeneous distribution of particles may hence lead to higher effective concentrations of total suspended solids with regards to the modelling of turbidity.

Regarding the long-term exposure to sunlight, sediment bedforms are decisive. As might have been anticipated from the findings of §6.5, the trapping phenomenon observed in the presence of ripples which buries contaminants under a thick layer of sediment is very efficient at shielding particle-attached bacteria from radiation. In fact, fig. 9.3(b) indicates that the exposed surface area is reduced to merely 20% during the first trapping cycle, and stays consistently around 30% in the long-term. In contrast to this, the exposure factor is found to stay within the range of 60–80% if the sediment bed is macroscopically flat with a tendency to decrease if the contaminants tend to get covered by the turbulent deposition of sediment particles. Irrespective of the details of the bedform, a tendency of contaminants to be entrained into the sediment bed appears to provide favourable conditions for the survival of particle-attached bacteria. It also serves as a plausible explanation of the spontaneous increase of FIB concentrations in rivers observed in the absence of immediate contamination events which is hypothesised to be caused by storages of bacteria associated to river beds (McDonald et al. 1982, Sherer et al. 1988, 1992, Wilkinson et al. 1995).

9.3 Lateral dispersion barriers during settling

It was observed in §6.7 that the mean-squared displacement of contaminants during the settling does not increase indefinitely with the distance travelled, but rather saturates during the settling phase. Moreover, when the converged value is inspected across the various cases, it appears to level off past a certain value of the relative turbulence intensity. A similar trend is observed in a simplified proxy system where particles settle within an exact coherent state of the fluid flow, and for this system it was indeed demonstrated that the lateral displacement during the settling does not considerably increase anymore if the relative turbulence intensity exceeds a value of $I_\tau \gtrapprox 2$, cf. §8.5.

It has been hypothesised that the saturation in displacement is linked to the tendency of contaminants to accumulate in large-scale high-speed streaks of the streamwise fluid velocity, which appears plausible as the maximum standard deviation of the lateral displacement is approximately $H_f/2$ while the spanwise wavelength of the large-scale streak is in the range of 1–$2H_f$, cf. §5.5. A potential physical mechanism which gives rise to this phenomenon may then be outlined as follows: In the beginning of the settling process, contaminants are driven towards an adjacent high-momentum region due to the preferential sampling of coherent vortices which presumably occurs on multiple scales. The focusing of particles in these regions can be considered as stable since on average the fluid flow only diverges from it laterally close to the sediment bed, cf. §5.1. As a consequence contaminants do not tend to exhibit a net spanwise movement once they are focused until they are close to deposition.

On the contrary, the diverging averaged flow near the sediment bed encourages the spanwise movement towards a large-scale low-speed streak, and thus, drives contaminants away from their spanwise position which they obtained during the initial settling, which tends to increase dispersion. In the bedload transport phase, contaminants are observed to accumulate near sediment ridges which thus constitute the region of convergence in the near bed region, cf. §6.5. Here, it is important to note that no dispersion barrier phenomenon occurs in the phase of near-bed transport: While the sediment ridges are a stable position on the bed, contaminants will occasionally be uplifted by resuspension events which predominantly occur in low-speed streaks. However, due to the symmetry breaking effect of gravity, the upward movement is unstable unlike the initial downward movement, and thus encourages spanwise

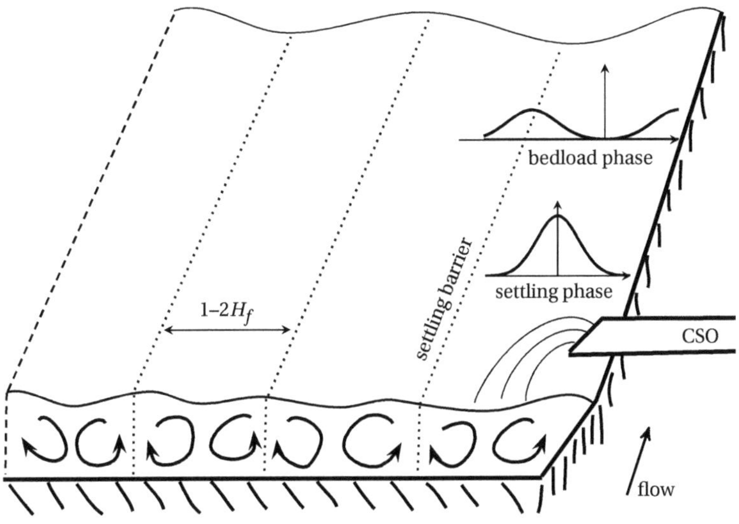

Figure 9.4: Conceptual sketch of the conjectured lateral dispersion barrier during the transient settling phase. The turbulent channel flow contains long streamwise-aligned streaks of streamwise velocity fluctuations with alternating signs which are associated to vortex clusters or packets. In the conditionally-averaged picture, these structures are separated by large-scale rollers which cause contaminants to accumulate in one of the large-scale high-speed streaks/downwellings which are closest to their initial position by means of multiscale interactions. Moreover, the analysis of lateral dispersion indicates that contaminants rarely cross low-speed streaks (······) during the settling phase. It may hence be speculated that contaminants which are introduced near the bank of a river stay contained in a region with a the width of 1–2H_f as long as they are suspended, and only spread further in lateral direction once they are deposited onto the bed. Since the streamwise contaminant flux is considerably higher during the settling phase, this implies that contaminants might stay confined to the near-shore region over significant downstream distances from the source.

dispersion. Contaminants may then again be transport towards a neighbouring high-speed streak, and so forth. In this idealised depiction, the spanwise transport of an individual contaminant particle thus consists of discrete jumps between flow features with spanwise dimensions of the order of the clear fluid height.

It may be speculated that the attenuation of lateral dispersion during the initial settling of contaminants has interesting consequences on the spreading of bacteria in rivers. Regarding combined sewer overflow events, the relief system from which the waste water is discharged is usually located near the river bank. As discussed in §4.6, the streamwise contaminant flux is considerably higher during the settling phase than at later stages, which implies that contaminants may be transported over a considerable streamwise distance until they start to disperse away from the shoreline. This recognition might therefore have significant impact on the near-shore water quality, as well as the prediction thereof, as it suggests that the spanwise distribution of contaminants remains inhomogeneous even at considerable distances from the source of pollution. In terms of modelling, this conjectured phenomenon especially casts doubt on the common assumption of well-mixedness of pollutants in the lateral direction if they enter the stream from a point source. A conceptual sketch of the transport barrier mechanism is provided in fig. 9.4.

In order to assess the relevance of this phenomenon in the real-world application, one might also speculate on the implications of a variation in the parameter point regarding the spreading, especially in view of particles smaller or lighter than the ones investigated in this thesis. A reduction in particle size or density implies that the terminal velocity of the contaminant particle reduces, which in turn increases the relative turbulence intensity $I_\tau = u_\tau / v_{p,\infty}$ and the characteristic settling time $t_{p,s} = H_f / v_{p,\infty}$. Since evidence was presented that the limitation in lateral dispersion becomes independent of the relative turbulence intensity for $I_\tau \gg 1$ and only depends on the depth of the water body, and particularly not directly on the particle diameter and presumably only weakly on the density, the upper bound of the

Table 9.1: Details on the initial conditions and boundary conditions on the particles' surface for the concentration fields which are employed to mimic the spreading of freely-suspended bacteria. In all cases the boundary conditions at the free-surface as well as at the bottom bounding wall are zero flux conditions.

scalar	initial condition	BC contaminants	BC sediment	Sc
S1	layer	$C_{SB} = 1$	$C_{SB} = 0$	1
S2	layer	$\frac{\partial C_{SB}}{\partial n} = 0$	$\frac{\partial C_{SB}}{\partial n} = 0$	1
SB	layer	–	$\frac{\partial C_{SB}}{\partial n} = 0$	1

mean-squared displacement remains approximately unmodified in terms of H_f. At the same time, the increase in the characteristic settling time in conjunction with the observation of §6.6 that the ratio between the actual deposition time and $t_{p,s}$ is not considerably dependent on I_τ implies that smaller particles will be transported over longer streamwise distances before they enter the near-bed region. As a consequence, the ratio between the streamwise-travelled distance (which was argued to increases with decreasing particle size/density) and the spanwise-travelled distance (which is speculated to be independent of the particle size/density at high relative turbulence intensities) before deposition might increase for smaller/lighter particles which further emphasises the point that contaminants might remain confined to the near-shore region over considerable downstream distances. Unfortunately an assessment of this conjecture requires a parameter sweep with respect to the Galileo number in a range which is outside the scope of this thesis and for which $I_\tau \gg 1$ in order to be in the presumed asymptotic regime of lateral dispersion. It should also be emphasised that the lateral dispersion in the near-bed region is expected to behave qualitatively different as the dispersion coefficient is conjectured to be proportional to the Shields number in this regime, cf. §6.7, and thus, the contamination will not stay confined as soon as the initial settling phase is concluded.

Self-evidently, the description of the mechanism is in a sense idealised, and in reality the alleged barrier should be not be understood as an absolute fact, but rather in a statistically-averaged sense. Influencing factors which need to be considered when assessing the relevance of this phenomenon in the bigger picture are the lifetime and evolution of LSMs and VLSMs, as the cyclic burst of large-scale streaks or their meandering characteristics at very-large scales might weaken their ability to obstruct dispersion. Moreover, the transport of contaminants near the bed in the later stages is subject to a great variety of phenomena aside from the ridge accumulation, such as the interaction with ripples or other trapping effects (§6.5) which require further investigations since they might play a leading role in the spanwise homogenisation of contaminant concentrations.

9.4 Effect of particles on freely-suspended bacteria

The content of this thesis was focused on the transport of bacteria in the particle-attached mode due to its beneficial characteristics regarding the survival of bacteria on one hand, and the complexity of the particle-laden flow problem which serves as an abstraction on the other hand. Most bacteria transport models which explicitly treat the particle-attached transport also consider the transport of suspended bacteria, although regularly both modes remain uncoupled which is justified by irreversible bonding. In this section, a brief primer is provided on the coupled transport of particle-attached and freely-suspended bacteria.

The accurate modelling of the transport of suspended bacteria is an intricate task even in the absence of particles. This is mainly caused by the low diffusivity of bacteria in comparison to the viscosity which results in high values of the Schmidt number, and consequently, in fluctuations in the concentration field significantly smaller than the Kolmogorov scales (Pinelli 2022). This intricacy poses high demands on the numerical method and the mesh resolution, and is avoided in the following by setting $Sc = 1$. The combined particle and scalar transport problem is treated using the method described in §3.2.3. The concentration of suspended bacteria is assumed to have no effect on the specific weight of the fluid, and thus, its transport is treated equivalently to that of a passive scalar. In fact, the tracking of suspended bacteria was included in the HR simulation series presented regularly throughout this thesis, where the pure particle-laden flow problem could be discussed independently of the mass transfer problem due to its passivity which implies that the evolution of the fluid flow and the particle trajectories are completely unaffected by the transport of suspended bacteria under the current assumptions. In contrast, the effect of particles on suspended bacteria is two-fold as particles may modify the concentration of suspended bacteria directly by means of their boundary condition for the scalar transport problem, but also modify the fluid velocity field which affects the advective term in the advection-diffusion equation even in the case of no mass transfer (irreversible attachment). These two effects are examined systematically by imposing two distinct types of scalar boundary conditions at the surface of the contaminants in the following. In the first case, a Dirichlet boundary condition of $C_{SB}^* = 1$ is imposed to emulate a simple detachment process where particles act as a source of suspended bacteria ("S1"), where C_{SB}^* denotes a normalised concentration. Here, it is assumed that the particle-attached and freely-suspended modes are in equilibrium in the discharged sewage, and that the dilution with freshwater causes bacteria to be primarily detached from contaminant particles. Furthermore, it is assumed that the number of attached bacteria remains large compared to the number of bacteria which are detached during the time interval such that no balance equation for the bacteria concentration on the particles' surface needs to be solved. In the second case, a zero-flux Neumann boundary condition of $\frac{\partial C_{SB}}{\partial n} = 0$ is imposed which makes the particles inert with respect to the scalar field ("S2"). In both cases the initial condition for the concentration field is a layer that embeds the contaminant particles which is given by the equation

$$C_{SB}^*(y,0) = \frac{1}{2}\left(\mathrm{erf}\left(\sqrt{\pi}\frac{y - y_{SB}}{\varsigma_{SB}}\right) + 1\right) \tag{9.8}$$

and constitutes a sigmoid function where $C_{SB}^* = 1$ for $y \gg y_{SB}$ and $C_{SB}^* = 0$ for $y \ll y_{SB}$. In particular, $\mathrm{erf}(x) = \frac{2}{\sqrt{\pi}} \int_0^x \exp(-t^2)\,\mathrm{d}t$ denotes the Gauss error function. In order to avoid a discontinuity at the edge of the layer, the transition occurs within a finite vertical region whose extent is determined by ς_{SB}. Henceforth, these two parameters are set to $y_{SB} = 0.9H_f$ and $\varsigma_{SB} = 0.025H_f$. The domain is hermetic and the sediment particle boundary conditions are $C_{SB}^* = 0$ for S1 and $\frac{\partial C_{SB}}{\partial n} = 0$ for S2. To characterise the effect of the presence of settling particles on the transport of suspended bacteria in the case without mass transfer, a reference simulation "SB" is conducted which does not contain any contaminant particles, but has the same initial condition for the scalar field. A summary of the imposed boundary conditions is provided in table 9.1.

Figure 9.5 provides the vertical profile of the xz-averaged concentrations of suspended bacteria at two instances in time in presence and absence of contaminant particles. If bacteria may detach from the particles (S1), the mean concentration is observed to be enhanced over the entire flow depth. In particular, the rapid rise in C_{SB}^* in the lower parts of the channel can be entirely attributed to the detachment of bacteria from particles as a comparison with the evolution of S2 immediately suggests. The rate of downward propagation is thereby mainly determined by the settling velocity of the contaminant particles as a comparison between G30-D15-HR and G10-D10-HR suggests. If bacteria are bonded irreversibly to particles (S2) the profile of the scalar field remains much closer to the reference curve for the spreading in absence of contaminant particles. However, the inert particles clearly lead to a faster mixing of the freely-suspended layer by manipulating the fluid velocity field. Regarding this effect, the

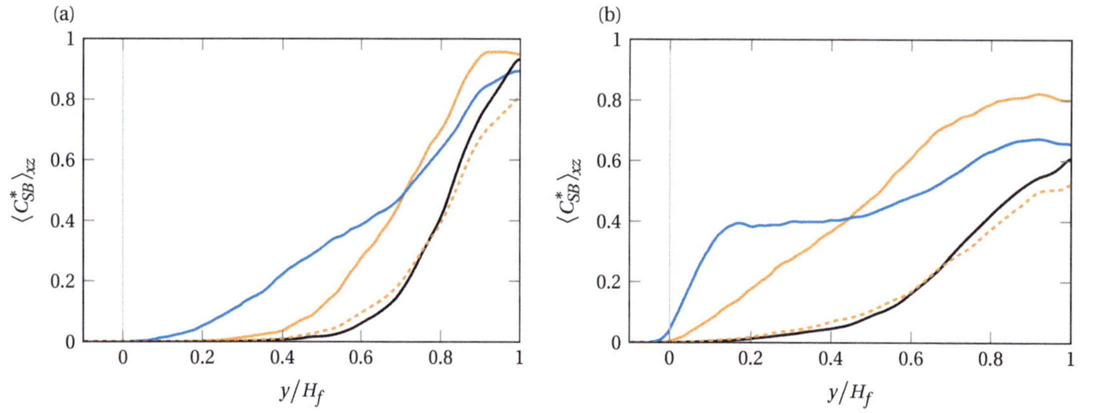

Figure 9.5: Mean concentration of suspended bacteria averaged in wall-parallel planes as a function of the vertical position. Two instances in time are displayed, namely (a) $t = 7.2t_b$ and (b) $t = 12t_b$. The line colours and styles denote different cases and boundary/initial conditions for the concentration field. Legend: G30-D15-HR-S1 (—), G10-D10-HR-S1 (—), G10-D10-HR-S2 (- - -), and HR-SB (—).

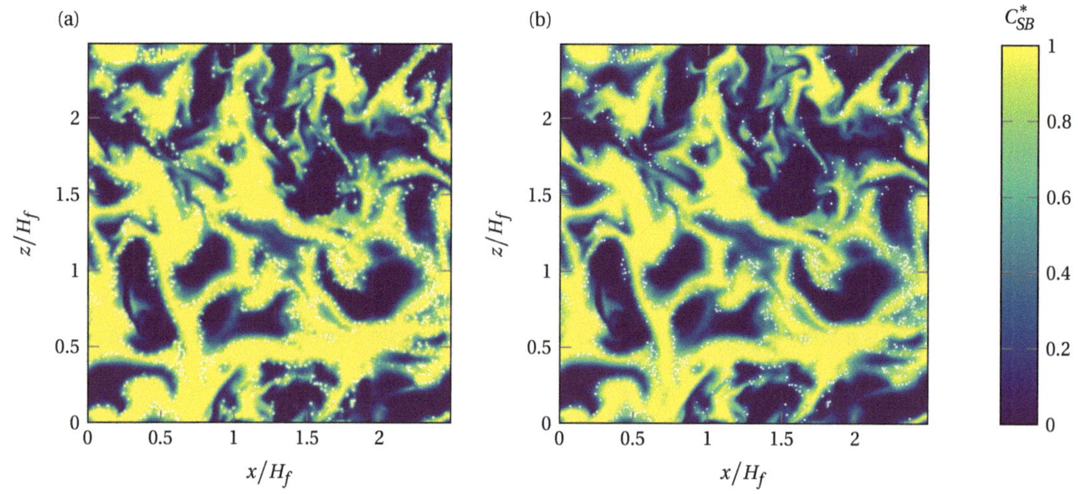

Figure 9.6: Exemplary visualisation of the concentration of suspended bacteria in the wall-normal plane at $y \approx 0.92H_f$ and $t \approx 2.8t_b$. Contaminant particles which are currently located in this plane are displayed as white dots. (a) G10-D10-HR-S1 showing detaching bacteria, (b) G10-D10-HR-S2 showing irreversibly attached bacteria.

difference between the two particle parameter points investigated is miniscule, and thus, the curve for G30-D15-HR-S2 has been omitted in the figure. It should be noted that while the disparity in the observed concentration profiles for S1 and S2 appears drastic, these profiles certainly dependent on the actual rate of detachment of bacteria which has only been modelled crudely and is likely to be overpredicted. Therefore, the recognitions just made should be understood in a qualitative sense.

An exemplary depiction of the instantaneous structure of C_{SB}^{*} is provided in fig. 9.6. The wall-parallel plane which is located at a position inside the layer of initialisation highlights the intermittent nature of the scalar field. In fact, both the freely-suspended and particle-attached bacteria are transported away from the free surface in fluid downwellings, although it can be noted that contaminant particle are located near the edges of high concentration regions, which is in accordance with the recognitions of §6.3.3. For this reason, the detachment of bacteria may be relevant even at early stages of the settling phase due to the early segregation between the two modes of transport.

9.5 Summary and concluding remarks

The purpose of this chapter was to transfer the newly obtained knowledge from the predominantly hydrodynamic analyses of this thesis to bacteria transport modelling, and to speculate on emerging phenomena which may have received only little attention in the water quality community so far.

To begin with, a simple but popular one-dimensional model for the streamwise spreading of bacteria was analysed with respect to its predictions on the transport of particle-attached bacteria. It was demonstrated by means of juxtaposition with the DNS data that the streamwise contaminant flux is underpredicted by the model, whereas the rate of dispersion and deposition are generally overpredicted. Practical modifications have been proposed based on the recognitions of §6 which likely eliminate some of the flaws of the model with only a marginal increase in complexity.

The organisation of particles into spatially intermittent structures has been discussed in the context of bacteria inactivation due to the exposure to ultraviolet radiation. The formation of clusters while particles are still in suspension was determined to reduce the exposure to sunlight due to shading effects. For this reason, the attenuation of sunlight is not only a function of the concentration of total suspended solids, but also depends on the degree of spatial inhomogeneity exhibited by the solid phase, which in turn depends primarily on the relative turbulence intensity. A similar phenomenon occurs due to the interaction of contaminants with the sediment bed. The trapping mechanism which immobilises contaminants under a thick layer of sediment in the presence of ripples was determined to reduce the exposure to solar irradiance dramatically. A similar effect, but to a weaker degree, can be observed for macroscopically flat beds if the parameter point enables an entrainment of contaminants into the sediment bed. The shielding from UV irradiance likely provides a favourable environment for the survival of bacteria, and it was hypothesised that the observed burying process is one of the storage mechanisms which is related to intermittent bursts in FIB concentrations observed in rivers even in the absence of immediate contamination events.

The tendency of contaminant particles to accumulate in large-scale streaks of high streamwise fluid momentum was speculated to lead to transport barriers which inhibit the lateral dispersion during the initial settling phase. This supposition is substantiated by the observed stagnation in lateral mean-squared displacement and the saturation of the ultimate displacement in regard to increasing relative turbulence intensity. As contaminants are introduced near river banks in combined sewer overflow events, this finding suggests that a strong spanwise inhomogeneity in FIB concentration might persist even at significant downstream distances from the source of pollution, which again has important implications on water quality near the shoreline of a river if this phenomenon proves to be relevant. Once the contaminants are deposited or close enough to the sediment bed to experience flow conditions which diverge laterally on average, the dispersion mechanism changes its nature due to the symmetry breaking effect of gravity, and consequently, no transport barriers are to be expected in this phase.

Finally, the effect of the presence of particles on the spreading of freely-suspended bacteria was outlined for the cases of detaching bacteria and the modification of the flow field by inert particles. If bacteria are not bound irreversibly, the tendency of particles to move downwards at faster velocities than the fluid causes mass transfer which increases the concentration of freely-suspended bacteria more rapidly at the bottom of the channel than in absence of settling particles. Moreover, it was found that the mere presence of particles facilitates the vertical dispersion of suspended bacteria, since the settling particles enhance vertical fluid momentum even if they are inert with respect to the scalar field.

10 Conclusion and outlook

The mechanistic modelling of bacteria fate and transport typically operates by successively decomposing the complex problem into separate processes, which are modelled individually based on empirical and theoretical knowledge, and are refined in steps according to the shortcomings in their predictions. While this procedure has proven to be successful for the description of tangible water-pollution problems and promotes the understanding of the empirical observations due to its decomposed nature, it offers little possibility for inductive reasoning as the model parameters and assumptions are generally tuned to the problem of interest. The transport components of these frameworks are either low-dimensional or based on recent advances in the modelling of flows on a geographic scale, i.e. on the scale of entire estuaries for instance, and thus, rely substantially on the modelling of hydrodynamical processes which occur on length scales of the order of the water depth and below. This is especially problematic in the context of bacterial contaminations as it is known that bacteria tend to attach to particles, which is of great significance as they provide a favourable environment for their survival, and which couples their spreading to the dynamics of particles. Indeed, particle-attached transport is thought to be vital in the explanation of various emerging phenomena such as the delayed release of enteric bacteria stored in river sediment even in absence of immediate contamination events (McDonald et al. 1982, Wilkinson et al. 1995).

There is a strong discrepancy between the sizes of the particles of interest (around 10–100 μm) and the smallest length scales which are resolved by any higher-dimensional transport model (several meters) that could be identified in the literature. It is therefore inevitable to rely on models for the description of the dynamics of particles. At the same time, from a hydrodynamics point of view, the interaction between particles and turbulence is not sufficiently understood to reliably capture all of its peculiarities. This is especially true in the case of the transient settling problem of particles in a turbulent horizontal channel flow, which can be thought to be an abstraction of the settling processes occurring after a combined sewer overflow event, and which has hitherto received only little attention in the hydrodynamic literature, especially in regard to high-fidelity simulation methods from first principles. Moreover, despite the recent advances in experimental and numerical methods in the study of sediment-laden flows, the mixing of contaminants with mobile river beds has so far been overlooked. There is thus a knowledge gap to be filled to promote the fundamental understanding of the abstracted particle-laden flow problem, as well as to assess the impact of the specific features of particle transport at the unresolved scales on the understanding of the spreading of bacteria as a whole.

In regard to advancing the understanding of bacteria transport, fully-resolved direct numerical simulations constitute a novel approach to the problem which describes the abstracted contamination scenario originating from the smallest relevant scales. In particular, the current approach isolates the subproblem of particle-bound bacteria transport, and formulates it based on fundamental physical principles from the bottom up under the avoidance of modelling assumptions until it contains all elements required to emulate the river contamination problem (settling particles, sediment bed, turbulence). Using this approach, the interaction between particles and turbulence can be described from first principles which gives the simulation results similar credibility to laboratory experiments with the advantage that the full state of the system is known at all times. Consequently, the simulated system provides an opportunity for inductive reasoning, and should be regarded as complementary to the traditional approaches to bacteria transport modelling with the aim of raising awareness on the relevance of typically unresolved phenomena.

In the following, the major findings of this thesis will be summarised from a hydrodynamic standpoint and in the context of bacteria transport. Subsequently, the thesis is concluded by a statement of the limitations of the current work and suggestions on aspects which require further research.

10.1 Summary of hydrodynamic findings

For this thesis, eight distinct parameter points have been investigated concerning the settling of particles from the free surface of turbulent horizontal open channel flows in the presence of a pre-existing sediment bed consisting of $\mathcal{O}(10^5)$ individual particles. The parameter space includes two distinct bed topologies (ripples, ridges), two values of the bulk Reynolds number ($Re_b = \{3009, 9483\}$), and a sweep in the contaminant parameters ($Ga = 10$–30, $\rho_p/\rho_f = 1.5$–2.5, $d_p^+ = 8$–18). The contaminant parameters have been selected to cover a wide range of relative turbulence intensities which is given by $I_\tau = 0.47$–2.88. The simulation method is based on the immersed boundary method in the formulation of Uhlmann (2005), and every relevant scale of turbulence, as well as the microscopic flow around every contaminant and sediment particle is fully resolved by the method. Similar setups have only been investigated experimentally (Cuthbertson and Ervine 2007) or numerically for unresolved particles and in absence of a sediment bed (Bragg et al. 2021a) before, and thus, the simulations of this thesis are first to simulate this problem from first principles, even if the context of bacteria transport is disregarded.

The instantaneous settling velocity of the contaminant particles was found to be enhanced on average in comparison to the terminal velocity of a corresponding isolated particle for most parameter points under investigation, which is in accordance with the experimental findings of Cuthbertson and Ervine (2007) over a similar parameter range. It was determined that the prevalent scaling parameter which describes this enhancement is the relative turbulence intensity, and that enhancement generally occurs when its value is of the order of unity or above. Here, the term "instantaneous settling velocity" can be interpreted in two ways, with the first one being the ensemble-average over all contaminant particles at a given instance in time—an indicator of how fast the system approaches its equilibrium state—and the second one being the ensemble-average over all contaminants at their respective time of crossing a given wall-parallel plane, which in general does not occur simultaneously. The latter is also denoted as the "spatial-average" within this thesis and provides a way to assess the role of wall-normal inhomogeneity in the channel. It was found that the excess in settling velocity is well described by the local standard deviation of the vertical fluid velocity once the initial acceleration phase of the contaminants has concluded.

The streamwise component of the contaminant was also found to be increased with respect to the local mean fluid velocity. A quadrant analysis of particle velocity fluctuations with respect to the fluid mean and the ambient terminal velocity suggested that positive fluctuations in u_c regularly go hand in hand with negative fluctuations in v_c. An evaluation of the fluid velocities in the immediate surrounding of the contaminants indicated that this behaviour is caused by the preferential sampling of turbulent sweep events—a phenomenon which has already been postulated by Cuthbertson and Ervine (2007) and Breugem and Uijttewaal (2007), but has now been confirmed by direct measurements of the fluid velocity.

In contrast to the vertical particle velocity, the enhancement in the streamwise component was found to be explicitly dependent on the details of the flow, i.e. it cannot be solely parametrised by the relative turbulence intensity, but rather also depends on the flow Reynolds number and the topology of the sediment bed. A detailed investigation on the cause of the enhancement revealed that contaminants tend to accumulate in large-scale high-speed streaks of the streamwise fluid velocity, which are associated to the LSMs and VLSMs in wall-bounded turbulence. It is known that the details of these coherent structures vary in the flow Reynolds number range under investigation, which explains the difference in scaling behaviour among cases with a different background flow. In comparison, the enhancement in v_c is predominantly driven by coherent fluctuations in v_f at significantly smaller scales, which is in accordance with the anisotropy in length scales between u_f' and v_f as is suggested by the premultiplied spectra of these quantities. In fact, a novel scale decomposition method based on the Gaussian band-pass filtering proposed by Motoori and Goto (2021) and Motoori et al. (2022) enabled a quantification of the contributions of fluid velocity fluctuations at a given scale to the velocity experienced by the contaminant particles. The decomposition indicated that structures with spatial dimensions of $(1.2, 0.4, 0.6)^T H_f$ to $(2.4, 0.8, 1.2)^T H_f$ in $(x, y, z)^T$ are most significant regarding the enhancement of the streamwise particle velocity, whereas for the vertical particle velocity the characteristic size range lies between $(0.3, 0.1, 0.15)^T H_f$ and $(0.6, 0.2, 0.3)^T H_f$. These results coincide approximately with the length scales which are the most active for the respective fluid velocity component in terms of their root-mean-square as obtained from the Fourier analysis.

It was hypothesised that the mechanism which transports contaminant particles into sweeps is related to the preferential sweeping mechanism of Maxey and Corrsin (1986). Indeed, shortly after the introduction of contaminants, this effect was demonstrated to be active and to immediately produce a significant spatial intermittency in the distribution of contaminants which are initially randomly seeded within a wall-parallel plane close to the free surface of the open channel. Regarding the arrangement into large-scale streaks, a determination of the primary mechanism which leads to the organisation is more intricate. It was conjectured that it is the result of the interactions of particles with the vortical structures which give rise to LSMs and VLSMs with prime candidates being vortex clusters (del Álamo et al. 2006) or packets of hairpin vortices (Kim and Adrian 1999).

The preferential sampling phenomenon also leads to the formation of particle clusters. The tendency of particles to accumulate increases with the relative turbulence intensity. Although it could be excluded that clusters are the cause of the enhanced settling velocity, they were determined to play a major role for the avoidance of a collective hindrance effect similar to the effect described by Richardson and Zaki (1954). By investigating non-turbulent proxy simulations of collective settling, it was found that this effect may substantially reduce the ensemble-averaged settling velocity if particles cannot escape the horizontal alignment imposed by their initial arrangement, and that a formation of clusters reduces its relevance due to the reduction in the cross-sectional area blocked by the ensemble. By investigating the conditionally averaged particle concentration in the vicinity of the settling contaminants, it was detected that hindrance is indeed a factor which needs to be considered during turbulent settling when the relative turbulence intensity is insufficient to efficiently break up the layer arrangement. In fact, the parameter points for which this phenomenon occurs are the only ones which exhibit a reduction in settling velocity and a perceptibly altered temporal and spatial evolution thereof, which both can be attributed to this collective effect.

While clusters are relevant in the period immediately after the settling commences, the contaminants tend to disperse in the vertical direction as the transient progresses such that their instantaneous distance grows with time. However, pattern formation can still be detected when the preferential concentration of crossing locations in wall-parallel planes is examined in a suitable frame of reference. A

Fourier analysis of the preferential concentration thereby reveals that the successive organisation of particles into large-scale streaks is not primarily controlled by the inertia of particles, but is rather driven by the spatial inhomogeneity of turbulence as these streaks only become dominant in the lower half of the channel. Nonetheless, a lag in the evolution of the dominant mode can be perceived which suggests a finite response time of the contaminants to changes in the flow environment.

The wall-normal inhomogeneity of wall-bounded turbulence also affects the averaged wake of the settling particles. For the current parameter points a mean wake which resembles the respective undisturbed state is only observed for low relative turbulence intensities. For a set of particle parameters, the mean wake changes substantially depending on the vertical position which could be ascribed to the spatial variation in $\langle v_f^2 \rangle_{xzt}$. In fact, when the length of the wake is quantified and evaluated as a function of a local relative turbulence intensity based on the root-mean-square fluid velocity in vertical direction, a clear trend can be observed which quantitatively matches the data reported by Homann et al. (2013) for a sphere towed through isotropic turbulence at a similar particle Reynolds number. The downstream length of the wake was found to saturate at a value of $\mathcal{O}(d_p)$ when the relative turbulence intensity exceeds unity. Although the flow in the channel is sheared on average, the effect of shear on the wake was deemed to be negligible in the bulk of the channel. At vertical positions closer to the sediment bed, the mean wake exhibits signs of distortion. A comparison with a particle settling in laminar Couette flow under comparable conditions to the turbulent mean suggested that although the distortion is considerable, the settling velocity is only altered marginally. Regarding the streamwise particle velocity, the rapid decrease in the mean streamwise fluid velocity encountered by the contaminants manifests itself in strong positive values for the difference between u_c and $\langle u_f \rangle_{xzt}$ due to the fact that particles sustain part of their streamwise momentum from higher positions. In fact, this phenomenon becomes the dominant cause for the enhancement in streamwise particle momentum near the wall as preferential sampling loses its relevance in the same region.

Regarding the deposition of contaminants, it was found that the time until deposition is log-normally distributed. The mean of the distribution is increased compared to the naïve estimate given by the ratio of the clear fluid height and the ambient terminal velocity. This implies that contaminants settle slower on average over the full settling distance despite moving at significantly faster speeds compared to the ambient terminal velocity instantaneously, which is a result of the fact that at high turbulence intensities particles also have an increased tendency to move upwards. The concept of deposition time can be generalised by evaluating the distribution of crossing times through any wall-parallel plane. In this case, a generalised Gamma distribution was found to provide the best fit, although a log-normal distribution is likely to be the preferred choice at least close to the sediment bed due to the principle of parsimony. This generalisation enables the definition of a velocity which quantifies the propagation speed of the ensemble of contaminants denoted as the group velocity, and indicates that the ensembles propagates to the centre of the open channel at a rate faster than the terminal velocity at sufficiently high relative turbulence intensities, but significantly slows down shortly before deposition. Similar to the deposition times, the streamwise distance covered before deposition also follows a log-normal distribution, as was previously suggested by Li and Shen (1975) based on a simple random-walk settling model. Interestingly, the variance of the settling length obtained from the DNS data follows an empirical fit based on experimental data provided by the same authors remarkably well, which suggests that the mechanisms responsible for the distribution are rather robust.

The spanwise displacement follows a Gaussian distribution around a zero mean at virtually all wall-normal positions. Its variance increases initially but saturates in the lower half of the channel, which implies that the distribution remains constant past a certain vertical position and no net dispersion occurs afterwards. This intriguing behaviour was linked to the accumulation of contaminants in the large-scale streaks which essentially determines the ultimate lateral position before deposition occurs. In the near-bed region contaminants are found to disperse laterally again, but the rate of dispersion is significantly slower than during the settling presumably due to frequent interparticle interactions. A dispersion coefficient was derived from the approximately linear increase of the mean-squared displacement in time during this phase, and its value is hypothesised to scale with the value of the Shields parameter, albeit too few data points are available to obtain conclusive evidence.

The interaction between contaminants and the sediment bed heavily depends on the topology of the bed. In case of a macroscopically flat bed, contaminants are transported onto the pre-existing sediment ridges by the footprints of the large-scale turbulent structures such that they position themselves below a large-scale low-speed streak (Scherer et al. 2022). Despite the fact that most of this transport occurs on a thin interface layer of particles, the contaminants may become buried by other particles such that they are immobilised. The efficiency of this process was found to be increased for smaller particles. In addition, the combined outcome of this partial immobilisation and the cyclic bursting and reformation or meandering of streaks appears to lead to a randomisation of the horizontal positions of the contaminants despite their tendency to accumulate in certain regions. When ripple features are considered, the contaminants become trapped in the trough of the ripple where they are subsequently buried beneath a thick sediment layer due to the propagation of the feature. This immobilises the contaminants until the ripple has been displaced far enough downstream that they may be eroded again on the windward side. Immediately after the initial settling, nearly all contaminants are buried at the same time such that the number of suspended contaminants is severely depleted. The ensuing erosion releases numerous contaminants at similar times which manifests itself in a burst in suspended contaminants. This phenomenon repeats itself cyclically, however, due to the increasing spread in the times of immobilisation, the oscillations in contaminant position and streamwise particle flux become attenuated in the long term.

Finally, motivated by the pursuit of a system of minimal complexity which can be utilised to study the particle-vortex interactions during the settling phase of contaminants, the settling of particles in a travelling wave solution to the Navier–Stokes equations was investigated. This travelling wave belongs to the group of exact coherent states which are non-trivial flow states which derive from the dynamical systems approach to turbulence and constitute attractors or saddles in the phase space. These exact coherent states have been demonstrated to be of relevance for turbulent flows, and are hypothesised to be blueprints for the coherent structures observed in fully turbulent flows. The particular travelling wave which is investigated in this thesis exhibits topological resemblance to a hairpin vortex (Shekar and Graham 2018), and indeed, the choice of this solution was partially motivated by the presumed relevance of hairpin vortices for LSMs. Moreover, the wall-normal evolution of the flow statistics of this flow are in good agreement with that obtained for turbulent flows, especially with regard to the vertical velocity fluctuations, and despite the fact that the travelling wave is time-invariant in a suitable frame of reference. When the ensemble-averaged settling velocity is assessed for contaminants settling in this simplified system, an extraordinary agreement to the turbulent evolution is obtained if the relative turbulence intensity and the density ratio are matched. An advantage of this novel approach is that the computational cost is significantly reduced such that a finely meshed parameter sweep can be performed. From this parameter sweep it was deduced that for $I_\tau \lessapprox 0.25$ the ensemble-averaged settling velocity is reduced due to the loitering effect where particles are delayed because of their inability to avoid counteracting flows. Meanwhile, when $I_\tau \gg 1$, the excess in settling velocity due to preferential sweeping converges towards a constant which is of $\mathcal{O}(u_\tau)$ and whose exact value depends on the density ratio. This implies that the settling velocity normalised by its ambient value varies linearly with I_τ at sufficiently high relative turbulence intensities. It was moreover ascertained that the mean-squared lateral displacement indeed levels off at high relative turbulence intensities, which implies that particles seldom cross multiple roller structures during the initial settling phase, as has been hypothesised for the turbulent scenario.

10.2 Summary of implications on bacteria transport

As has been stated in the research objectives in the introduction to this thesis, the goals of this study in the context of bacteria transport were multifarious and can, broadly speaking, be summarised by 1) the examination of quantities which directly enter existing models, 2) the detailed analysis of phenomena which are known to be relevant, and 3) the detection of emergent phenomena which have not been considered in this context hitherto.

Concerning the first point on the list, the evaluation of the deposition time and the settling length are two major factors which could directly be related to the popular one-dimensional stream model by Jamieson et al. (2005). A comparison between the DNS data and the model revealed that the streamwise contaminant flux is generally underpredicted by the model, whereas the rate of dispersion and deposition are overpredicted. Moreover, the nature of the deposition term was shown to result in a wrong distribution of deposition times, and practical modifications have been proposed to improve the simple model with only a marginal increase in complexity.

Regarding the second point, the major phenomenon to be discussed is the interaction between contaminant particles and the sediment bed. One of the justifications for the (almost) exclusive consideration of particle-attached bacteria in this thesis was the fact that solid matter provides a beneficial environment for bacteria to survive even under harsh conditions, which increases the relevance of particle-bound bacteria from the long-term perspective on the spreading. A significant factor of influence for the fate of bacteria is the exposure to ultraviolet irradiance. While the opaque nature of particles or their surface characteristics may themselves be advantageous, a more dynamical aspect to consider is the shade which is provided by other suspended or deposited particles. The direct numerical simulations provided evidence that contaminated particles may be efficiently removed from suspension by being buried underneath a layer of sediment. This entrainment is especially pronounced in the presence of ripple features due to the capturing mechanism previously described. A simple quantification method regarding the exposure to sunlight estimated that in case of ripples, the surface area of contaminant particles which is exposed to sunlight is drastically reduced to around 30% of the value observed at the beginning of the transient in the long-term. Even in the absence of bed features, shading was detected to be a significant factor since contaminants still exhibit a tendency to become buried by other particles, especially if they are small. Nonetheless, for macroscopically flat beds the exposure factor was found to be closer to 60%, albeit it might further reduce for longer observation intervals. The immobilisation of contaminants by river sediment and their simultaneous shielding from inactivation by sunlight may be speculated to be related to the occurrence of bacteria storages in river beds, which are thought to be responsible for surges in the concentration of enteric bacteria even in absence of immediate contamination events (McDonald et al. 1982, Wilkinson et al. 1995). If this is indeed the case, the data obtained within this thesis might be helpful to develop a better understanding of this process.

Shading effects were also found to play a role during the initial settling of contaminants due to their tendency to form clusters, albeit in this case the effect is less pronounced compared to the deposition-induced shading. Nonetheless, this phenomenon has the interesting implication that turbidity effects are not merely a function of the concentration of suspended solids, but rather also dependent on the relative turbulence intensity as this quantity is the primary parameter which governs the spatial inhomogeneity of the particle distribution which in turn manipulates the effective turbidity locally.

To address the third point, one of the striking recognitions of this thesis is that there appears to be a fundamental distinction in the ability of particle-attached bacteria to spread in lateral direction between the initial settling phase and the phase of long-term transport in the bedload or by resuspension. The tendency of settling particles to accumulate in large-scale flow features with a spacing of 1–2 H_f during the settling limits the mean-squared spanwise displacement to values which are smaller than this distance. At the same time, the initial settling phase exhibits a streamwise contaminant flux which is considerably larger than that in the near-bed region due to the exposure of contaminants to faster streamwise fluid velocities. The consequence is that a contamination which is introduced at the free surface near the river bank, such as in the case of combined sewer overflow events, may be transported along the shoreline over a considerable downstream distance with limited dispersion in the lateral direction. In particular, there is an argument to be made that the lateral confinement becomes more significant the smaller the contaminant particles are, or more precisely, the lower the corresponding value of the Galileo number is for a given background flow. This is due to the reason that the upper-bound observed for the mean-squared lateral displacement is hypothesised to be independent of the relative turbulence intensity, whereas the settling length is not. Nevertheless, it should be acknowledged that these alleged transport barriers are only active during the initial settling phase as they are a result of the symmetry breaking effect of gravity. Regarding the spanwise dispersion in the near-bed region, it was found that the lateral dispersion coefficient scales with the value of the Shields number of the contaminant particles, and is in particular not subject to any confinement effect.

Another point which is to the best of the author's knowledge not yet fully addressed in bacteria transport models is the effect of contaminant particles on the spreading of freely-suspended bacteria. The interaction between the two modes of transport was determined to be twofold: contaminant particles may act as a source of freely-suspended bacteria if the attachment is not irreversible, but also affect the transport characteristics by modifying the fluid velocity field in the irreversibly-attached case. If a simple detachment model is considered based on the concentration gradient of bacteria between the surface of the particle and the fluid phase, a high detachment rate can be expected even if the contaminant particles are initially in equilibrium with their surrounding, since the settling process tends to separate the two phases which leads to steeper gradients. However, this effect becomes less relevant as the relative turbulent velocity increases. The particle-induced modification of the velocity field increases the vertical spreading rate of freely-suspended bacteria as compared to the reference case in absence of contaminant particles. Here, the significance of the effect was not found to be strongly dependent on the particle parameter point for the cases considered.

10.3 Limitations of the present work

Naturally, the approach employed in this thesis suffers from several constraints, the most significant of which will be summarised in the following. To begin with, the use of direct numerical simulations and the aspiration to simulate the smallest scales of the contamination problem from first principles comes at the cost of limitations for the largest scales which can be included in the simulations due to the immense computational effort and the availability of computing resources at the time. However, the separation of scales is not unimportant for the current work as it restricts the friction Reynolds number (which is considerably smaller than the parameter point found in actual rivers), and arguably more importantly, the relative size of contaminant particles in comparison to the water depth. While the ratio

H_f/d_p may easily reach values beyond $\mathcal{O}(10^4)$ when particles with diameters in the upper micrometre range and water depths of the order of metres are considered, a maximum value of $H_f/d_p \approx 70$ is studied in this thesis. While the inferences can be argued to be relatively robust with respect to this ratio—at least for the range considered in this thesis and corroborated by the good agreement for the settling in the travelling wave solution where this ratio is minimal—it may be worthwhile to test this assumption or the inferences made on large fluvial systems using a different approach.

Moreover, the simulation domain can also be considered small in the horizontal directions even for DNS standards. This shortcoming was mainly owed to the desire of studying the direct influence of sediment bed topologies by suppressing the ripple bedform instability using a truncated domain, but is also a result of the vast computational costs of the method. However, as has been discussed in the main text, the short streamwise extent also truncates the size of LSMs and VLSMs which were in hindsight found to be important phenomena for the current treatise. Hence, the consequences of these structures exhibiting a finite length stay unknown and uncertainties remain concerning the accurate description of their dynamics.

While the original plan for this thesis included the incorporation of bacteria fate and a more tightly coupled analysis of the interaction between the particle-attached and freely-suspended bacteria transport modes, this ambition was discarded in favour of a more universal setup of the numerical simulations. One of the major motives for this decision was the lack of availability of well-founded models for the detachment behaviour of bacteria as well as inactivation models which take inhomogeneities at the scales considered into account. At the same time a more detailed analysis would eventually need to consider initial temperature differences between the discharged effluent and the river (Pinelli 2022) which leads to scalar-induced buoyancy effects which in turn might alter the transport characteristics of embedded particles. Due to the knowledge gap on the settling problem even in absence of this effect, the active coupling between scalar fields and the fluid velocity field was rejected in order to increase the relevancy of this work for the hydrodynamical literature.

A ubiquitous limitation on any research which is often not explicitly addressed is the limitation in time that a researcher can allocate to post-process the data. In regard to this thesis, the interaction between particles and scalar fields ought to be more thoroughly investigated then what was finally included in the main text. In fact, an appreciable amount of research time was invested to set up this problem and to manage the excessive volume of data generated by the dual mesh approach. Still, the majority of this data is yet to be assessed, and hopefully, further inferences can be made from its empirical content in the future.

10.4 Recommendations for future research

To conclude this thesis, some recommendations for future research will be suggested which go beyond the improvement of the limitations discussed in the preceding section.

As has been elaborated numerous times at this point, the conjectured lateral transport barriers might exhibit a relevant macroscopic effect for which the awareness in the bacteria transport community is still low. However, due to the limitations in the parameter points which were able to be investigated, the upscaling towards large fluvial systems remains speculative. For this reason, purposeful laboratory experiments should be conducted to look for empirical evidence of this phenomenon. These experiments need to be conducted at relative turbulence intensities significantly above unity and in a channel with a streamwise length which is sufficient for the particles to settle into the near-wall region. Moreover, it would be beneficial to be able to measure the spanwise distribution at any given vertical position since at high relative turbulence intensities particles may be displaced considerably in lateral direction immediately before deposition. While the possibility of simultaneously measuring the fluid flow would be beneficial to correlate the observations to LSMs and VLSMs, it does not appear to be a requirement for a first assessment of the transport barrier effect. In order to reduce the uncertainty in the instantaneous lateral position of the large-scale streak it might furthermore be helpful to design the bottom wall of the channel such that the position of the mean secondary flow becomes locked in place (e.g. Stroh et al. 2020).

If empirical evidence for the transport barriers can be found, it would be interesting to examine the role of secondary effects, such as the finite lifetime or meandering motion of these structures. Such an analysis could again be conducted in laboratory experiments, but may also be tackled by simpler simulation methods. In particular, it seems worthwhile to assess whether this effect can be reproduced using unresolved particles in order to benefit from the use of spectral flow solvers. If this is the case, an adequate parameter space appears to be accessible at a reasonable computational effort.

In regard to lateral dispersion in the near-bed region, a detailed study stating dispersion coefficients could not be identified apart from the study of Seizilles et al. (2014) which, however, considers the laminar flow regime within which the dispersion mechanism is likely to be dominated by particle collisions rather than by the interaction with flow structures. A larger number of parameter points would be helpful to draw conclusions on the scaling law, and to possibly confirm the Shields number scaling postulated in this thesis. For this purpose, it may be sufficient to analyse existing DNS data for sediment transport in terms of the displacement of the interface particles to assess the displacement in the monodisperse scenario. However, an investigation of bidisperse particle size distributions is also highly relevant, and may contribute to the knowledge on the insertion of contaminants into deeper layers of the bed by the burying mechanisms described within the current work. Furthermore, the details of the contaminant-sediment interaction are likely influenced by further influencing factors which have not been considered so far, such as cohesive effects or non-spherical particle shapes. An elaborate study on this topic seems to be justified by the fact that this interaction is also relevant in the more general context of pollutants, for instance regarding the deposition of microplastics and its long-term consequences.

Returning to the original overarching research question of bacteria transport in river systems, it appears to be required to fill the gap in intermediate length scales between traditional models and the current approach by a method which accurately resolves turbulent structures of the order of the fluid depth, but is also capable of depicting the entire cross-section of a river and to account for particle transport. A potential candidate for a simulation technique which is suitable for this investigation is the large eddy simulation coupled to a transport model for discrete particles (e.g. Elghannay and Tafti 2018). Yet, this method comes at the cost of the need to model the hydrodynamic forces on the particles, and it remains to be demonstrated that this does not lead to inadequate simplifications in the current context. The availability of data for the spreading on intermediate scales benefits the transfer of knowledge from the observations made using DNS to the coarse-grained traditional models.

From the hydrodynamical point of view, an in-depth analysis of the interactions which transport particles into high-speed streaks during the settling would benefit the understanding of this phenomenon. Such an investigation would likely require a suitable criterion to identify the coherent structures with which the contaminants interact, and conditioned statistics of said interactions. The analysis is complicated by the facts that the settling is transient, that the structure of turbulence is inhomogeneous in wall-normal direction, and that the interactions presumably occur on multiple scales and may even be non-local.

Speaking of non-local interactions, it was demonstrated in this thesis that exact coherent structures are an excellent tool for the investigation of particle-vortex interactions. An interesting question which could be tackled directly with the available dataset is the role of non-locality in these interactions and that of the role of particle response time. This examination is considerably simpler for the travelling wave compared to fully turbulent flows due to the lack of temporal dynamics in the underlying flow state and the complete knowledge of the undisturbed flow field even at the positions were a particle is currently located. The investigation of particle dynamics using exact coherent states is still in its infancy, and it is exciting to see what contributions this novel approach will bring to the field of multiphase flows.

A Spatial-averaging procedure

The time-dependent ensemble average can be expressed using the Dirac delta function as

$$\langle\varphi\rangle_{\mathbb{T}}(t) := \frac{\sum_i^{N_{c\Sigma}} \int_0^\infty \varphi^{(i)}(\tilde{t})\delta(\tilde{t}-t)\,\mathrm{d}\tilde{t}}{\sum_i^{N_{c\Sigma}} \int_0^\infty \delta(\tilde{t}-t)\,\mathrm{d}\tilde{t}} \tag{A.1}$$

where $\delta(t)$ is the Dirac delta function. It is important to note that the equation is dimensionally consistent due to the scaling property $\delta(ax) = \delta(x)/|a|$, and thus, the dimension of $\delta(t)$ is one over time. Using the definition of the Dirac delta function, the integral in the denominator evaluates to

$$\int_0^\infty \delta(\tilde{t}-t)\,\mathrm{d}\tilde{t} = 1 \tag{A.2}$$

for $t \in [0,\infty)$. Furthermore, using the mean value theorem, the integral in the numerator is equivalent to

$$\int_0^\infty \varphi^{(i)}(\tilde{t})\delta(\tilde{t}-t)\,\mathrm{d}\tilde{t} = \varphi^{(i)}(t), \tag{A.3}$$

and hence, the simpler definition of the ensemble average is recovered, viz.

$$\langle\varphi\rangle_{\mathbb{T}}(t) = \frac{1}{N_{c\Sigma}} \sum_i^{N_{c\Sigma}} \varphi^{(i)}(t). \tag{A.4}$$

From this, a discrete-in-time formula can readily be derived.

Similarly, the spatial ensemble average defined in eq. (4.5) can be expressed in terms of the Dirac delta function as

$$\langle\varphi\rangle_{\mathbb{X}\uparrow\downarrow}(y) := \frac{\sum_i^{N_{c\Sigma}} \int_0^\infty \varphi^{(i)}(\tilde{t})\,\delta\left(\frac{y_p^{(i)}(\tilde{t})-y}{v_p^{(i)}(\tilde{t})}\right)\mathrm{d}\tilde{t}}{\sum_i^{N_{c\Sigma}} \int_0^\infty \delta\left(\frac{y_p^{(i)}(\tilde{t})-y}{v_p^{(i)}(\tilde{t})}\right)\mathrm{d}\tilde{t}}, \tag{A.5}$$

given an infinite observation time.[1] Here, the argument of the Dirac delta function has been scaled such that the temporal integration produces a unit time interval. Since the scaling is only relevant at $y_p = y$, the scaling property of the Dirac delta function can be exploited without loss of generality, i.e.

$$\langle\varphi\rangle_{\mathbb{X}\uparrow\downarrow}(y) = \frac{\sum_i^{N_{c\Sigma}} \int_0^\infty \varphi^{(i)}(\tilde{t})\,|v_p^{(i)}(\tilde{t})|\,\delta\big(y_p^{(i)}(\tilde{t})-y\big)\,\mathrm{d}\tilde{t}}{\sum_i^{N_{c\Sigma}} \int_0^\infty |v_p^{(i)}(\tilde{t})|\,\delta\big(y_p^{(i)}(\tilde{t})-y\big)\,\mathrm{d}\tilde{t}}. \tag{A.6}$$

Possible methods of evaluating the ensemble average are presented in the following.

[1] Strictly speaking, the definition given in eq. (4.5) restricts itself to the first crossing of a given particle, whereas eq. A.5 considers all crossings. However, it is straightforward to impose a similar restriction by adjusting the upper bound of the integral.

A.1 Trajectory interpolation

In general, the function $y_p(t)$ is not invertible. However, that set of times where the (i)-th particle crosses the plane located at y is given by

$$T^{(i)} := \{t \in \mathbb{R} : y_p^{(i)}(t) = y\} \tag{A.7}$$

and the members of it are the cross-through times of the (i)-th particle. Using this set, eq. (A.6) can be rewritten as

$$\langle \varphi \rangle_{\mathbb{X}\uparrow\downarrow}(y) = \frac{\sum_i^{N_{c\Sigma}} \sum_{t_{c,\mathbb{X}} \in T^{(i)}} \varphi^{(i)}(t_{c,\mathbb{X}})}{\sum_i^{N_{c\Sigma}} |T^{(i)}|}, \tag{A.8}$$

where $|T^{(i)}|$ is the cardinality of the set, i.e. the number of crossings of the (i)-th particle. The cross-through times can be determined from the zero-crossings of the discrete trajectory data. With the knowledge of $T^{(i)}$, the value of $\varphi^{(i)}(t_{c,\mathbb{X}})$ is readily determined using linear interpolation.

A.2 Binned approximation

When the time resolution of the discrete data is insufficient for trajectory interpolation—as is the case for the more coarsely sampled fluid data in this thesis—equation A.6 can be approximated by a weighted-binning approach. For this purpose, the Dirac delta function is approximated using a rectangle function, viz.

$$\delta(y_p^{(i)}(t) - y) \approx \frac{H\big(y_p^{(i)}(t) - y + \frac{\delta y}{2}\big) - H\big(y_p^{(i)}(t) - y - \frac{\delta y}{2}\big)}{\delta y} =: \frac{\Delta_{\mathbb{X}\uparrow\downarrow}^{(i)}(y)}{\delta y}, \tag{A.9}$$

where $H(y)$ is the Heaviside step function and $\Delta_{\mathbb{X}\uparrow\downarrow}(y)$ is the binning function defined as

$$\Delta_{\mathbb{X}\uparrow\downarrow}(y) := \begin{cases} 1 & \text{if} \quad y_c^{(i)} \in \big[y - \frac{\delta y}{2}, y + \frac{\delta y}{2}\big) \\ 0 & \text{otherwise} \end{cases}. \tag{A.10}$$

Here, the parameter δy denotes the width of the bin. Substitution of the Dirac delta function by this approximation in eq. (A.6) yields

$$\langle \varphi \rangle_{\mathbb{X}\uparrow\downarrow}(y) \approx \frac{\sum_i^{N_{c\Sigma}} \int_0^\infty \Delta_{\mathbb{X}\uparrow\downarrow}^{(i)}(y)\, |v_p^{(i)}(\tilde{t})|\, \varphi^{(i)}(\tilde{t})\, \mathrm{d}\tilde{t}}{\sum_i^{N_{c\Sigma}} \int_0^\infty \Delta_{\mathbb{X}\uparrow\downarrow}^{(i)}(y)\, |v_p^{(i)}(\tilde{t})|\, \mathrm{d}\tilde{t}}. \tag{A.11}$$

A variation of this formulation is given in (5.3) and is used to approximate the spatial average for fluid data where the time resolution is significantly coarser than for the particle trajectories.

B Gamma distribution fit of cross-through times

The three-parameter generalised Gamma distribution for the probability density is given by

$$\mathrm{pdf}\left(t_{c,\mathbb{X}}\right) := \frac{c b^{a/c}}{\Gamma\left(\frac{a}{c}\right)} t_{c,\mathbb{X}}^{a-1} \exp\left(-b t_{c,\mathbb{X}}^{c}\right), \tag{B.1}$$

where $a, b, c > 0$ are the three parameters to be fitted to the data and

$$\Gamma(s) = \int_0^\infty t^{s-1} e^{-t} \,\mathrm{d}t \tag{B.2}$$

is the gamma function. Eq. (B.1) is formulated such that normalization properties of a probability density function are fulfilled automatically, i.e. $\int_0^\infty \mathrm{pdf}\left(t_{c,\mathbb{X}}\right) \mathrm{d}t_{c,\mathbb{X}} = 1$ for any valid set of parameters.

The free parameters are determined from an empirical PDF with a bin size of approximately $0.4 t_b$ by finding a minimum of the loss function

$$R_j(a,b,c) = \sum_i \left(\mathrm{pdf}_j(\tau_i) - \mathrm{cdf}_j(t_{obs})\,\mathrm{epdf}_j(\tau_i)\right)^2 \Delta\tau, \tag{B.3}$$

where τ_i is the center of the i-th sampling bin, $\Delta\tau$ the (constant) width of the bins and the subscript j denotes evaluation at the discrete vertical position y_j of the crossing plane. Note that the prefactor $\mathrm{cdf}_j(t_{obs})$ to the empirical data depends on the free parameters and provides a progressive renormalization such that

$$\sum_i \mathrm{cdf}(t_{obs})\,\mathrm{epdf}(\tau_i)\,\Delta\tau \approx \mathrm{cdf}(t_{obs}) \tag{B.4}$$

after optimization. The prefactor corresponds to the value of the cumulative distribution function evaluated at the observation time. The distribution has a closed form and can be evaluated from the current values of the free parameters by

$$\mathrm{cdf}\left(\tau\right) = \frac{\gamma\left(\frac{a}{c}, b\tau^c\right)}{\Gamma\left(\frac{a}{c}\right)} \tag{B.5}$$

where

$$\gamma(s,x) = \int_0^x t^{s-1} e^{-t} \,\mathrm{d}t \tag{B.6}$$

is the lower incomplete gamma function.

For the given empirical data, multiple local minima of R_j exist, and thus, without a proper choice for an initial guess for a, b, c, the resulting coefficients and renormalization constants fluctuate substantially with wallnormal position. In order to reduce noise, initial guesses are generated based on the coefficients obtained for a neighboring crossing plane. For this, a regularised version of the loss function is minimised,

$$R_j^*(a,b,c) = \sum_i \left(\mathrm{pdf}(\tau_i) - \mathrm{epdf}(\tau_i)\,\mathrm{cdf}(t_{obs})\right)^2 \Delta\tau$$
$$+\theta\left(\left(\frac{a_j}{a_{j-1}}-1\right)^2+\left(\frac{b_j}{b_{j-1}}-1\right)^2+\left(\frac{c_j}{c_{j-1}}-1\right)^2\right), \tag{B.7}$$

where large deviations of a_j, b_j or c_j at the current position y_j from its corresponding values at the previously evaluated position y_{j-1} are penalised. A value of $\theta = 1$ has been found to provide satisfactory results.

The full algorithm is summarised in the following:

0. Using an arbitrary initial guess, fit a_j, b_j, c_j for the first crossing plane to be processed by minimizing $R_j(a,b,c)$.
1. Proceed to the next plane.
2. Using $a_{j-1}, b_{j-1}, c_{j-1}$ as starting values, minimise $R_j^*(a,b,c)$ to obtain a new initial guess.
3. Minimise $R_j(a,b,c)$ to obtain a_j, b_j, c_j.
4. If this was the last plane, stop. Otherwise repeat from step 1.

It is more suitable to commence the algorithm from a plane close to the plane of initialization of the contaminants rather than from the bottom of the domain. This is because there it is reasonable to assume that the transient is fully captured, and thus, the constraint $\mathrm{cdf}_j(t_{obs}) = 1$ may be enforced. Consequently, the algorithm has been employed from top to bottom in this work.

This simple algorithm yielded satisfactory results for all cases and vertical positions with the notable exceptions of G30-D10-LR for $y > 0.8H_f$ where large discontinuities for the fit coefficients were obtained presumably due to the large width of the sampling bins and narrow peak of the Gamma distribution, and G10-D10-HR fot $y < 0.1H_f$ where difficulties in determining the rescaling factor occured presumably due to the significant truncation of the empirical data. The obtained parameters of the generalised Gamma distribution as well as the residuals are summarised in fig. B.1. A comparison of the empirical PDF and Gamma distributions for various cases and heights is shown in fig. B.2

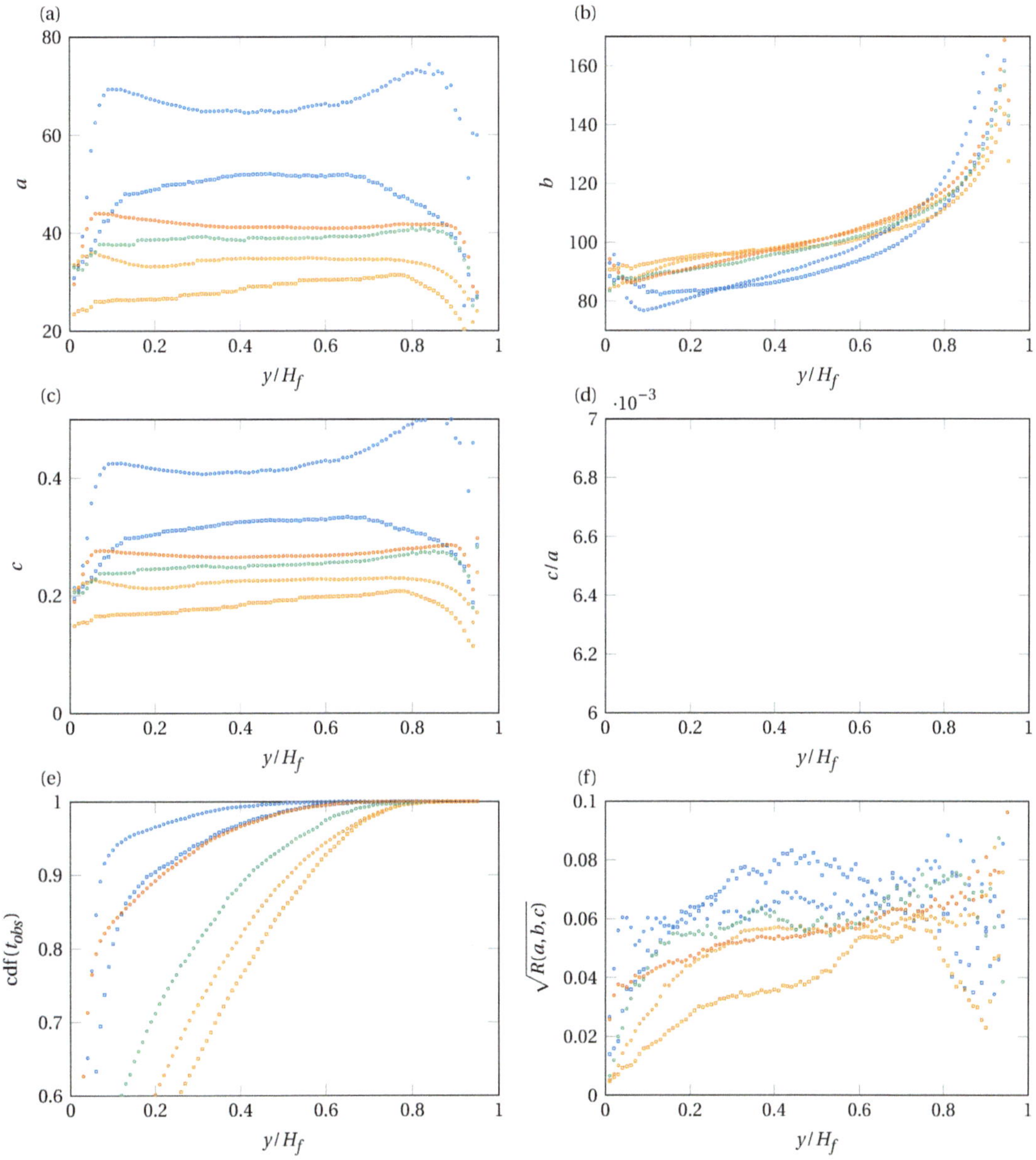

Figure B.1: Summary of the parameters of the generalised Gamma distribution obtained by minimizing eq. (B.3). (a-c) The set of parameters obtained for each case and crossing plane, (d) the ratio c/a, (e) the rescaling factor and (f) the residual, i.e. value of the loss function for the resulting parameters. Legend: G30-D15-HR (□), G10-D10-HR (□), G30-D15-LR (○), G10-D6-LR (○), G15-D10-LR (○) and G15-D6-LR (○).

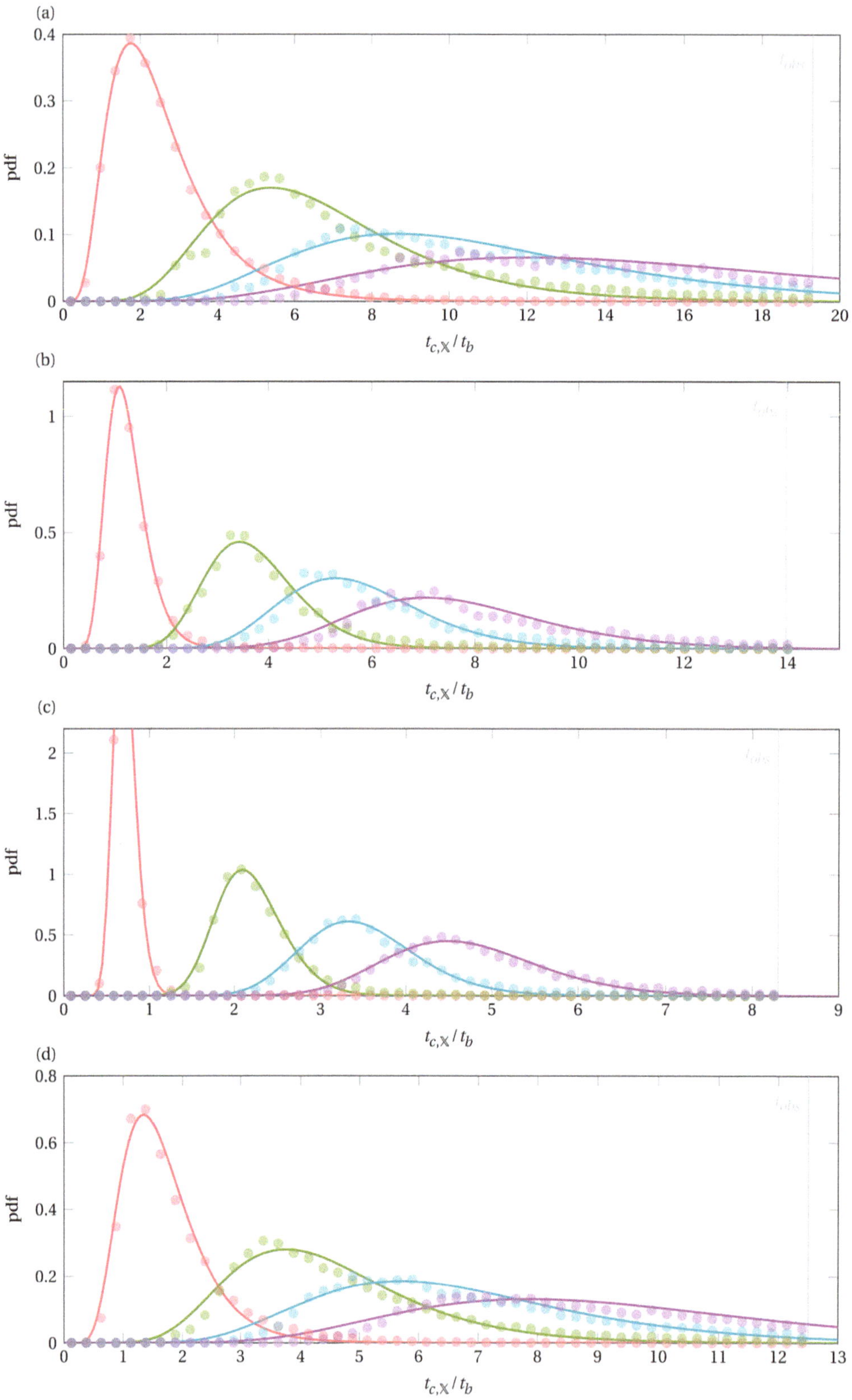

Figure B.2: Generalised gamma distributions fitted to the rescaled empirical data for the cases (a) G10-D10-HR, (b) G30-D15-HR, (c) G10-D6-LR, (d) G30-D10-LR. The data corresponds to the vertical positions $y = 0.9H_f$ (—), $y = 0.7H_f$ (—), $y = 0.5H_f$ (—) and $y = 0.3H_f$ (—).

C Gaussian filtering

C.1 Low-pass filter

The one-dimensional Gaussian filter kernel is commonly defined as

$$K_\sigma^L = \frac{1}{\sqrt{2\pi}\sigma} \exp\left(-\frac{x^2}{2\sigma^2}\right), \quad \sigma > 0 \tag{C.1}$$

Applying this filter to a function $\phi(x)$ in the following way

$$\mathcal{L}\{\phi\}(x) = K_\sigma^L(x) * \phi(x) := \int_{-\infty}^{\infty} K_\sigma^L(x - x') \cdot \phi(x') \, \mathrm{d}x' \tag{C.2}$$

Applying the Fourier transformation and utilizing the convolution theorem yields

$$\mathcal{L}\{\phi\}(k) = \hat{K}_\sigma^L(k) \cdot \phi(k) \tag{C.3}$$

where

$$\hat{K}_\sigma^L(k) := \mathcal{F}\left[K_\sigma^L(x)\right], \tag{C.4}$$

$$\hat{\phi}(k) := \mathcal{F}\left[\phi(x)\right] \tag{C.5}$$

are the Fourier transforms of the filter kernel and the function to be filtered, respectively, which is here defined using the transform pair

$$g(k) = \mathcal{F}\left[f(x)\right] := \int_{-\infty}^{\infty} f(x) \exp(-ikx) \, \mathrm{d}x, \tag{C.6}$$

$$f(x) = \mathcal{F}^{-1}\left[g(k)\right] := \frac{1}{2\pi} \int_{-\infty}^{\infty} g(k) \exp(ikx) \, \mathrm{d}k, \tag{C.7}$$

$$\tag{C.8}$$

and the angular wavenumber $k = 2\pi/\lambda$ with λ being the wavelength.

The Fourier transform of the filter kernel is again a Gaussian in spectral space and reads

$$\hat{K}_\sigma^L(k) = \exp\left(-\frac{\sigma^2 k^2}{2}\right). \tag{C.9}$$

In this work, the characteristic wavenumber of the filter is defined as the wavenumber at which the power spectrum is attenuated by half, i.e.

$$\mathcal{L}\{\phi\}(k_c)/\phi(k_c) = K_{\sigma,c}^L \triangleq 1/\sqrt{2}, \tag{C.10}$$

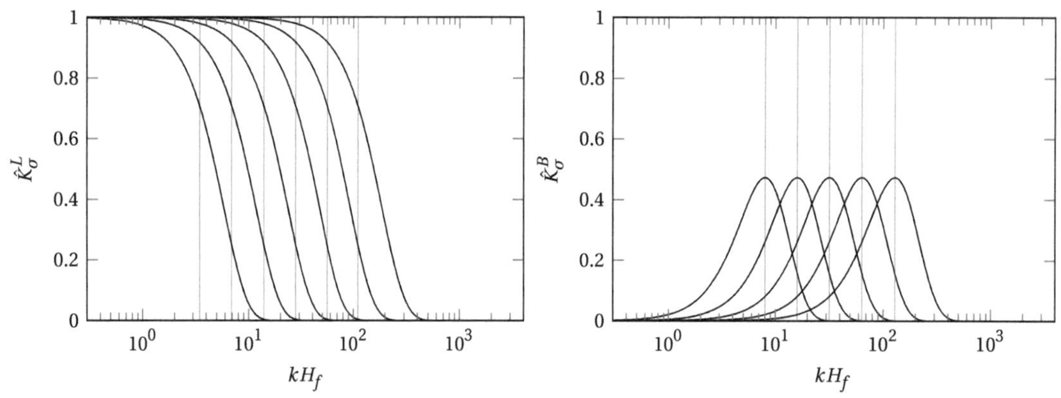

Figure C.1: Spectral support of the low-pass filters (left) and the derived band-pass filters (right).

and thus

$$k_c = \sqrt{\frac{2\ln\sqrt{2}}{\sigma^2}} \approx 0.833/\sigma. \tag{C.11}$$

C.2 Band-pass filter

It was suggested by Motoori and Goto (2021) to educe turbulent structures of a given size range using a Gaussian bandpass filter constructed by the difference of two Gaussian lowpass filter, i.e.

$$\mathcal{B}\{\phi\}(x) := K^B_\sigma(x) * \phi(x) = (K^L_\sigma(x) - K^L_{2\sigma}(x)) * \phi(x), \tag{C.12}$$

where $K^L_{2\sigma}$ is a lowpass filter with filter width 2σ. The spectral properties of the bandpass kernel can be analyzed analogous to the kernel of the lowpass filter. The Fourier transform yields

$$\hat{K}^B_\sigma(k) = \exp\left(-\sigma^2 k^2/2\right) - \exp\left(-2\sigma^2 k^2\right). \tag{C.13}$$

For each band, a characteristic wavenumber can be defined as the wavenumber for which $\hat{K}^B_\sigma$ reaches its maximum, i.e.

$$\left.\frac{\mathrm{d}\hat{K}^B_\sigma}{\mathrm{d}k}\right|_{k_c} = 4\sigma^2 k_c \exp\left(-2\sigma^2 k_c^2\right) - \sigma^2 k_c \exp\left(-\sigma^2 k_c^2/2\right) \triangleq 0. \tag{C.14}$$

which leads to

$$k_c = \sqrt{\frac{2\ln 4}{3\sigma^2}} \approx 0.961/\sigma \tag{C.15}$$

and a corresponding peak value of $K^B_{\sigma,c} = 3\sqrt[3]{2}/8 \approx 0.472$.

Due to the infinite support of the Gaussian kernel, the width of the spectral band is also infinite in principle. It decays rapidly however and is generally of the order of k_c for attenuation values of interest. Figure C.2 shows the normalised bandwidth as a function of attenuation w.r.t. the peak transmission.

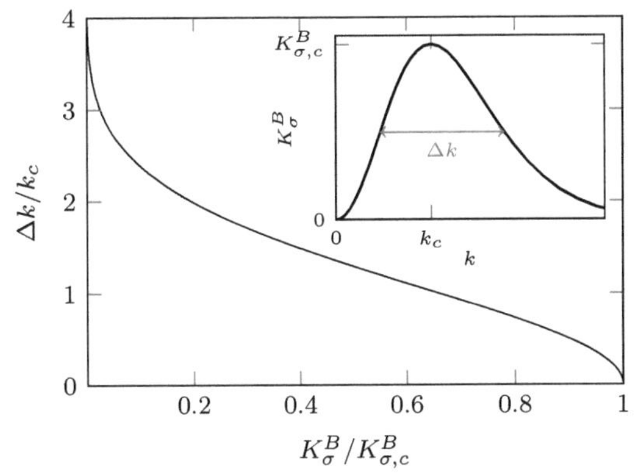

Figure C.2: Normalised bandwidth of the band-pass filter as a function of the attenuation with respect to the peak transmission.

It is certainly possible to construct bands using other differences in filter widths as factor 2 used in this work and Motoori and Goto (2021). By choosing a larger spacing between the two lowpass filters, the characteristic frequency gets less attenuated at the cost of a larger bandwidth, i.e. the filter is less selective in the scales to be extracted, while the opposite is true for narrower bands.

C.3 High-pass filter

A high-pass filter can be constructed by subtracting the low-pass filtered field from the full field, i.e.

$$\mathcal{H}\{\phi\}(x) = \phi(x) - K^L_\sigma(x) * \phi(x) = (1 - K^L_\sigma(x)) * \phi(x) := K^H_\sigma(x) * \phi(x) \tag{C.16}$$

and thus the Fourier transform of the kernel is given by

$$\hat{K}^H_\sigma(k) = 1 - \exp\left(-\frac{\sigma^2 k^2}{2}\right). \tag{C.17}$$

The characteristic wavenumber is defined in the same way as for the low-pass filter and is given by

$$k_c = \sqrt{\frac{2\ln(1-\sqrt{2}/2)^{-1}}{\sigma^2}} \approx 1.567/\sigma. \tag{C.18}$$

D Stability of the exact coherent state

The exact coherent state primarily discussed in §8 is unstable, even with respect to infinitesimal perturbations. Nonetheless, the analysis of the characteristics of settling particles greatly relies on the presumption that the flow state is not considerably altered by the presence of fully resolved particles, although their presence naturally introduces finite-amplitude perturbations. The purpose of this chapter is to explore this hypothesis and to quantify the disturbance for examplary parameter points.

In order to assess a typical time scale over which the flow state diverges from its exact coherent state under ideal conditions, direct numerical simulations have been performed using the same pseudo-spectral solver which has been used to obtain the ECS. No initial perturbations were added to the flow, and hence, the instability solely occurs due to round-off errors in the numerical description of the ECS and numerical errors in its time integration. Figure D.1(a) shows the temporal evolution of the instability using the friction Reynolds number as an observable under the conditions previously described. The value of Re_τ indicates that the ECS is maintained for around $250\,t_b$ before relaminarisation can be observed. This time significantly exceeds the observation times of the multiphase simulations, which are $t_{obs} = \{15, 30, 45\}\,t_b$ depending on the parameter point. However, as this characterisation does not take into account finite-amplitude perturbations, further assessments have to be made.

In fig. D.1(b), the temporal evolution of the change in turbulent kinetic energy (TKE)—defined using the variance based on the instantaneous domain-average—relative to the TKE of the ECS is presented for a singlephase DNS and three multiphase simulations at $Ga = 10$, $\rho_p/\rho_f = 1.75$ with a varying number of particles per run ($N_p = \{1, 16, 32\}$). In this analysis, all simulations have been conducted using the immersed boundary method DNS code for comparability. Within the entire observation interval of $t_{obs} \approx 220\,t_b$, the change in TKE for the singlephase run is only of $\mathcal{O}(1\%)$, which indicates that the time until relaminarisation does not reduce significantly when interpolation errors and changes in the numerical integration method are introduced. When particles are introduced to the flow, the variation in TKE is still less than 1% for all investigated values of N_p in the relevant observation interval. This suggests that no rapid destabilisation is triggered even if multiple particles are added at once. In general, the deviation from singlephase behaviour grows with increasing number of particles per simulation. However, this does not necessarily imply an alteration in the destabilising mechanism, but might rather be attributed to local flow modifications in the vicinity of particles.

In order to get a more intuitive perspective on the perturbation of the ECS, fig. D.2 displays the vortical structure of the final state of three multiphase simulations, which have been previously included in the discussion in §8 and contain $N_p = 16$ particles per run. These cases (fig. D.2(b-d)) have been selected such that they cover a wide range of values of the Galileo number and the density ratio, and a singlephase simulation using the same method is shown alongside as a reference (fig. D.2(a)). There are deviations from the ECS observable besides the boundary layers introduced by the particles themselves, such as the introduction of spanwise asymmetry presumably caused by the asymmetric initial distribution of the

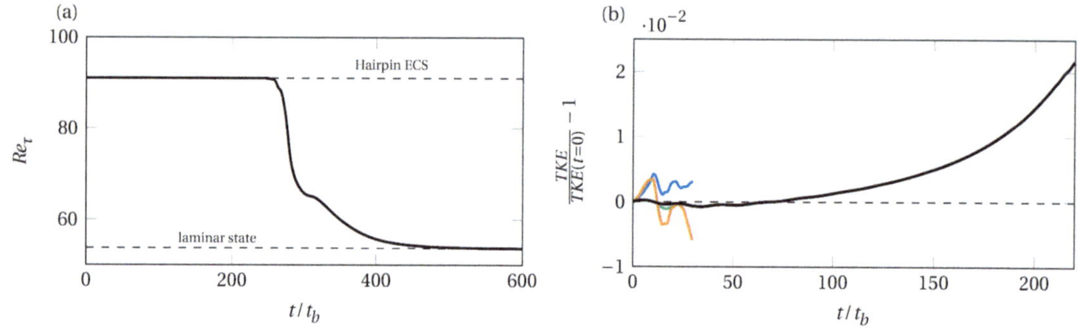

Figure D.1: (a) Instantaneous friction Reynolds number as a function of simulation time (—). Dashed lines (- - -) indicate the values expected for the exact coherent state ($Re_\tau \approx 91$) and the corresponding laminar state at the same value of the bulk Reynolds number ($Re_\tau \approx 54$). The data is shown for a singlephase simulation using the pseudo-spectral solver CHANNELFLOW (Gibson et al. 2008, Gibson 2014) under ideal conditions, i.e. the initial perturbation solely arises from finite precision arithmetic. (b) Relative change in turbulent kinetic energy (TKE) as a function of time for a singlephase simulation (—), and three multiphase simulations at $Ga = 10$, $\rho_p/\rho_f = 1.75$ with varying numbers of particles per simulation run: $N_p = 1$ (—), $N_p = 16$ (—), $N_p = 32$ (—). The simulations were conducted using the immersed boundary DNS code.

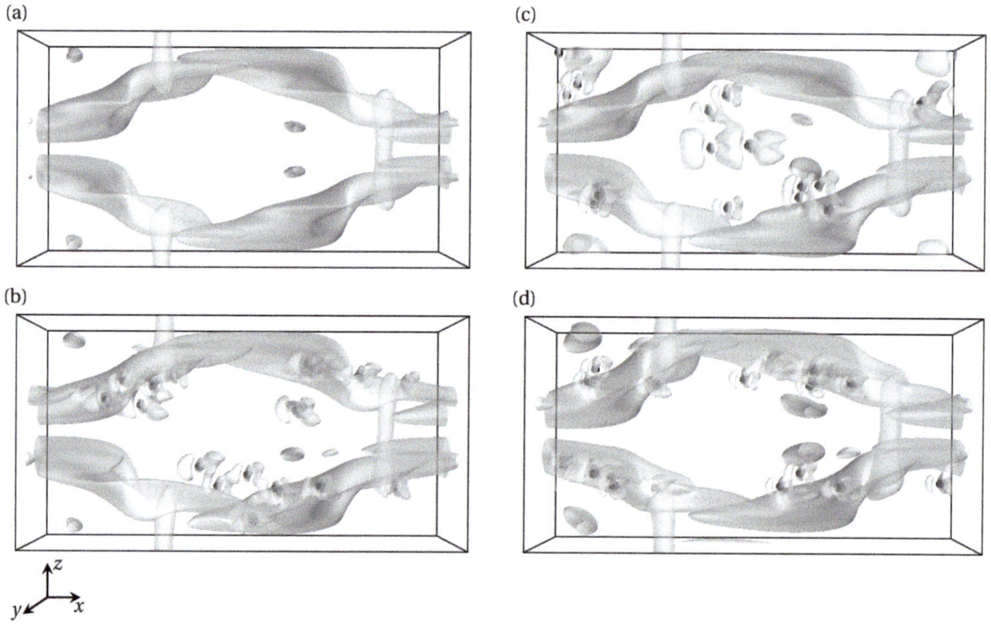

Figure D.2: Visualisation of vortical structures using the Q-criterion defined in eq. (2.51). (a) Unperturbed singlephase simulation using the immersed boundary DNS code as a reference. (b) $Ga = 10$, $\rho_p/\rho_f = 1.75$. (c) $Ga = 10$, $\rho_p/\rho_f = 10$. (d) $Ga = 50$, $\rho_p/\rho_f = 1.75$. Isocontours of $Q^+ = 2.7 \cdot 10^{-3}$ are shown by translucent gray surfaces for $t = 30t_b$. For the multiphase simulations, 16 particles were introduced to the flow which are depicted as black spheres.

settling particles. Nonetheless, the overall flow topology is reasonably conserved. The initial hypothesis of negligible deviations from the ECS during the observation period is thus acceptable.

Since most of the quantities presented in §8 are associated to particle trajectories—and thus only indirectly associated to the flow state of the fluid—it is worthwhile to further explore their sensitivity on the number of particles per run. Figure D.3(a) shows the ensemble-averaged particle velocity for a single physical parameter point, but for a varying number of particles per run. Within the investigated range of $N_p \in [1, 32]$ only few deviations can be observed between the simulation setups. This is especially true in the early phase when most particles are still settling, i.e. when they did not make contact with the bottom boundary yet. A similar conclusion can be drawn for the deviation of the trajectories from the reference trajectory ($N_p = 1$), which is quantified in fig. D.3(b) by the instantenous root-mean square

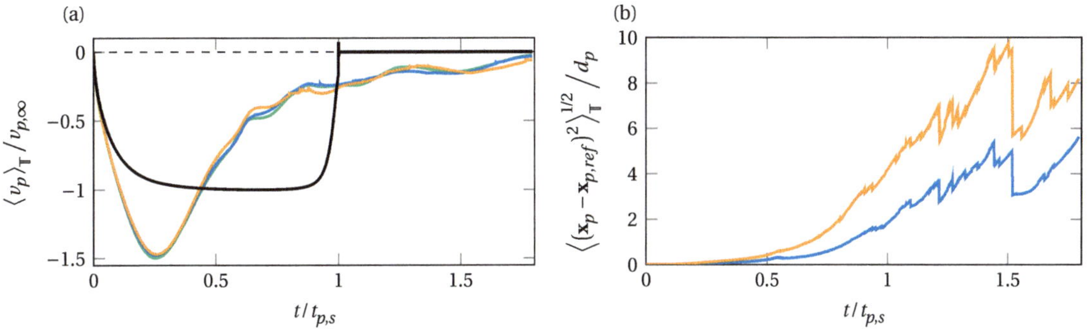

Figure D.3: (a) Ensemble-averaged particle settling velocity, as defined in eq. (4.2), as a function of time. The coloured lines show $\langle v_p \rangle_{\mathbb{T}}$ for varying number of particles per realisation (—, $N_p = 1$; —, $N_p = 16$; —, $N_p = 32$) with the total number of simulated particles kept constant at $N_{p\Sigma} = 288$, i.e. the smaller the value of N_p, the more realisations had to be performed. The solid black line indicates a reference for v_p for a single particle in the same domain in absence of a background flow. (b) Ensemble-averaged difference between the observed particle positions $\mathbf{x}_p$ for multi-particle setups (—, $N_p = 16$; —, $N_p = 32$) and the reference position $\mathbf{x}_{p,ref}$ for the setup with a single particle per run ($N_p = 1$) at a given time. The physical parameter point is the same as the one shown in fig. D.1 ($Ga = 10$, $\rho_p / \rho_f = 1.75$).

of the difference in observed particle positions $\mathbf{x}_p^{(i)}$ (for $N_p = \{16.32\}$) and the reference position $\mathbf{x}_{p,ref}^{(i)}$ (for $N_p = 1$). As is typical for chaotic systems, small perturbations amplify substantially over time, but statistical properties appear to stay mostly unaffected. In fact, the flucatuations in $\langle v_p \rangle_{\mathbb{T}}$ at times where fig. D.3(b) indicates considerable divergences in trajectories may well be attributed to a low total sample size $N_{p\Sigma}$.

List of Figures

List of Tables

Bibliography

R. J. Adrian. Hairpin vortex organization in wall turbulence. *Physics of Fluids*, 19(4):041301, Apr. 2007. doi: 10.1063/1.2717527. (cited on pages 25, 26, 28, 174, and 184)

R. J. Adrian and I. Marusic. Coherent structures in flow over hydraulic engineering surfaces. *Journal of Hydraulic Research*, 50(5):451–464, Oct. 2012. doi: 10.1080/00221686.2012.729540. (cited on pages 28 and 37)

R. J. Adrian, C. D. Meinhart, and C. D. Tomkins. Vortex organization in the outer region of the turbulent boundary layer. *Journal of Fluid Mechanics*, 422:1–54, Nov. 2000. doi: 10.1017/S0022112000001580. (cited on page 26)

J. C. D. Álamo, J. Jiménez, P. Zandonade, and R. D. Moser. Scaling of the energy spectra of turbulent channels. *Journal of Fluid Mechanics*, 500:135–144, Jan. 2004. doi: 10.1017/S002211200300733X. (cited on page 121)

A. Aliseda, A. Cartellier, F. Hainaux, and J. C. Lasheras. Effect of preferential concentration on the settling velocity of heavy particles in homogeneous isotropic turbulence. *Journal of Fluid Mechanics*, 468:77–105, Oct. 2002. doi: 10.1017/S0022112002001593. (cited on page 40)

Z. Amoura, V. Roig, F. Risso, and A.-M. Billet. Attenuation of the wake of a sphere in an intense incident turbulence with large length scales. *Physics of Fluids*, 22(5):055105, May 2010. doi: 10.1063/1.3425628. (cited on pages 41, 113, 114, and 124)

J. L. Aragón, G. G. Naumis, M. Bai, M. Torres, and P. K. Maini. Turbulent Luminance in Impassioned van Gogh Paintings. *Journal of Mathematical Imaging and Vision*, 30(3):275–283, Mar. 2008. doi: 10.1007/s10851-007-0055-0. (cited on page 1)

R. Aris. On the dispersion of a solute in a fluid flowing through a tube. *Proceedings of the Royal Society of London. Series A. Mathematical and Physical Sciences*, 235(1200):67–77, Apr. 1956. doi: 10.1098/rspa.1956.0065. (cited on page 43)

M. T. Auer and S. L. Niehaus. Modeling fecal coliform bacteria—I. Field and laboratory determination of loss kinetics. *Water Research*, 27(4):693–701, Apr. 1993. doi: 10.1016/0043-1354(93)90179-L. (cited on pages 3, 4, and 6)

D. Auerbach, P. Cvitanović, J.-P. Eckmann, G. Gunaratne, and I. Procaccia. Exploring chaotic motion through periodic orbits. *Physical Review Letters*, 58(23):2387–2389, June 1987. doi: 10.1103/PhysRevLett.58.2387. (cited on page 31)

P. Bagchi and S. Balachandar. Effect of free rotation on the motion of a solid sphere in linear shear flow at moderate Re. *Physics of Fluids*, 14(8):2719–2737, July 2002. doi: 10.1063/1.1487378. (cited on page 115)

P. Bagchi and S. Balachandar. Response of the wake of an isolated particle to an isotropic turbulent flow. *Journal of Fluid Mechanics*, 518:95–123, 2004. (cited on pages 41 and 111)

R. A. Bagnold. The flow of cohesionless grains in fluids. *Philosophical Transactions of the Royal Society of London. Series A, Mathematical and Physical Sciences*, 249(964):235–297, Dec. 1956. doi: 10.1098/rsta.1956.0020. (cited on page 35)

S. C. C. Bailey and A. J. Smits. Experimental investigation of the structure of large- and very-large-scale motions in turbulent pipe flow. *Journal of Fluid Mechanics*, 651:339–356, May 2010. doi: 10.1017/S0022112009993983. (cited on page 26)

L. Baker, A. Frankel, A. Mani, and F. Coletti. Coherent clusters of inertial particles in homogeneous turbulence. *Journal of Fluid Mechanics*, 833:364–398, Dec. 2017. doi: 10.1017/jfm.2017.700. (cited on page 40)

S. Balachandar. A scaling analysis for point–particle approaches to turbulent multiphase flows. *International Journal of Multiphase Flow*, 35(9):801–810, Sept. 2009. doi: 10.1016/j.ijmultiphaseflow.2009.02.013. (cited on page 39)

S. Balachandar and J. K. Eaton. Turbulent Dispersed Multiphase Flow. *Annual Review of Fluid Mechanics*, 42, 2010. (cited on page 39)

B. Balakumar and R. Adrian. Large- and very-large-scale motions in channel and boundary-layer flows. *Philosophical Transactions of the Royal Society A: Mathematical, Physical and Engineering Sciences*, 365(1852):665–681, Mar. 2007. doi: 10.1098/rsta.2006.1940. (cited on page 26)

P. R. Bandyopadhyay and R. D. Watson. Structure of rough-wall turbulent boundary layers. *The Physics of Fluids*, 31(7):1877–1883, July 1988. doi: 10.1063/1.866686. (cited on page 35)

G. K. Batchelor, A. A. Townsend, and H. Jeffreys. The nature of turbulent motion at large wave-numbers. *Proceedings of the Royal Society of London. Series A. Mathematical and Physical Sciences*, 199(1057):238–255, Oct. 1949. doi: 10.1098/rspa.1949.0136. (cited on page 18)

C. Bauer. *Direct Numerical Simulation of Very-Large-Scale Motions in Turbulent Pipe Flow*. PhD thesis, Universität Illmenau, Göttingen, 2020. (cited on page 26)

C. Bauer, Y. Sakai, and M. Uhlmann. DNS of plane open channel flow: Scaling near the free-slip boundary and evidence of very-large-scale motions. *in preparation*, 2022. (cited on pages 20, 23, 28, 121, 122, 177, and 184)

M. Bayazit. Random Walk Model For Motion Of A Solid Particle In Turbulent Open-Channel Flow. *Journal of Hydraulic Research*, 10(1):1–14, Jan. 1972. doi: 10.1080/00221687209500016. (cited on pages 155 and 156)

J. Bec. Fractal clustering of inertial particles in random flows. *Physics of Fluids*, 15(11):L81–L84, Nov. 2003. doi: 10.1063/1.1612500. (cited on page 38)

A. Beer. Bestimmung der Absorption des rothen Lichts in farbigen Flüssigkeiten. *Annalen der Physik*, 162(5):78–88, 1852. doi: 10.1002/andp.18521620505. (cited on page 191)

J. B. Belanger. *Essai Sur La Solution Numerique de Quelques Problemes Relatifs Au Mouvement Permanent Des Eaux Courantes*. Carilian-Goeury, Paris, 1828. (cited on page 15)

E. Ben-Jacob, O. Schochet, A. Tenenbaum, I. Cohen, A. Czirók, and T. Vicsek. Generic modelling of cooperative growth patterns in bacterial colonies. *Nature*, 368(6466):46–49, Mar. 1994. doi: 10.1038/368046a0. (cited on page 14)

J. Bertrand. Sur l'homogénéité dans les formules de physique. *Cahiers de recherche de l'Academie de Sciences*, 86:916–920, 1878. (cited on page 14)

J. Best. The fluid dynamics of river dunes: A review and some future research directions. *Journal of Geophysical Research: Earth Surface*, 110(F4), 2005. doi: 10.1029/2004JF000218. (cited on page 36)

E. W. Bijker, E. van Hijum, and P. Vellinga. Sand transport by waves. In *Coastal Engineering Proceedings*, pages 67–67, Honolulu, Hawaii, United States, Jan. 1976. doi: 10.9753/icce.v15.67. (cited on page 40)

R. F. Blackwelder and H. Eckelmann. Streamwise vortices associated with the bursting phenomenon. *Journal of Fluid Mechanics*, 94(3):577–594, Oct. 1979. doi: 10.1017/S0022112079001191. (cited on page 24)

A. F. Blumberg and G. L. Mellor. A Description of a Three-Dimensional Coastal Ocean Circulation Model. In *Three-Dimensional Coastal Ocean Models*, pages 1–16. American Geophysical Union (AGU), 1987. ISBN 978-1-118-66504-6. doi: 10.1029/CO004p0001. (cited on page 8)

A. B. Boehm. Model of Microbial Transport and Inactivation in the Surf Zone and Application to Field Measurements of Total Coliform in Northern Orange County, California. *Environmental Science & Technology*, 37(24):5511–5517, Dec. 2003. doi: 10.1021/es034321x. (cited on page 6)

S. Bottin and H. Chaté. Statistical analysis of the transition to turbulence in plane Couette flow. *The European Physical Journal B - Condensed Matter and Complex Systems*, 6(1):143–155, Nov. 1998. doi: 10.1007/s100510050536. (cited on page 29)

J. Boussinesq. *Théorie de l'écoulement Tourbillonnant et Tumultueux Des Liquides Dans Les Lits Rectilignes a Grande Section*. Gauthier-Villars et fils, 1897. ISBN 978-1-289-85832-2. (cited on page 15)

P. Bradshaw. A note on "critical roughness height" and "transitional roughness". *Physics of Fluids*, 12(6): 1611–1614, June 2000. doi: 10.1063/1.870410. (cited on page 34)

A. Bragg, D. C. Swailes, and R. Skartlien. Particle transport in a turbulent boundary layer: Non-local closures for particle dispersion tensors accounting for particle-wall interactions. *Physics of Fluids*, 24 (10):103304, Oct. 2012. doi: 10.1063/1.4757657. (cited on page 43)

A. D. Bragg and L. R. Collins. New insights from comparing statistical theories for inertial particles in turbulence: I. Spatial distribution of particles. *New Journal of Physics*, 16(5):055013, May 2014. doi: 10.1088/1367-2630/16/5/055013. (cited on page 38)

A. D. Bragg, P. J. Ireland, and L. R. Collins. On the relationship between the non-local clustering mechanism and preferential concentration. *Journal of Fluid Mechanics*, 780:327–343, Oct. 2015. doi: 10.1017/jfm.2015.474. (cited on page 38)

A. D. Bragg, D. H. Richter, and G. Wang. Mechanisms governing the settling velocities and spatial distributions of inertial particles in wall-bounded turbulence. *Physical Review Fluids*, 6(6):064302, June 2021a. doi: 10.1103/PhysRevFluids.6.064302. (cited on pages 35, 43, 44, 45, 71, and 200)

A. D. Bragg, D. H. Richter, and G. Wang. When is settling important for particle concentrations in wall-bounded turbulent flows? *arXiv:2101.04607 [physics]*, Sept. 2021b. (cited on page 40)

W. A. Breugem and W. S. J. Uijttewaal. Sediment transport by coherent structures in a turbulent open channel flow experiment. In B. J. Geurts, H. Clercx, and W. Uijttewaal, editors, *Particle-Laden Flow*, ERCOFTAC Series, pages 43–55. Springer Netherlands, 2007. ISBN 978-1-4020-6218-6. (cited on pages 44 and 200)

W. Brooks, S. Corsi, M. Fienen, and R. Carvin. Predicting recreational water quality advisories: A comparison of statistical methods. *Environmental Modelling & Software*, 76:81–94, Feb. 2016. doi: 10.1016/j.envsoft.2015.10.012. (cited on page 5)

G. L. Brown and A. S. W. Thomas. Large structure in a turbulent boundary layer. *The Physics of Fluids*, 20(10):S243–S252, Oct. 1977. doi: 10.1063/1.861737. (cited on page 26)

P. C. Brown, E. Borowska, T. Schwartz, and H. Horn. Impact of the particulate matter from wastewater discharge on the abundance of antibiotic resistance genes and facultative pathogenic bacteria in downstream river sediments. *Science of The Total Environment*, 649:1171–1178, Feb. 2019. doi: 10.1016/j.scitotenv.2018.08.394. (cited on page 3)

P. C. Brown, E. Borowska, R. Peschke, T. Schwartz, and H. Horn. Decay of elevated antibiotic resistance genes in natural river sediments after sedimentation of wastewater particles. *Science of The Total Environment*, 705:135861, Feb. 2020. doi: 10.1016/j.scitotenv.2019.135861. (cited on page 3)

E. Buckingham. On Physically Similar Systems; Illustrations of the Use of Dimensional Equations. *Physical Review*, 4(4):345–376, Oct. 1914. doi: 10.1103/PhysRev.4.345. (cited on page 14)

N. B. Budanur, K. Y. Short, M. Farazmand, A. P. Willis, and P. Cvitanović. Relative periodic orbits form the backbone of turbulent pipe flow. *Journal of Fluid Mechanics*, 833:274–301, Dec. 2017. doi: 10.1017/jfm.2017.699. (cited on page 30)

D. Butler and J. W. Davies. *Urban Drainage*. CRC Press, 2 edition, 2004. ISBN 978-0-429-21735-7. (cited on pages 2 and 3)

I. Calmet and J. Magnaudet. Statistical structure of high-Reynolds-number turbulence close to the free surface of an open-channel flow. *Journal of Fluid Mechanics*, 474:355–378, Jan. 2003. doi: 10.1017/S0022112002002793. (cited on page 20)

S. M. Cameron, V. I. Nikora, and M. J. Witz. Entrainment of sediment particles by very large-scale motions. *Journal of Fluid Mechanics*, 888, Apr. 2020. doi: 10.1017/jfm.2020.24. (cited on page 36)

B. J. Cantwell. Organized Motion in Turbulent Flow. *Annual Review of Fluid Mechanics*, 13(1):457–515, Jan. 1981. doi: 10.1146/annurev.fl.13.010181.002325. (cited on page 28)

J. H. Cartwright and H. Nakamura. What kind of a wave is Hokusai's Great wave off Kanagawa? *Notes and Records of the Royal Society*, 63(2):119–135, June 2009. doi: 10.1098/rsnr.2007.0039. (cited on page 1)

H. J. Casey. *Über Geschiebebewegung*, volume 19. Preußische Versuchsanstalt für Wasserbau und Schiffbau, 1935. (cited on page 36)

M. Cellino and W. Graf. Experiments on suspension flow in open channels with bed forms. *Journal of Hydraulic Research*, 38(4):289–298, July 2000. doi: 10.1080/00221680009498328. (cited on page 37)

M. Cellino and U. Lemmin. Influence of Coherent Flow Structures on the Dynamics of Suspended Sediment Transport in Open-Channel Flow. *Journal of Hydraulic Engineering*, 130(11):1077–1088, Nov. 2004. doi: 10.1061/(ASCE)0733-9429(2004)130:11(1077). (cited on page 36)

C. Chan-Braun, M. García-Villalba, and M. Uhlmann. Force and torque acting on particles in a transitionally rough open channel flow. *Journal of Fluid Mechanics*, 684:441–474, Oct. 2011. doi: 10.1017/jfm.2011.311. (cited on pages 34, 35, and 58)

G. J. Chandler and R. R. Kerswell. Invariant recurrent solutions embedded in a turbulent two-dimensional Kolmogorov flow. *Journal of Fluid Mechanics*, 722:554–595, May 2013. doi: 10.1017/jfm.2013.122. (cited on pages 30 and 31)

S. J. Chapman. Subcritical transition in channel flows. *Journal of Fluid Mechanics*, 451:35–97, Jan. 2002. doi: 10.1017/S0022112001006255. (cited on page 21)

G. W. Characklis, M. J. Dilts, O. D. Simmons, C. A. Likirdopulos, L.-A. H. Krometis, and M. D. Sobsey. Microbial partitioning to settleable particles in stormwater. *Water Research*, 39(9):1773–1782, May 2005. doi: 10.1016/j.watres.2005.03.004. (cited on page 3)

F. Charru, B. Andreotti, and P. Claudin. Sand Ripples and Dunes. *Annual Review of Fluid Mechanics*, 45 (1):469–493, 2013. doi: 10.1146/annurev-fluid-011212-140806. (cited on page 36)

L. Chen, S. Goto, and J. C. Vassilicos. Turbulent clustering of stagnation points and inertial particles. *Journal of Fluid Mechanics*, 553:143–154, Apr. 2006. doi: 10.1017/S0022112006009177. (cited on page 38)

W.-B. Chen and W.-C. Liu. Investigating the fate and transport of fecal coliform contamination in a tidal estuarine system using a three-dimensional model. *Marine Pollution Bulletin*, 116(1):365–384, Mar. 2017. doi: 10.1016/j.marpolbul.2017.01.031. (cited on page 9)

H. Chick. The Process of Disinfection by Chemical Agencies and Hot Water. *Epidemiology & Infection*, 10(2):237–286, Aug. 1910. doi: 10.1017/S0022172400042959. (cited on page 6)

A. Chouippe and M. Uhlmann. Forcing homogeneous turbulence in direct numerical simulation of particulate flow with interface resolution and gravity. *Physics of Fluids*, 27(12):123301, Dec. 2015. doi: 10.1063/1.4936274. (cited on page 70)

A. Chouippe and M. Uhlmann. On the influence of forced homogeneous-isotropic turbulence on the settling and clustering of finite-size particles. *Acta Mechanica*, 230(2):387–412, Feb. 2019. doi: 10.1007/s00707-018-2271-7. (cited on pages 41 and 70)

A. Chouippe, M. Krayer, M. Uhlmann, J. Dušek, A. Kiselev, and T. Leisner. Heat and water vapor transfer in the wake of a falling ice sphere and its implication for secondary ice formation in clouds. *New Journal of Physics*, 21(4):043043, Apr. 2019. doi: 10.1088/1367-2630/ab0a94. (cited on page 42)

F. Christiansen, P. Cvitanovic, and V. Putkaradze. Spatiotemporal chaos in terms of unstable recurrent patterns. *Nonlinearity*, 10(1):55–70, Jan. 1997. doi: 10.1088/0951-7715/10/1/004. (cited on page 31)

City of Karlsruhe. The urban wastewater management of Karlsruhe, 2010. (cited on page 2)

R. Clift, J. R. Grace, and M. E. Weber. *Bubbles, Drops, and Particles*. Academic Press, Inc., New York, 1978. (cited on page 42)

A. Colagrossi, S. Marrone, P. Colagrossi, and D. Le Touzé. Da Vinci's observation of turbulence: A French-Italian study aiming at numerically reproducing the physics behind one of his drawings, 500 years later. *Physics of Fluids*, 33(11):115122, Nov. 2021. doi: 10.1063/5.0070984. (cited on page 1)

S. W. Coleman and J. C. Vassilicos. A unified sweep-stick mechanism to explain particle clustering in two- and three-dimensional homogeneous, isotropic turbulence. *Physics of Fluids*, 21(11):113301, Nov. 2009. doi: 10.1063/1.3257638. (cited on page 38)

D. Coles. Transition in circular Couette flow. *Journal of Fluid Mechanics*, 21(3):385–425, Mar. 1965. doi: 10.1017/S0022112065000241. (cited on page 32)

M. Colombini. Turbulence-driven secondary flows and formation of sand ridges. *Journal of Fluid Mechanics*, 254:701–719, Sept. 1993. doi: 10.1017/S0022112093002319. (cited on page 36)

M. Colombini and G. Parker. Longitudinal streaks. *Journal of Fluid Mechanics*, 304:161–183, Dec. 1995. doi: 10.1017/S0022112095004381. (cited on page 36)

J. P. Connolly, A. F. Blumberg, and J. D. Quadrini. Modeling Fate of Pathogenic Organisms in Coastal Waters of Oahu, Hawaii. *Journal of Environmental Engineering*, 125(5):398–406, May 1999. doi: 10.1061/(ASCE)0733-9372(1999)125:5(398). (cited on page 8)

S. Corrsin. Turbulent flow. *American Scientist*, 49(3):300–325, 1961. (cited on page 18)

D. Craig, H. Fallowfield, and N. Cromar. Use of microcosms to determine persistence of Escherichia coli in recreational coastal water and sediment and validation with in situ measurements. *Journal of Applied Microbiology*, 96(5):922–930, 2004. doi: 10.1111/j.1365-2672.2004.02243.x. (cited on page 4)

J. Crank. *The Mathematics of Diffusion*. Dover Publications, 1975. (cited on pages 160 and 188)

C. Crowe, M. Sommerfeld, and Y. Tsuji. *Multiphase Flows with Particles and Droplets*. CRC Press, New York, 1998. (cited on page 52)

P. Cugier and P. Le Hir. Development of a 3D Hydrodynamic Model for Coastal Ecosystem Modelling. Application to the Plume of the Seine River (France). *Estuarine, Coastal and Shelf Science*, 55(5):673–695, Nov. 2002. doi: 10.1006/ecss.2001.0875. (cited on page 8)

A. J. Cuthbertson and D. A. Ervine. Experimental Study of Fine Sand Particle Settling in Turbulent Open Channel Flows over Rough Porous Beds. *Journal of Hydraulic Engineering*, 133(8):905–916, Aug. 2007. doi: 10.1061/(ASCE)0733-9429(2007)133:8(905). (cited on pages 40, 44, 45, 62, 71, 74, 75, 80, 81, 90, 200, 224, and 229)

A. J. S. Cuthbertson. *The Motion of Fine Sand Particles in Turbulent Open Channel Shear Flows over Porous Bed Conditions*. PhD thesis, University of Glasgow, 2001. (cited on page 44)

P. Cvitanović. Recurrent flows: The clockwork behind turbulence. *Journal of Fluid Mechanics*, 726:1–4, July 2013. doi: 10.1017/jfm.2013.198. (cited on page 31)

E. Darakas. E. Coli Kinetics - Effect of Temperature on the Maintenance and Respectively the Decay Phase. *Environmental Monitoring and Assessment*, 78(2):101–110, Sept. 2002. doi: 10.1023/A:1016313520396. (cited on page 6)

O. Darrigol. Between Hydrodynamics and Elasticity Theory: The First Five Births of the Navier-Stokes Equation. *Archive for History of Exact Sciences*, 56(2):95–150, 2002. (cited on page 1)

A. Davey. On the stability of plane Couette flow to infinitesimal disturbances. *Journal of Fluid Mechanics*, 57(2):369–380, Feb. 1973. doi: 10.1017/S0022112073001217. (cited on pages 28 and 115)

P. A. Davidson. *Turbulence: An Introduction for Scientists and Engineers*. Oxford University Press, 2015. ISBN 0-19-872258-3. (cited on pages 19, 23, and 25)

A. de Brauwere, N. K. Ouattara, and P. Servais. Modeling Fecal Indicator Bacteria Concentrations in Natural Surface Waters: A Review. *Critical Reviews in Environmental Science and Technology*, 44(21): 2380–2453, Nov. 2014. doi: 10.1080/10643389.2013.829978. (cited on page 5)

A. J. C. B. de Saint-Venant. Theorie du mouvement non-permanent des eaux avec application aux crues des rivers et a l'introduntion des Marees dans leur lit. *Academic de Sci. Comptes Redus*, 73(99):148–154, 1871. (cited on page 15)

S. De Toffol, C. Engelhard, and W. Rauch. Combined sewer system versus separate system – a comparison of ecological and economical performance indicators. *Water Science and Technology*, 55(4): 255–264, Feb. 2007. doi: 10.2166/wst.2007.116. (cited on page 3)

R. B. Dean. Reynolds Number Dependence of Skin Friction and Other Bulk Flow Variables in Two-Dimensional Rectangular Duct Flow. *Journal of Fluids Engineering*, 100(2):215–223, June 1978. doi: 10.1115/1.3448633. (cited on page 176)

J. C. del Álamo and J. Jiménez. Spectra of the very large anisotropic scales in turbulent channels. *Physics of Fluids*, 15(6):L41–L44, June 2003. doi: 10.1063/1.1570830. (cited on page 121)

J. C. del Álamo, J. Jiménez, P. Zandonade, and R. D. Moser. Self-similar vortex clusters in the turbulent logarithmic region. *Journal of Fluid Mechanics*, 561:329–358, Aug. 2006. doi: 10.1017/S0022112006000814. (cited on pages 27, 123, 124, 148, 162, 165, and 201)

R. J. Dewey and M. D. Palmer. St. John's Harbour water quality model. *Canadian Journal of Civil Engineering*, 11(1):65–73, Mar. 1984. doi: 10.1139/l84-008. (cited on page 7)

S. Dey. *Fluvial Hydrodynamics*, volume 818. Springer, 2014. (cited on page 37)

S. Dey and S. Z. Ali. Fluvial instabilities. *Physics of Fluids*, 32(6):061301, June 2020. doi: 10.1063/5.0010038. (cited on page 37)

S. Dey, S. Zeeshan Ali, and E. Padhi. Terminal fall velocity: The legacy of Stokes from the perspective of fluvial hydraulics. *Proceedings of the Royal Society A: Mathematical, Physical and Engineering Sciences*, 475(2228):20190277, Aug. 2019. doi: 10.1098/rspa.2019.0277. (cited on page 41)

J. A. Dickenson and J. J. Sansalone. Distribution and disinfection of bacterial loadings associated with particulate matter fractions transported in urban wet weather flows. *Water Research*, 46(20):6704–6714, Dec. 2012. doi: 10.1016/j.watres.2011.12.039. (cited on page 3)

U. Dittmer, A. Bachmann-Machnik, and M. A. Launay. Impact of Combined Sewer Systems on the Quality of Urban Streams: Frequency and Duration of Elevated Micropollutant Concentrations. *Water*, 12 (3):850, Mar. 2020. doi: 10.3390/w12030850. (cited on page 3)

C. D. Dritselis and N. S. Vlachos. Numerical investigation of momentum exchange between particles and coherent structures in low Re turbulent channel flow. *Physics of Fluids*, 23(2):025103, Feb. 2011. doi: 10.1063/1.3553292. (cited on page 41)

I. G. Droppo, S. N. Liss, D. Williams, T. Nelson, C. Jaskot, and B. Trapp. Dynamic Existence of Waterborne Pathogens within River Sediment Compartments. Implications for Water Quality Regulatory Affairs. *Environmental Science & Technology*, 43(6):1737–1743, Mar. 2009. doi: 10.1021/es802321w. (cited on page 4)

Y. Duan, Q. Chen, D. Li, and Q. Zhong. Contributions of very large-scale motions to turbulence statistics in open channel flows. *Journal of Fluid Mechanics*, 892, June 2020. doi: 10.1017/jfm.2020.174. (cited on page 27)

K. Duncan, B. Mehlig, S. Östlund, and M. Wilkinson. Clustering by Mixing Flows. *Physical Review Letters*, 95(24):240602, Dec. 2005. doi: 10.1103/PhysRevLett.95.240602. (cited on page 38)

J. K. Eaton and J. R. Fessler. Preferential concentration of particles by turbulence. *International Journal of Multiphase Flow*, 20:169–209, Aug. 1994. doi: 10.1016/0301-9322(94)90072-8. (cited on page 37)

ECCC. Wastewater Systems Effluent Regulations Reported Data. Technical report, Environment and Climate Change Canada, 2019. (cited on page 2)

B. Eckhardt, T. M. Schneider, B. Hof, and J. Westerweel. Turbulence Transition in Pipe Flow. *Annual Review of Fluid Mechanics*, 39(1):447–468, 2007. doi: 10.1146/annurev.fluid.39.050905.110308. (cited on pages 29 and 30)

H. Edelsbrunner and E. P. Mücke. Three-dimensional Alpha Shapes. *ACM Trans. Graph.*, 13(1):43–72, Jan. 1994. doi: 10.1145/174462.156635. (cited on page 56)

J. E. Edinger and E. M. Buchak. Numerical hydrodynamics of estuaries. In *Estuarine and Wetland Processes*, pages 115–146. Springer, 1980. (cited on page 8)

J. W. Elder. The dispersion of marked fluid in turbulent shear flow. *Journal of Fluid Mechanics*, 5(4): 544–560, May 1959. doi: 10.1017/S0022112059000374. (cited on page 43)

H. Elghannay and D. Tafti. LES-DEM simulations of sediment transport. *International Journal of Sediment Research*, 33(2):137–148, June 2018. doi: 10.1016/j.ijsrc.2017.09.006. (cited on page 207)

S. Elghobashi and G. C. Truesdell. Direct simulation of particle dispersion in a decaying isotropic turbulence. *Journal of Fluid Mechanics*, 242:655–700, Sept. 1992. doi: 10.1017/S0022112092002532. (cited on page 37)

S. Elghobashi and G. C. Truesdell. On the two-way interaction between homogeneous turbulence and dispersed solid particles. I: Turbulence modification. *Physics of Fluids A: Fluid Dynamics*, 5(7):1790–1801, July 1993. doi: 10.1063/1.858854. (cited on page 41)

J. B. Ellis. Pollutional Aspects of Urban Runoff. In H. C. Torno, J. Marsalek, and M. Desbordes, editors, *Urban Runoff Pollution*, NATO ASI Series, pages 1–38, Berlin, Heidelberg, 1986. Springer. ISBN 978-3-642-70889-3. doi: 10.1007/978-3-642-70889-3_1. (cited on page 3)

A. Enciso, M. Á. García-Ferrero, and D. Peralta-Salas. The Biot–Savart operator of a bounded domain. *Journal de Mathématiques Pures et Appliquées*, 119:85–113, Nov. 2018. doi: 10.1016/j.matpur.2017.11.004. (cited on page 18)

F. Engelund and J. Fredsoe. Sediment Ripples and Dunes. *Annual Review of Fluid Mechanics*, 14(1): 13–37, 1982. doi: 10.1146/annurev.fl.14.010182.000305. (cited on page 35)

C. W. Erkenbrecher. Sediment bacterial indicators in an urban shellfishing subestuary of the lower Chesapeake Bay. *Applied and Environmental Microbiology*, 42(3):484–492, Sept. 1981. doi: 10.1128/aem.42.3.484-492.1981. (cited on page 4)

J. H. Ferziger, M. Perić, and R. L. Street. *Computational Methods for Fluid Dynamics*, volume 3. Springer, 2002. ISBN 3-540-42074-6. (cited on page 53)

J. R. Fessler, J. D. Kulick, and J. K. Eaton. Preferential concentration of heavy particles in a turbulent channel flow. *Physics of Fluids*, 6(11):3742–3749, Nov. 1994. doi: 10.1063/1.868445. (cited on pages 45, 138, and 146)

L. Fiabane, R. Zimmermann, R. Volk, J.-F. Pinton, and M. Bourgoin. Clustering of finite-size particles in turbulence. *Physical Review E*, 86(3):035301, Sept. 2012. doi: 10.1103/PhysRevE.86.035301. (cited on page 39)

L. Fiabane, R. Volk, J.-F. Pinton, R. Monchaux, A. Cartellier, and M. Bourgoin. Do finite-size neutrally buoyant particles cluster? *Physica Scripta*, 2013(T155):014056, 2013. doi: 10.1088/0031-8949/2013/T155/014056. (cited on page 39)

H. B. Fischer. Longitudinal Dispersion and Turbulent Mixing in Open-Channel Flow. *Annual Review of Fluid Mechanics*, 5(1):59–78, Jan. 1973. doi: 10.1146/annurev.fl.05.010173.000423. (cited on pages 158 and 189)

K. A. Flack and M. P. Schultz. Roughness effects on wall-bounded turbulent flows. *Physics of Fluids*, 26 (10):101305, Oct. 2014. doi: 10.1063/1.4896280. (cited on page 35)

K. A. Flack, M. P. Schultz, and T. A. Shapiro. Experimental support for Townsend's Reynolds number similarity hypothesis on rough walls. *Physics of Fluids*, 17(3):035102, Mar. 2005. doi: 10.1063/1.1843135. (cited on page 35)

O. Flores and J. Jiménez. Effect of wall-boundary disturbances on turbulent channel flows. *Journal of Fluid Mechanics*, 566:357–376, Nov. 2006. doi: 10.1017/S0022112006001534. (cited on pages 27 and 123)

O. Flores and J. Jiménez. Hierarchy of minimal flow units in the logarithmic layer. *Physics of Fluids*, 22 (7):071704, July 2010. doi: 10.1063/1.3464157. (cited on page 123)

O. Flores, J. Jiménez, and J. C. D. Álamo. Vorticity organization in the outer layer of turbulent channels with disturbed walls. *Journal of Fluid Mechanics*, 591:145–154, Nov. 2007. doi: 10.1017/S0022112007008506. (cited on pages 27 and 123)

W. Fornari, F. Picano, G. Sardina, and L. Brandt. Reduced particle settling speed in turbulence. *Journal of Fluid Mechanics*, 808:153–167, Dec. 2016. doi: 10.1017/jfm.2016.648. (cited on pages 70, 118, and 124)

W. Fornari, S. Zade, L. Brandt, and F. Picano. Settling of finite-size particles in turbulence at different volume fractions. *Acta Mechanica*, 230(2):413–430, Feb. 2019. doi: 10.1007/s00707-018-2269-1. (cited on page 43)

A. F. Fortes, D. D. Joseph, and T. S. Lundgren. Nonlinear mechanics of fluidization of beds of spherical particles. *Journal of Fluid Mechanics*, 177:467–483, Apr. 1987. doi: 10.1017/S0022112087001046. (cited on page 42)

C. Fortier and A. Mailhot. Climate Change Impact on Combined Sewer Overflows. *Journal of Water Resources Planning and Management*, 141(5):04014073, May 2015. doi: 10.1061/(ASCE)WR.1943-5452.0000468. (cited on page 3)

W. Froude. On experiments with H.M.S. "Greyhound". *Transactions of the Institution of Naval Architects*, 15(1):36–73, 1874. (cited on page 15)

B. Ganapathisubramani, E. K. Longmire, and I. Marusic. Characteristics of vortex packets in turbulent boundary layers. *Journal of Fluid Mechanics*, 478:35–46, Mar. 2003. doi: 10.1017/S0022112002003270. (cited on page 26)

J. J. Gannon, M. K. Busse, and J. E. Schillinger. Fecal coliform disappearance in a river impoundment. *Water Research*, 17(11):1595–1601, Jan. 1983. doi: 10.1016/0043-1354(83)90017-9. (cited on pages 3 and 4)

G. Gao, R. A. Falconer, and B. Lin. Numerical modelling of sediment–bacteria interaction processes in surface waters. *Water Research*, 45(5):1951–1960, Feb. 2011. doi: 10.1016/j.watres.2010.12.030. (cited on page 7)

M. H. García. Sediment transport and morphodynamics. In *Sedimentation Engineering: Processes, Measurements, Modeling, and Practice*, pages 21–163. American Society of Civil Engineers, 2008. (cited on page 37)

T. Garcia-Armisen and P. Servais. Partitioning and Fate of Particle-Associated E. coli in River Waters. *Water Environment Research*, 81(1):21–28, 2009. doi: 10.2175/106143008X304613. (cited on page 4)

T. Garcia-Armisen, B. Thouvenin, and P. Servais. Modelling faecal coliforms dynamics in the Seine estuary, France. *Water Science and Technology*, 54(3):177–184, Aug. 2006. doi: 10.2166/wst.2006.466. (cited on page 8)

M. García-Villalba, A. G. Kidanemariam, and M. Uhlmann. DNS of vertical plane channel flow with finite-size particles: Voronoi analysis, acceleration statistics and particle-conditioned averaging. *International Journal of Multiphase Flow*, 46:54–74, 2012. doi: 10.1016/j.ijmultiphaseflow.2012.05.007. (cited on pages 70 and 118)

J. Gasperi, S. Garnaud, V. Rocher, and R. Moilleron. Priority pollutants in wastewater and combined sewer overflow. *Science of The Total Environment*, 407(1):263–272, Dec. 2008. doi: 10.1016/j.scitotenv.2008.08.015. (cited on page 2)

GCS. Finite-size Particle Dynamics and Scalar Transport in Turbulent Open Channel Flow Over a Mobile Sediment Bed. https://www.gauss-centre.eu/results/computational-and-scientific-engineering/finite-size-particle-dynamics-and-scalar-transport-in-turbulent-open-channel-flow-over-a-mobile-sediment-bed, June 2021. (cited on page 58)

E. Gedenk. Simulations of Sediment Movement Could Help Protect Waterways. *InSiDE, Innovatives Supercomputing in Deutschland*, 20(1):18–20, 2022. (cited on page 58)

J. Gibson, J. Drake, and B. Karney. UV Disinfection of Wastewater and Combined Sewer Overflows. In S. I. Ahmad, editor, *Ultraviolet Light in Human Health, Diseases and Environment*, Advances in Experimental Medicine and Biology, pages 267–275. Springer International Publishing, Cham, 2017. ISBN 978-3-319-56017-5. doi: 10.1007/978-3-319-56017-5_22. (cited on page 3)

J. F. Gibson. Channelflow: A spectral Navier-Stokes simulator in C++. Technical report, J. F. Gibson, 2014. (cited on pages 174 and 220)

J. F. Gibson, J. Halcrow, and P. Cvitanović. Visualizing the geometry of state space in plane Couette flow. *Journal of Fluid Mechanics*, 611:107–130, Sept. 2008. doi: 10.1017/S002211200800267X. (cited on pages 174 and 220)

J. P. Gollub and S. V. Benson. Many routes to turbulent convection. *Journal of Fluid Mechanics*, 100(3): 449–470, Oct. 1980. doi: 10.1017/S0022112080001243. (cited on page 29)

J. P. Gollub and H. L. Swinney. Onset of Turbulence in a Rotating Fluid. *Physical Review Letters*, 35(14): 927–930, Oct. 1975. doi: 10.1103/PhysRevLett.35.927. (cited on page 32)

S. Gomé, L. S. Tuckerman, and D. Barkley. Statistical transition to turbulence in plane channel flow. *Physical Review Fluids*, 5(8):083905, Aug. 2020. doi: 10.1103/PhysRevFluids.5.083905. (cited on page 29)

G. H. Good, S. Gerashchenko, and Z. Warhaft. Intermittency and inertial particle entrainment at a turbulent interface: The effect of the large-scale eddies. *Journal of Fluid Mechanics*, 694:371–398, Mar. 2012. doi: 10.1017/jfm.2011.552. (cited on page 41)

R. A. Gore and C. T. Crowe. Effect of particle size on modulating turbulent intensity. *International Journal of Multiphase Flow*, 15(2):279–285, Apr. 1989. doi: 10.1016/0301-9322(89)90076-1. (cited on page 41)

S. Goto and J. C. Vassilicos. Self-similar clustering of inertial particles and zero-acceleration points in fully developed two-dimensional turbulence. *Physics of Fluids*, 18(11):115103, Nov. 2006. doi: 10.1063/1.2364263. (cited on page 38)

S. Goto and J. C. Vassilicos. Sweep-Stick Mechanism of Heavy Particle Clustering in Fluid Turbulence. *Physical Review Letters*, 100(5):054503, Feb. 2008. doi: 10.1103/PhysRevLett.100.054503. (cited on page 38)

M. Guala, S. E. Hommema, and R. J. Adrian. Large-scale and very-large-scale motions in turbulent pipe flow. *Journal of Fluid Mechanics*, 554:521–542, May 2006. doi: 10.1017/S0022112006008871. (cited on pages 26 and 120)

J. Guckenheimer. Strange Attractors in Fluids: Another View. *Annual Review of Fluid Mechanics*, 18(1): 15–29, 1986. doi: 10.1146/annurev.fl.18.010186.000311. (cited on page 30)

J. Guckenheimer and P. Holmes. *Nonlinear Oscillations, Dynamical Systems, and Bifurcations of Vector Fields*, volume 42. Springer Science & Business Media, 1st edition, 1983. ISBN 1-4612-1140-9. (cited on page 29)

K. Gustavsson and B. Mehlig. Statistical models for spatial patterns of heavy particles in turbulence. *Advances in Physics*, 65(1):1–57, Jan. 2016. doi: 10.1080/00018732.2016.1164490. (cited on page 39)

W. H. Hager and O. Castro-Orgaz. William Froude and the Froude Number. *Journal of Hydraulic Engineering*, 143(4):02516005, Apr. 2017. doi: 10.1061/(ASCE)HY.1943-7900.0001213. (cited on page 15)

J. Halcrow, J. F. Gibson, P. Cvitanović, and D. Viswanath. Heteroclinic connections in plane Couette flow. *Journal of Fluid Mechanics*, 621:365–376, Feb. 2009. doi: 10.1017/S0022112008005065. (cited on page 31)

Y.-S. Ham, H. Kobori, and M. Takasago. Effects of combined sewer overflow and stormwater on indicator bacteria concentrations in the Tama River due to the high population density of Tokyo Metropolitan area. *Environmental Monitoring and Assessment*, 152(1):459, May 2008. doi: 10.1007/s10661-008-0330-6. (cited on page 2)

J. M. Hamilton, J. Kim, and F. Waleffe. Regeneration mechanisms of near-wall turbulence structures. *Journal of Fluid Mechanics*, 287:317–348, 1995. (cited on page 24)

M. R. Head and P. Bandyopadhyay. New aspects of turbulent boundary-layer structure. *Journal of Fluid Mechanics*, 107:297–338, June 1981. doi: 10.1017/S0022112081001791. (cited on pages 25 and 26)

B. Hof, C. W. H. van Doorne, J. Westerweel, F. T. M. Nieuwstadt, H. Faisst, B. Eckhardt, H. Wedin, R. R. Kerswell, and F. Waleffe. Experimental Observation of Nonlinear Traveling Waves in Turbulent Pipe Flow. *Science*, 305(5690):1594–1598, Sept. 2004. doi: 10.1126/science.1100393. (cited on page 31)

H. Homann, J. Bec, and R. Grauer. Effect of turbulent fluctuations on the drag and lift forces on a towed sphere and its boundary layer. *Journal of Fluid Mechanics*, 721:155–179, Apr. 2013. doi: 10.1017/jfm.2013.66. (cited on pages xi, 111, 113, 114, 118, 124, and 202)

E. Hopf. A mathematical example displaying features of turbulence. *Communications on Pure and Applied Mathematics*, 1(4):303–322, 1948. doi: 10.1002/cpa.3160010401. (cited on page 28)

MA. House, JB. Ellis, EE. Herricks, T. Hvitved-Jacobsen, J. Seager, L. Lijklema, H. Aalderink, and IT. Cliforde. Urban drainage-impacts on receiving water quality. *Water Science and technology*, 27(12):117, 1993. (cited on page 3)

D. Hoyal, J. Atkinson, J. Depinto, and S. Taylor. The effect of turbulence on sediment deposition. *Journal of Hydraulic Research*, 33(3):349–360, May 1995. doi: 10.1080/00221689509498576. (cited on page 44)

S. G. Huisman, T. Barois, M. Bourgoin, A. Chouippe, T. Doychev, P. Huck, C. E. B. Morales, M. Uhlmann, and R. Volk. Columnar structure formation of a dilute suspension of settling spherical particles in a quiescent fluid. *Physical Review Fluids*, 1(7):074204, Nov. 2016. doi: 10.1103/PhysRevFluids.1.074204. (cited on page 42)

J. C. R. Hunt, A. A. Wray, and P. Moin. Eddies, streams, and convergence zones in turbulent flows. In *Studying Turbulence Using Numerical Simulation Databases, 2. Proceedings of the 1988 Summer Program*, Dec. 1988. (cited on pages 31 and 38)

S. Ikeda. Self-Formed Straight Channels in Sandy Beds. *Journal of the Hydraulics Division*, 107(4):389–406, Apr. 1981. doi: 10.1061/JYCEAJ.0005644. (cited on page 36)

D. L. Inman, R. J. Tait, and C. E. Nordstrom. Mixing in the surf zone. *Journal of Geophysical Research (1896-1977)*, 76(15):3493–3514, 1971. doi: 10.1029/JC076i015p03493. (cited on page 6)

P. J. Ireland and L. R. Collins. Direct numerical simulation of inertial particle entrainment in a shearless mixing layer. *Journal of Fluid Mechanics*, 704:301–332, Aug. 2012. doi: 10.1017/jfm.2012.241. (cited on page 41)

T. Itano and S. C. Generalis. Hairpin Vortex Solution in Planar Couette Flow: A Tapestry of Knotted Vortices. *Physical Review Letters*, 102(11):114501, Mar. 2009. doi: 10.1103/PhysRevLett.102.114501. (cited on page 30)

R. Jamieson, R. Gordon, D. Joy, and H. Lee. Assessing microbial pollution of rural surface waters: A review of current watershed scale modeling approaches. *Agricultural Water Management*, 70(1):1–17, Oct. 2004. doi: 10.1016/j.agwat.2004.05.006. (cited on page 4)

R. Jamieson, D. M. Joy, H. Lee, R. Kostaschuk, and R. Gordon. Transport and deposition of sediment-associated Escherichia coli in natural streams. *Water Research*, 39(12):2665–2675, July 2005. doi: 10.1016/j.watres.2005.04.040. (cited on pages xiii, 4, 5, 7, 43, 62, 187, 188, 189, 190, 204, and 226)

A. Jenkins, M. J. Kirkby, A. McDonald, P. Naden, and D. Kay. A Process Based Model of Faecal Bacterial Levels in Upland Catchments. *Water Science and Technology*, 16(5-7):453–462, May 1984. doi: 10.2166/wst.1984.0151. (cited on page 6)

M. Jenny, J. Dušek, and G. Bouchet. Instabilities and transition of a sphere falling or ascending freely in a Newtonian fluid. *Journal of Fluid Mechanics*, 508:201–239, 2004. doi: 10.1017/S0022112004009164. (cited on pages 36, 41, and 42)

J. Jiménez. Turbulent flows over rough walls. *Annu. Rev. Fluid Mech.*, 36:173–196, 2004. (cited on pages 34, 35, and 36)

J. Jiménez. Cascades in wall-bounded turbulence. *Annual Review of Fluid Mechanics*, 44(1):27–45, 2012. doi: 10.1146/annurev-fluid-120710-101039. (cited on page 123)

J. Jiménez. Near-wall turbulence. *Physics of Fluids*, 25(10):101302, Oct. 2013. doi: 10.1063/1.4824988. (cited on pages 27 and 28)

J. Jiménez. Coherent structures in wall-bounded turbulence. *Journal of Fluid Mechanics*, 842, May 2018. doi: 10.1017/jfm.2018.144. (cited on pages 24 and 28)

J. Jiménez and P. Moin. The minimal flow unit in near-wall turbulence. *Journal of Fluid Mechanics*, 225: 213–240, 1991. (cited on page 24)

J. Jiménez and A. Pinelli. The autonomous cycle of near-wall turbulence. *Journal of Fluid Mechanics*, 389:335–359, June 1999. doi: 10.1017/S0022112099005066. (cited on page 24)

J. Jiménez, A. A. Wray, P. G. Saffman, and R. S. Rogallo. The structure of intense vorticity in isotropic turbulence. *Journal of Fluid Mechanics*, 255:65–90, 1993. (cited on page 18)

J. Jiménez, M. Uhlmann, A. Pinelli, and G. Kawahara. Turbulent shear flow over active and passive porous surfaces. *Journal of Fluid Mechanics*, 442:89–117, Sept. 2001. doi: 10.1017/S0022112001004888. (cited on page 35)

H. E. Jobson and W. W. Sayre. Vertical transfer in open channel flow. *Journal of the Hydraulics Division*, 96(3):703–724, 1970. (cited on page 44)

A. Kadota and I. Nezu. Three-dimensional structure of space-time correlation on coherent vortices generated behind dune crest. *Journal of Hydraulic Research*, 37(1):59–80, Jan. 1999. doi: 10.1080/00221689909498532. (cited on page 37)

D. Kaftori, G. Hetsroni, and S. Banerjee. Particle behavior in the turbulent boundary layer. I. Motion, deposition, and entrainment. *Physics of Fluids*, 7(5):1095–1106, May 1995a. doi: 10.1063/1.868551. (cited on page 44)

D. Kaftori, G. Hetsroni, and S. Banerjee. Particle behavior in the turbulent boundary layer. II. Velocity and distribution profiles. *Physics of Fluids*, 7(5):1107–1121, May 1995b. doi: 10.1063/1.868552. (cited on page 44)

T. Kajishima. Influence of particle rotation on the interaction between particle clusters and particle-induced turbulence. *International Journal of Heat and Fluid Flow*, 25(5):721–728, Oct. 2004a. doi: 10.1016/j.ijheatfluidflow.2004.05.007. (cited on page 42)

T. Kajishima. Numerical investigation of collective behavior of gravitationally-settling particles in a homogeneous field. In *Proc. 5th International Conferences on Multiphase Flow, Yokohama, 2004*, 2004b. (cited on page 129)

T. Kajishima and S. Takiguchi. Interaction between particle clusters and particle-induced turbulence. *International Journal of Heat and Fluid Flow*, 23(5):639–646, Oct. 2002. doi: 10.1016/S0142-727X(02)00159-5. (cited on page 42)

S. M. Kashefipour, B. Lin, and R. A. Falconer. Dynamic modelling of bacterial concentrations in coastal waters: Effects of solar radiation on decay. In *Advances in Hydraulics and Water Engineering*, pages 993–998. World Scientific Publishing Company, July 2002. ISBN 978-981-238-090-6. doi: 10.1142/9789812776969_0183. (cited on page 7)

G. Kawahara and S. Kida. Periodic motion embedded in plane Couette turbulence: Regeneration cycle and burst. *Journal of Fluid Mechanics*, 449:291–300, Dec. 2001. doi: 10.1017/S0022112001006243. (cited on pages 31 and 32)

G. Kawahara, M. Uhlmann, and L. van Veen. The Significance of Simple Invariant Solutions in Turbulent Flows. *Annual Review of Fluid Mechanics*, 44(1):203–225, 2012. doi: 10.1146/annurev-fluid-120710-101228. (cited on page 30)

K. Kawanisi and R. Shiozaki. Turbulent effects on the settling velocity of suspended sediment. *Journal of hydraulic engineering*, 134(2):261–266, 2008. (cited on page 41)

N. K. Khullar, U. C. Kothyari, and K. G. Ranga Raju. Suspended Wash Load Transport of Nonuniform Sediments. *Journal of Hydraulic Engineering*, 136(8):534–543, Aug. 2010. doi: 10.1061/(ASCE)HY.1943-7900.0000223. (cited on page 35)

A. G. Kidanemariam. *The Formation of Patterns in Subaqueous Sediment.* KIT Scientific Publishing, 2016. ISBN 978-3-7315-0532-7. (cited on pages 37, 52, and 150)

A. G. Kidanemariam and M. Uhlmann. Direct numerical simulation of pattern formation in subaqueous sediment. *Journal of Fluid Mechanics*, 750:R2, 2014a. doi: 10.1017/jfm.2014.284. (cited on pages 55 and 60)

A. G. Kidanemariam and M. Uhlmann. Interface-resolved direct numerical simulation of the erosion of a sediment bed sheared by laminar channel flow. *International Journal of Multiphase Flow*, 67: 174–188, 2014b. doi: 10.1016/j.ijmultiphaseflow.2014.08.008. (cited on page 52)

A. G. Kidanemariam and M. Uhlmann. Formation of sediment patterns in channel flow: Minimal unstable systems and their temporal evolution. *Journal of Fluid Mechanics*, 818:716–743, May 2017. doi: 10.1017/jfm.2017.147. (cited on pages 57 and 60)

A. G. Kidanemariam, C. Chan-Braun, T. Doychev, and M. Uhlmann. Direct numerical simulation of horizontal open channel flow with finite-size, heavy particles at low solid volume fraction. *New Journal of Physics*, 15, 2013. doi: 10.1088/1367-2630/15/2/025031. (cited on pages 44, 62, and 102)

A. G. Kidanemariam, M. Scherer, and M. Uhlmann. Open-channel flow over evolving subaqueous ripples. *Journal of Fluid Mechanics*, 937:A26, Apr. 2022. doi: 10.1017/jfm.2022.113. (cited on page 150)

KT. Kiger and C. Pan. Suspension and turbulence modification effects of solid particulates on a horizontal turbulent channel flow. *J. Turbulence*, 3(19):1–17, 2002. (cited on page 44)

B. R. Kim, J. M. Higgins, and D. J. Bruggink. Reservoir Circulation Patterns and Water Quality. *Journal of Environmental Engineering*, 109(6):1284–1294, Dec. 1983. doi: 10.1061/(ASCE)0733-9372(1983)109:6(1284). (cited on page 8)

J. Kim and P. Moin. Application of a fractional-step method to incompressible Navier-Stokes equations. *Journal of Computational Physics*, 59(2):308–323, June 1985. doi: 10.1016/0021-9991(85)90148-2. (cited on page 50)

J. Kim, P. Moin, and R. Moser. Turbulence statistics in fully developed channel flow at low Reynolds number. *Journal of Fluid Mechanics*, 177:133–166, Apr. 1987. doi: 10.1017/S0022112087000892. (cited on pages 22 and 24)

K. C. Kim and R. J. Adrian. Very large-scale motion in the outer layer. *Physics of Fluids*, 11(2):417–422, Jan. 1999. doi: 10.1063/1.869889. (cited on pages 26, 148, 162, and 201)

M. Kleidorfer, M. Möderl, R. Sitzenfrei, C. Urich, and W. Rauch. A case independent approach on the impact of climate change effects on combined sewer system performance. *Water Science and Technology*, 60(6):1555–1564, Sept. 2009. doi: 10.2166/wst.2009.520. (cited on page 3)

S. J. Kline, W. C. Reynolds, F. A. Schraub, and P. W. Runstadler. The structure of turbulent boundary layers. *Journal of Fluid Mechanics*, 30(4):741–773, Dec. 1967. doi: 10.1017/S0022112067001740. (cited on page 24)

A. N. Kolmogorov. The local structure of turbulence in incompressible viscous fluid for very large Reynolds numbers. *Dokl. Akad. Nauk SSSR*, 30:301–305, 1941. (cited on page 16)

B. Kubrak, H. Herlina, F. Greve, and J. G. Wissink. Low-diffusivity scalar transport using a WENO scheme and dual meshing. *Journal of Computational Physics*, 240:158–173, May 2013. doi: 10.1016/j.jcp.2012.12.039. (cited on page 53)

J. D. Kulick, J. R. Fessler, and J. K. Eaton. Particle response and turbulence modification in fully developed channel flow. *Journal of Fluid Mechanics*, 277:109–134, Oct. 1994. doi: 10.1017/S0022112094002703. (cited on page 41)

L. D. Landau. On the problem of turbulence. In *Dokl. Akad. Nauk USSR*, volume 44, page 311, 1944. (cited on page 28)

O. E. Lanford. The Strange Attractor Theory of Turbulence. *Annual Review of Fluid Mechanics*, 14(1): 347–364, 1982. doi: 10.1146/annurev.fl.14.010182.002023. (cited on pages 29 and 30)

J. J. Leendertse and A. B. Nelson. *A Water-Quality Simulation Model for Well Mixed Estuaries and Coastal Seas: Volume IX, the Computer Program.* RAND Corporation, Santa Monica, CA, 1978. (cited on page 7)

D. Legendre, A. Merle, and J. Magnaudet. Wake of a spherical bubble or a solid sphere set fixed in a turbulent environment. *Physics of Fluids*, 18(4):048102, Apr. 2006. doi: 10.1063/1.2191885. (cited on pages 41 and 113)

S. H. Lehnigk. A class of conservative diffusion processes with delta function initial conditions. *Journal of Mathematical Physics*, 17(6):973–976, June 1976. doi: 10.1063/1.522990. (cited on page 188)

R. M. Li and H. W. Shen. Solid particle settlement in open-channel flow. *Journal of the Hydraulics Division*, 101(HY7), July 1975. (cited on pages 44, 155, 156, 157, 162, and 202)

T. Li, Q. Tan, and S. Zhu. Characteristics of combined sewer overflows in Shanghai and selection of drainage systems. *Water and Environment Journal*, 24(1):74–82, 2010. doi: 10.1111/j.1747-6593.2008.00141.x. (cited on page 2)

H.-K. Liu. Mechanics of Sediment-Ripple Formation. *Journal of the Hydraulics Division*, 83(2):1197–23, Apr. 1957. doi: 10.1061/JYCEAJ.0000075. (cited on page 37)

W.-C. Liu and W.-C. Huang. Modeling the transport and distribution of fecal coliform in a tidal estuary. *Science of The Total Environment*, 431:1–8, Aug. 2012. doi: 10.1016/j.scitotenv.2012.05.016. (cited on page 8)

W.-C. Liu, W.-T. Chan, and C.-C. Young. Modeling fecal coliform contamination in a tidal Danshuei River estuarine system. *Science of The Total Environment*, 502:632–640, Jan. 2015. doi: 10.1016/j.scitotenv.2014.09.065. (cited on page 9)

E. N. Lorenz. Deterministic Nonperiodic Flow. *Journal of the Atmospheric Sciences*, 20(2):130–141, Mar. 1963. doi: 10.1175/1520-0469(1963)020<0130:DNF>2.0.CO;2. (cited on page 30)

A. Lozano-Durán and J. Jiménez. Effect of the computational domain on direct simulations of turbulent channels up to $Re_\tau = 4200$. *Physics of Fluids*, 26(1):011702, Jan. 2014a. doi: 10.1063/1.4862918. (cited on pages 26, 120, 121, and 124)

A. Lozano-Durán and J. Jiménez. Time-resolved evolution of coherent structures in turbulent channels: Characterization of eddies and cascades. *Journal of Fluid Mechanics*, 759:432–471, Nov. 2014b. doi: 10.1017/jfm.2014.575. (cited on page 27)

A. Lozano-Durán, O. Flores, and J. Jiménez. The three-dimensional structure of momentum transfer in turbulent channels. *Journal of Fluid Mechanics*, 694:100–130, Mar. 2012a. doi: 10.1017/jfm.2011.524. (cited on page 27)

A. Lozano-Durán, O. Flores, and J. Jiménez. The three-dimensional structure of momentum transfer in turbulent channels. *Journal of Fluid Mechanics*, 694:100–130, Mar. 2012b. doi: 10.1017/jfm.2011.524. (cited on page 120)

A. Lozano-Durán, M. Holzner, and J. Jiménez. Multiscale analysis of the topological invariants in the logarithmic region of turbulent channels at a friction Reynolds number of 932. *Journal of Fluid Mechanics*, 803:356–394, Sept. 2016. doi: 10.1017/jfm.2016.504. (cited on pages 163 and 164)

LUBW. Deutsches Gewässerkundliches Jahrbuch. Rheingebiet, Teil I. Hoch- und Oberrhein. Technical report, Landesanstalt für Umwelt, Messungen und Naturschutz, Karlsruhe, 2009. (cited on page 2)

F. Lucci, A. Ferrante, and S. Elghobashi. Is Stokes number an appropriate indicator for turbulence modulation by particles of Taylor-length-scale size? *Physics of Fluids*, 23(2):025101, Feb. 2011. doi: 10.1063/1.3553279. (cited on page 39)

B. A. Madge and J. N. Jensen. Ultraviolet Disinfection of Fecal Coliform in Municipal Wastewater: Effects of Particle Size. *Water Environment Research*, 78(3):294–304, 2006. doi: 10.2175/106143005X94385. (cited on page 3)

R. Manning. *On the Flow of Water in Open Channels and Pipes*, volume 20 of *Transactions of the Institution of Civil Engineers of Ireland.* Institution of Civil Engineers of Ireland, 1890. (cited on page 6)

M. J. Manton. On the motion of a small particle in the atmosphere. *Boundary-Layer Meteorology*, 6(3): 487–504, May 1974. doi: 10.1007/BF02137681. (cited on page 40)

C. Marchioli and A. Soldati. Mechanisms for particle transfer and segregation in a turbulent boundary layer. *Journal of Fluid Mechanics*, 468:283–315, Oct. 2002. doi: 10.1017/S0022112002001738. (cited on page 44)

I. Marusic and S. Broomhall. Leonardo da Vinci and Fluid Mechanics. *Annual Review of Fluid Mechanics*, 53(1):1–25, 2021. doi: 10.1146/annurev-fluid-022620-122816. (cited on page 1)

I. Marusic and J. P. Monty. Attached Eddy Model of Wall Turbulence. *Annual Review of Fluid Mechanics*, 51(1):49–74, Jan. 2019. doi: 10.1146/annurev-fluid-010518-040427. (cited on page 28)

I. Marusic, B. J. McKeon, P. A. Monkewitz, H. M. Nagib, A. J. Smits, and K. R. Sreenivasan. Wall-bounded turbulent flows at high Reynolds numbers: Recent advances and key issues. *Physics of Fluids*, 22(6): 065103, June 2010. doi: 10.1063/1.3453711. (cited on page 35)

J. Mathieu and J. Scott. *An Introduction to Turbulent Flow.* Cambridge University Press, June 2000. ISBN 978-0-521-77538-0. doi: 10.1017/CBO9781316529850. (cited on page 18)

M. Maxey. Simulation Methods for Particulate Flows and Concentrated Suspensions. *Annual Review of Fluid Mechanics*, 49(1):171–193, Jan. 2017. doi: 10.1146/annurev-fluid-122414-034408. (cited on page 49)

M. R. Maxey. The gravitational settling of aerosol particles in homogeneous turbulence and random flow fields. *Journal of Fluid Mechanics*, 174:441–465, Jan. 1987. doi: 10.1017/S0022112087000193. (cited on pages 37, 38, and 40)

M. R. Maxey and S. Corrsin. Gravitational Settling of Aerosol Particles in Randomly Oriented Cellular Flow Fields. *Journal of the Atmospheric Sciences*, 43(11):1112–1134, June 1986. doi: 10.1175/1520-0469(1986)043<1112:GSOAPI>2.0.CO;2. (cited on pages 37, 40, and 201)

A. McDonald and D. Kay. Enteric bacterial concentrations in reservoir feeder streams: Baseflow characteristics and response to hydrograph events. *Water Research*, 15(8):961–968, Jan. 1981. doi: 10.1016/0043-1354(81)90203-7. (cited on page 5)

A. McDonald, D. Kay, and A. Jenkins. Generation of fecal and total coliform surges by stream flow manipulation in the absence of normal hydrometeorological stimuli. *Applied and Environmental Microbiology*, 44(2):292–300, Aug. 1982. doi: 10.1128/aem.44.2.292-300.1982. (cited on pages 5, 62, 192, 199, and 204)

B. J. McKeon. The engine behind (wall) turbulence: Perspectives on scale interactions. *Journal of Fluid Mechanics*, 817, Apr. 2017. doi: 10.1017/jfm.2017.115. (cited on page 28)

J. B. McLaughlin. Numerical computation of particles-turbulence interaction. *International Journal of Multiphase Flow*, 20:211–232, Aug. 1994. doi: 10.1016/0301-9322(94)90073-6. (cited on page 37)

R. Mei. Effect of turbulence on the particle settling velocity in the nonlinear drag range. *International Journal of Multiphase Flow*, 20(2):273–284, Apr. 1994. doi: 10.1016/0301-9322(94)90082-5. (cited on page 41)

E. Meyer-Peter and R. Müller. Formulas for bed-load transport. In *IAHSR 2nd Meeting, Stockholm, Appendix 2*. IAHR, 1948. (cited on pages 59, 61, and 190)

J. S. Mill. *A System of Logic, Ratiocinative and Inductive: Being a Connected View of the Principles of Evidence, and the Methods of Scientific Investigation*. John W. Parker, 1856. (cited on page 157)

R. Mittal and G. Iaccarino. Immersed Boundary Methods. *Annual Review of Fluid Mechanics*, 37(1): 239–261, 2005. doi: 10.1146/annurev.fluid.37.061903.175743. (cited on page 49)

R. Monchaux, M. Bourgoin, and A. Cartellier. Analyzing preferential concentration and clustering of inertial particles in turbulence. *International Journal of Multiphase Flow*, 40:1–18, Apr. 2012. doi: 10.1016/j.ijmultiphaseflow.2011.12.001. (cited on pages 39 and 127)

J. P. Monty, J. A. Stewart, R. C. Williams, and M. S. Chong. Large-scale features in turbulent pipe and channel flows. *Journal of Fluid Mechanics*, 589:147–156, Oct. 2007. doi: 10.1017/S002211200700777X. (cited on page 26)

J. P. Monty, N. Hutchins, H. C. H. Ng, I. Marusic, and M. S. Chong. A comparison of turbulent pipe, channel and boundary layer flows. *Journal of Fluid Mechanics*, 632:431–442, Aug. 2009. doi: 10.1017/S0022112009007423. (cited on page 26)

S. Motoki, K. Tsugawa, M. Shimizu, and G. Kawahara. The ultimate state of turbulent permeable-channel flow. *Journal of Fluid Mechanics*, 931, Jan. 2022. doi: 10.1017/jfm.2021.937. (cited on page 35)

Y. Motoori and S. Goto. Hierarchy of coherent structures and real-space energy transfer in turbulent channel flow. *Journal of Fluid Mechanics*, 911, Mar. 2021. doi: 10.1017/jfm.2020.1025. (cited on pages 163, 201, 216, and 217)

Y. Motoori, C. Wong, and S. Goto. Role of the hierarchy of coherent structures in the transport of heavy small particles in turbulent channel flow. *Journal of Fluid Mechanics*, 942, July 2022. doi: 10.1017/jfm.2022.327. (cited on pages 45, 163, 171, and 201)

S. P. Murray. Settling velocities and vertical diffusion of particles in turbulent water. *Journal of Geophysical Research (1896-1977)*, 75(9):1647–1654, 1970. doi: 10.1029/JC075i009p01647. (cited on page 41)

R. Nagaosa. Direct numerical simulation of vortex structures and turbulent scalar transfer across a free surface in a fully developed turbulence. *Physics of Fluids*, 11(6):1581–1595, June 1999. doi: 10.1063/1.870020. (cited on page 27)

R. Nagaosa and R. A. Handler. Statistical analysis of coherent vortices near a free surface in a fully developed turbulence. *Physics of Fluids*, 15(2):375–394, Jan. 2003. doi: 10.1063/1.1533071. (cited on page 27)

M. Nagata. Three-dimensional finite-amplitude solutions in plane Couette flow: Bifurcation from infinity. *Journal of Fluid Mechanics*, 217:519–527, Aug. 1990. doi: 10.1017/S0022112090000829. (cited on pages 30, 31, and 173)

C. Narayanan and D. Lakehal. Mechanisms of particle deposition in a fully developed turbulent open channel flow. *Physics of Fluids*, 15(3):763–775, Feb. 2003. doi: 10.1063/1.1545473. (cited on page 138)

CLMH. Navier. Mémoire sur les lois du mouvement des fluides. *Mémoires de l'Académie Royale des Sciences de l'Institut de France*, 6(1823):389–440, 1823. (cited on page 1)

M. Neiburger. Weather modification and smog. *Science (New York, N.Y.)*, 126(3275):637–645, 1957. (cited on page 1)

S. L. Neitsch, J. G. Arnold, J. R. Kiniry, and J. R. Williams. Soil and water assessment tool theoretical documentation version 2009. Technical report, Texas Water Resources Institute, 2011. (cited on page 6)

I. Nezu and H. Nakagawa. Cellular Secondary Currents in Straight Conduit. *Journal of Hydraulic Engineering*, 110(2):173–193, Feb. 1984. doi: 10.1061/(ASCE)0733-9429(1984)110:2(173). (cited on page 36)

P. Nielsen. On the motion of suspended sand particles. *Journal of Geophysical Research: Oceans*, 89(C1):616–626, 1984. doi: 10.1029/JC089iC01p00616. (cited on page 40)

P. Nielsen. *Coastal Bottom Boundary Layers and Sediment Transport*, volume 4. World Scientific, 1992. ISBN 981-02-0473-6. (cited on page 40)

P. Nielsen. Turbulence effects on the settling of suspended particles. *Journal of Sedimentary Research*, 63(5):835–838, Sept. 1993. doi: 10.1306/D4267C1C-2B26-11D7-8648000102C1865D. (cited on pages 40 and 41)

J. Nikuradse. *Strömungsgesetze in Rauhen Rohren.* VDI-Verl., Berlin, 1933. (cited on page 34)

Y. Niño, F. Lopez, and M. Garcia. Threshold for particle entrainment into suspension. *Sedimentology*, 50(2):247–263, 2003. doi: 10.1046/j.1365-3091.2003.00551.x. (cited on page 44)

Y. Nishiyama, Y. Kuwata, and K. Suga. Direct numerical simulation of turbulent heat transfer over fully resolved anisotropic porous structures. *International Journal of Heat and Fluid Flow*, 81:108515, Feb. 2020. doi: 10.1016/j.ijheatfluidflow.2019.108515. (cited on page 35)

W. of Ockham. *Quaestiones et Decisiones in Quattuor Libros Sententiarum Petri Lombardi: Centilogium Theologicum.* Johannes Trechsel, 1495. (cited on page 157)

S. Oka and S. Goto. Generalized sweep-stick mechanism of inertial-particle clustering in turbulence. *Physical Review Fluids*, 6(4):044605, Apr. 2021. doi: 10.1103/PhysRevFluids.6.044605. (cited on page 38)

S. A. Orszag. Accurate solution of the Orr–Sommerfeld stability equation. *Journal of Fluid Mechanics*, 50 (4):689–703, Dec. 1971. doi: 10.1017/S0022112071002842. (cited on page 21)

S. A. Orszag and L. C. Kells. Transition to turbulence in plane Poiseuille and plane Couette flow. *Journal of Fluid Mechanics*, 96(1):159–205, Jan. 1980. doi: 10.1017/S0022112080002066. (cited on page 115)

J. Page and R. R. Kerswell. Searching turbulence for periodic orbits with dynamic mode decomposition. *Journal of Fluid Mechanics*, 886, Mar. 2020. doi: 10.1017/jfm.2019.1074. (cited on page 32)

C. L. Palmer. The Pollutional Effects of Storm-Water Overflows from Combined Sewers. *Sewage and Industrial Wastes*, 22(2):154–165, 1950. (cited on page 3)

M. D. Palmer and R. J. Dewey. Beach fecal coliforms. *Canadian Journal of Civil Engineering*, 11(2): 217–224, June 1984. doi: 10.1139/l84-033. (cited on page 7)

J. S. Park and M. D. Graham. Exact coherent states and connections to turbulent dynamics in minimal channel flow. *Journal of Fluid Mechanics*, 782:430–454, Nov. 2015. doi: 10.1017/jfm.2015.554. (cited on pages 30 and 174)

V. C. Patel and M. R. Head. Some observations on skin friction and velocity profiles in fully developed pipe and channel flows. *Journal of Fluid Mechanics*, 38(1):181–201, Aug. 1969. doi: 10.1017/S0022112069000115. (cited on page 21)

A. E. Perry and M. S. Chong. On the mechanism of wall turbulence. *Journal of Fluid Mechanics*, 119: 173–217, June 1982. doi: 10.1017/S0022112082001311. (cited on page 25)

A. E. Perry, S. Henbest, and M. S. Chong. A theoretical and experimental study of wall turbulence. *Journal of Fluid Mechanics*, 165:163–199, Apr. 1986. doi: 10.1017/S002211208600304X. (cited on pages 25 and 121)

C. Peruzzi, D. Poggi, L. Ridolfi, and C. Manes. On the scaling of large-scale structures in smooth-bed turbulent open-channel flows. *Journal of Fluid Mechanics*, 889, Apr. 2020. doi: 10.1017/jfm.2020.73. (cited on page 36)

C. S. Peskin. Flow Patterns Around Heart Valves: A Numerical Method. *Journal of Computational Physics*, 10:252–271, 1972. (cited on pages 49 and 50)

C. S. Peskin. The immersed boundary method. *Acta Numerica*, 11(January 2002):479–517, 2002. doi: 10.1017/S0962492902000077. (cited on page 50)

T. Pestana, M. Uhlmann, and G. Kawahara. Can preferential concentration of finite-size particles in plane Couette turbulence be reproduced with the aid of equilibrium solutions? *Physical Review Fluids*, 5(3):034305, Mar. 2020. doi: 10.1103/PhysRevFluids.5.034305. (cited on page 173)

A. J. Petersen, L. Baker, and F. Coletti. Experimental study of inertial particles clustering and settling in homogeneous turbulence. *Journal of Fluid Mechanics*, 864:925–970, Apr. 2019. doi: 10.1017/jfm.2019.31. (cited on page 40)

P. J. Phillips, A. T. Chalmers, J. L. Gray, D. W. Kolpin, W. T. Foreman, and G. R. Wall. Combined Sewer Overflows: An Environmental Source of Hormones and Wastewater Micropollutants. *Environmental Science & Technology*, 46(10):5336–5343, May 2012. doi: 10.1021/es3001294. (cited on page 3)

M. Pinelli. *Direct numerical simulation of interfacial mass transfer and pollutant transport in turbulent open channel flow.* PhD thesis, Karlsruhe Institute of Technology, 2022. (cited on pages 195 and 206)

M. Pinelli, H. Herlina, J. G. Wissink, and M. Uhlmann. Direct numerical simulation of turbulent mass transfer at the surface of an open channel flow. *Journal of Fluid Mechanics*, 933, Feb. 2022. doi: 10.1017/jfm.2021.1080. (cited on page 27)

Y. Pomeau and P. Manneville. Intermittent transition to turbulence in dissipative dynamical systems. *Communications in Mathematical Physics*, 74(2):189–197, 1980. (cited on page 29)

S. B. Pope. *Turbulent Flows*. Cambridge University Press, 2000. ISBN 978-0-521-59886-6. doi: 10.1017/CBO9780511840531. (cited on pages 22 and 77)

K. R. Popper. *Logik Der Forschung*, volume 9 of *Schriften Zur Wissenschaftlichen Weltauffassung*. Springer-Verlag Wien, Wien, 1935. ISBN 978-3-7091-2021-7. (cited on page 157)

L. Prandtl. Bericht über die Entstehung der Turbulenz. *Z. Angew. Math. Mech*, 5:136–139, 1925. (cited on page 23)

R. L. Prentice. A Log Gamma Model and Its Maximum Likelihood Estimation. *Biometrika*, 61(3):539–544, 1974. doi: 10.2307/2334737. (cited on page 157)

A. Prosperetti and G. Tryggvason. *Computational Methods for Multiphase Flow*. Cambridge University Press, June 2009. ISBN 978-1-139-45990-7. doi: 10.1017/CBO9780511607486. (cited on page 49)

J. Qian, E. Walters, P. Rutschmann, M. Wagner, and H. Horn. Modelling the influence of total suspended solids on *E. coli* removal in river water. *Water Science and Technology*, 73(6):1320–1332, Mar. 2016. doi: 10.2166/wst.2015.605. (cited on pages 7 and 191)

N. M. Qureshi, M. Bourgoin, C. Baudet, A. Cartellier, and Y. Gagne. Turbulent Transport of Material Particles: An Experimental Study of Finite Size Effects. *Physical Review Letters*, 99(18):184502, Oct. 2007. doi: 10.1103/PhysRevLett.99.184502. (cited on page 39)

M. Rashidi, G. Hetsroni, and S. Banerjee. Particle-turbulence interaction in a boundary layer. *International Journal of Multiphase Flow*, 16(6):935–949, Nov. 1990. doi: 10.1016/0301-9322(90)90099-5. (cited on page 44)

A. L. Reda, P. Ferreira, M. Mendes, and M. B. Beck. Combined sewer overflows in Brazil: A 2014 situation report. In *Anais: SHEW'C2014-XIV Safety, Health and Environment World Congress. Cubatão, SP, Brazil*, volume 14, 2014. doi: DOI10.14684/SHEWC.14.2014.11-15. (cited on page 2)

M. W. Reeks. The transport of discrete particles in inhomogeneous turbulence. *Journal of Aerosol Science*, 14(6):729–739, Jan. 1983. doi: 10.1016/0021-8502(83)90055-1. (cited on page 43)

M. W. Reeks. On a kinetic equation for the transport of particles in turbulent flows. *Physics of Fluids A: Fluid Dynamics*, 3(3):446–456, Mar. 1991. doi: 10.1063/1.858101. (cited on page 43)

M. W. Reeks. On model equations for particle dispersion in inhomogeneous turbulence. *International Journal of Multiphase Flow*, 31(1):93–114, Jan. 2005. doi: 10.1016/j.ijmultiphaseflow.2004.08.002. (cited on page 43)

O. Reynolds. III. An experimental investigation of the circumstances which determine whether the motion of water shall be direct or sinuous, and of the law of resistance in parallel channels. *Proceedings of the Royal Society of London*, 35(224-226):84–99, Jan. 1883. doi: 10.1098/rspl.1883.0018. (cited on page 14)

O. Reynolds. IV. On the dynamical theory of incompressible viscous fluids and the determination of the criterion. *Philosophical Transactions of the Royal Society of London. (A.)*, 186:123–164, Jan. 1895. doi: 10.1098/rsta.1895.0004. (cited on page 14)

J. F. Richardson and W. N. Zaki. The sedimentation of a suspension of uniform spheres under conditions of viscous flow. *Chemical Engineering Science*, 3(2):65–73, Apr. 1954. doi: 10.1016/0009-2509(54)85015-9. (cited on pages 42, 132, 133, 161, and 201)

L. F. Richardson. *Weather Prediction by Numerical Process.* Cambridge University Press, Cambridge, 1922. (cited on page 16)

E. Rind and I. P. Castro. Direct numerical simulation of axisymmetric wakes embedded in turbulence. *Journal of Fluid Mechanics*, 710:482–504, Nov. 2012. doi: 10.1017/jfm.2012.374. (cited on page 41)

S. K. Robinson. Coherent motions in the turbulent boundary layer. *Annual Review of Fluid Mechanics*, 23(1):601–639, 1991. (cited on page 28)

M. Rodrigues, A. Oliveira, M. Guerreiro, A. B. Fortunato, J. Menaia, L. M. David, and A. Cravo. Modeling fecal contamination in the Aljezur coastal stream (Portugal). *Ocean Dynamics*, 61(6):841, Mar. 2011. doi: 10.1007/s10236-011-0392-9. (cited on page 9)

A. M. Roma, C. S. Peskin, and M. J. Berger. An Adaptive Version of the Immersed Boundary Method. *Journal of Computational Physics*, 153(2):509–534, Aug. 1999. doi: 10.1006/jcph.1999.6293. (cited on page 50)

V. A. Romanov. Stability of plane-parallel Couette flow. *Functional Analysis and Its Applications*, 7(2): 137–146, Apr. 1973. doi: 10.1007/BF01078886. (cited on pages 28 and 115)

M. M. Roper and K. C. Marshall. Modification of the interaction between Escherichia coli and bacteriophage in saline sediment. *Microbial Ecology*, 1(1):1–13, Dec. 1974. doi: 10.1007/BF02512375. (cited on page 4)

H. Rouse. Modern Conceptions of the Mechanics of Fluid Turbulence. *Transactions of the American Society of Civil Engineers*, 102(1):463–505, Jan. 1937. doi: 10.1061/TACEAT.0004872. (cited on page 35)

D. W. I. Rouson and J. K. Eaton. On the preferential concentration of solid particles in turbulent channel flow. *Journal of Fluid Mechanics*, 428:149–169, Feb. 2001. doi: 10.1017/S0022112000002627. (cited on page 45)

D. Ruelle and F. Takens. On the nature of turbulence. *Communications in Mathematical Physics*, 20(3): 167–192, 1971. (cited on pages 28, 30, and 31)

A. M. Sadeghi and J. G. Arnold. A SWAT/microbial sub-model for predicting pathogen loadings in surface and groundwater at watershed and basin scales. In *Total Maximum Daily Load (TMDL): Environmental Regulations, Proceedings of 2002 Conference*, page 56. American Society of Agricultural and Biological Engineers, 2002. ISBN 1-892769-24-7. (cited on page 6)

H. Salwen, F. W. Cotton, and C. E. Grosch. Linear stability of Poiseuille flow in a circular pipe. *Journal of Fluid Mechanics*, 98(2):273–284, May 1980. doi: 10.1017/S0022112080000146. (cited on page 28)

M. Sano and K. Tamai. A universal transition to turbulence in channel flow. *Nature Physics*, 12(3):249–253, Mar. 2016. doi: 10.1038/nphys3659. (cited on page 29)

W. W. Sayre. *Dispersion of Mass in Open-Channel Flow.* PhD thesis, Colorado State University. Libraries, 1967. (cited on page 43)

W. W. Sayre. Dispersion of Silt Particles in Open Channel Flow. *Journal of the Hydraulics Division*, 95(3): 1009–1038, May 1969. doi: 10.1061/JYCEAJ.0002077. (cited on page 43)

W. W. Sayre and D. W. Hubbell. *Transport and Dispersion of Labeled Bed Material, North Loup River, Nebraska.* U.S. Government Printing Office, 1965. (cited on page 43)

M. Scherer. *Turbulent Coherent Structures, Secondary Currents and Sediment Ridges.* PhD thesis, Karlsruhe Institute of Technology, 2022. (cited on pages 55 and 93)

M. Scherer, A. G. Kidanemariam, and M. Uhlmann. On the scaling of the instability of a flat sediment bed with respect to ripple-like patterns. *Journal of Fluid Mechanics*, 900, Oct. 2020. doi: 10.1017/jfm.2020.423. (cited on page 55)

M. Scherer, M. Uhlmann, A. G. Kidanemariam, and M. Krayer. On the role of turbulent large-scale streaks in generating sediment ridges. *Journal of Fluid Mechanics*, 930, Jan. 2022. doi: 10.1017/jfm.2021.891. (cited on pages 36, 41, 45, 55, 67, and 203)

L. Schiller and A. Naumann. über die grundlegenden Berechnungen bei der Schwerkraftaufbereitung. *Z. Ver. Dtsch. Ing*, 77(12):318–320, 1933. (cited on pages xi, 65, and 118)

I. Schnauder, B. Bockelmann-Evans, and B. Lin. Modelling faecal bacteria pathways in receiving waters. *Proceedings of the Institution of Civil Engineers - Maritime Engineering*, 160(4):143–153, Dec. 2007. doi: 10.1680/maen.2007.160.4.143. (cited on page 7)

C. Schultz-Fademrecht, M. Wichern, and H. Horn. The impact of sunlight on inactivation of indicator microorganisms both in river water and benthic biofilms. *Water Research*, 42(19):4771–4779, Dec. 2008. doi: 10.1016/j.watres.2008.08.022. (cited on page 4)

G. Seizilles, E. Lajeunesse, O. Devauchelle, and M. Bak. Cross-stream diffusion in bedload transport. *Physics of Fluids*, 26(1):013302, Jan. 2014. doi: 10.1063/1.4861001. (cited on page 207)

A. Semadeni-Davies, C. Hernebring, G. Svensson, and L.-G. Gustafsson. The impacts of climate change and urbanisation on drainage in Helsingborg, Sweden: Combined sewer system. *Journal of Hydrology*, 350(1):100–113, Feb. 2008. doi: 10.1016/j.jhydrol.2007.05.028. (cited on page 3)

G. Seminara. Fluvial Sedimentary Patterns. *Annual Review of Fluid Mechanics*, 42(1):43–66, Jan. 2010. doi: 10.1146/annurev-fluid-121108-145612. (cited on page 36)

A. Shekar and M. D. Graham. Exact coherent states with hairpin-like vortex structure in channel flow. *Journal of Fluid Mechanics*, 849:76–89, Aug. 2018. doi: 10.1017/jfm.2018.409. (cited on pages 30, 31, 174, 184, and 203)

B. M. Sherer, J. R. Miner, J. A. Moore, and J. C. Buckhouse. Resuspending Organisms from a Rangeland Stream Bottom. *Transactions of the ASAE*, 31(4):1217–1222, 1988. doi: 10.13031/2013.30847. (cited on pages 5 and 192)

B. M. Sherer, J. R. Miner, J. A. Moore, and J. C. Buckhouse. Indicator Bacterial Survival in Stream Sediments. *Journal of Environmental Quality*, 21(4):591–595, 1992. doi: 10.2134/jeq1992.00472425002100040011x. (cited on pages 5, 62, and 192)

A. Shields. Anwendung der Aehnlichkeitsmechanik und der Turbulenzforschung auf die Geschiebebewegung. *PhD Thesis Technical University Berlin*, 1936. (cited on page 36)

M. Shimizu and P. Manneville. Bifurcations to turbulence in transitional channel flow. *Physical Review Fluids*, 4(11):113903, Nov. 2019. doi: 10.1103/PhysRevFluids.4.113903. (cited on page 29)

M. A. Shockling, J. J. Allen, and A. J. Smits. Roughness effects in turbulent pipe flow. *Journal of Fluid Mechanics*, 564:267–285, Oct. 2006. doi: 10.1017/S0022112006001467. (cited on page 34)

E. D. Siggia. Numerical study of small-scale intermittency in three-dimensional turbulence. *Journal of Fluid Mechanics*, 107:375–406, June 1981. doi: 10.1017/S002211208100181X. (cited on page 18)

D. B. Simons and E. V. Richardson. Forms of bed roughness in alluvial channels. *Journal of the Hydraulics Division*, 87(3):87–105, 1961. (cited on page 37)

C. R. Smith and S. P. Metzler. The characteristics of low-speed streaks in the near-wall region of a turbulent boundary layer. *Journal of Fluid Mechanics*, 129:27–54, Apr. 1983. doi: 10.1017/S0022112083000634. (cited on page 24)

A. J. Smits, B. J. McKeon, and I. Marusic. High–Reynolds Number Wall Turbulence. *Annual Review of Fluid Mechanics*, 43(1):353–375, 2011. doi: 10.1146/annurev-fluid-122109-160753. (cited on pages 26 and 28)

E. Sober. *Simplicity*. Oxford : Clarendon Press, 1975. ISBN 978-0-19-824407-3. (cited on page 157)

K. D. Squires and J. K. Eaton. Particle response and turbulence modification in isotropic turbulence. *Physics of Fluids A: Fluid Dynamics*, 2(7):1191–1203, July 1990. doi: 10.1063/1.857620. (cited on page 41)

K. D. Squires and J. K. Eaton. Preferential concentration of particles by turbulence. *Physics of Fluids A: Fluid Dynamics*, 3(5):1169–1178, May 1991. doi: 10.1063/1.858045. (cited on pages 37 and 38)

E. W. Stacy. A Generalization of the Gamma Distribution. *The Annals of Mathematical Statistics*, 33(3): 1187–1192, 1962. (cited on page 83)

B. M. Steets and P. A. Holden. A mechanistic model of runoff-associated fecal coliform fate and transport through a coastal lagoon. *Water Research*, 37(3):589–608, Feb. 2003. doi: 10.1016/S0043-1354(02)00312-3. (cited on page 6)

G. G. Stokes. On the Theories of the Internal Friction of Fluids in Motion, and of the Equilibrium and Motion of Elastic Solids. *Transactions of the Cambridge Philosophical Society*, 8:287–305, 1845. doi: 10.1017/CBO9780511702242.005. (cited on page 1)

G. G. Stokes. On the effect of the internal friction of fluids on the motion of pendulums. *Transactions of the Cambridge Philosophical Society*, 9:8, 1851. (cited on page 1)

S. H. Strogatz. *Nonlinear Dynamics and Chaos: With Applications to Physics, Biology, Chemistry, and Engineering*. CRC Press, 2018. ISBN 0-429-49256-1. (cited on page 29)

A. Stroh, K. Schäfer, B. Frohnapfel, and P. Forooghi. Rearrangement of secondary flow over spanwise heterogeneous roughness. *Journal of Fluid Mechanics*, 885, Feb. 2020. doi: 10.1017/jfm.2019.1030. (cited on page 207)

M. Stumpf. Solutions to Exercises from Turbulent Flows by Stephen B. Pope: Solution to Exercise 7.8. https://pope.mae.cornell.edu/TurbulentFlows/solutions/, 2017. (cited on page 22)

K. Suga, Y. Nakagawa, and M. Kaneda. Spanwise turbulence structure over permeable walls. *Journal of Fluid Mechanics*, 822:186–201, July 2017. doi: 10.1017/jfm.2017.278. (cited on page 35)

K. Suga, Y. Okazaki, U. Ho, and Y. Kuwata. Anisotropic wall permeability effects on turbulent channel flows. *Journal of Fluid Mechanics*, 855:983–1016, Nov. 2018. doi: 10.1017/jfm.2018.666. (cited on page 35)

B. M. Sumer. Mean velocity and longitudinal dispersion of heavy particles in turbulent open-channel flow. *Journal of Fluid Mechanics*, 65(01):11–28, 1974. (cited on page 43)

B. M. Sumer. Settlement of solid particles in open-channel flow. *Journal of the Hydraulics Division*, 103 (HY11), Nov. 1977. (cited on page 43)

B. Suri, J. Tithof, R. O. Grigoriev, and M. F. Schatz. Forecasting Fluid Flows Using the Geometry of Turbulence. *Physical Review Letters*, 118(11):114501, Mar. 2017. doi: 10.1103/PhysRevLett.118.114501. (cited on page 30)

J. D. Swearingen and R. F. Blackwelder. The growth and breakdown of streamwise vortices in the presence of a wall. *Journal of Fluid Mechanics*, 182:255–290, Sept. 1987. doi: 10.1017/S0022112087002337. (cited on page 24)

G. Talamini, D. Shao, X. Su, X. Guo, and X. Ji. Combined sewer overflow in Shenzhen, China: The case study of Dasha River. In *SUSTAINABLE DEVELOPMENT AND PLANNING 2016*, pages 785–796, Penang, Malaysia, Dec. 2016. doi: 10.2495/SDP160661. (cited on page 2)

S. Taneda. Experimental Investigation of the Wake behind a Sphere at Low Reynolds Numbers. *Journal of the Physical Society of Japan*, 11(10):1104–1108, Oct. 1956. doi: 10.1143/JPSJ.11.1104. (cited on page 42)

M. Tanemura. Statistical distributions of Poisson Voronoi cells in two and three dimensions. *Forma*, 18 (4):221–247, 2003. (cited on page 142)

G. I. Taylor. Diffusion by continuous movements. *Proceedings of the london mathematical society*, 2(1): 196–212, 1922. (cited on page 43)

G. I. Taylor. Dispersion of soluble matter in solvent flowing slowly through a tube. *Proceedings of the Royal Society of London. Series A. Mathematical and Physical Sciences*, 219(1137):186–203, Aug. 1953. doi: 10.1098/rspa.1953.0139. (cited on page 43)

G. I. Taylor. The dispersion of matter in turbulent flow through a pipe. *Proceedings of the Royal Society of London. Series A. Mathematical and Physical Sciences*, 223(1155):446–468, May 1954. doi: 10.1098/ rspa.1954.0130. (cited on page 43)

T. Theodorsen. Mechanisms of turbulence. In *Proceedings of the 2nd Midwestern Conference on Fluid Mechanics*, Ohio State University, 1952. (cited on page 25)

R. V. Thomann and J. A. Mueller. *Principles of Surface Water Quality Modeling and Control.* Harper & Row, 1987. ISBN 978-0-06-350728-9. (cited on page 187)

W. H. Thomas and C. H. Gibson. Effects of small-scale turbulence on microalgae. *Journal of Applied Phycology*, 2(1):71–77, Mar. 1990. doi: 10.1007/BF02179771. (cited on page 1)

S. Toh and T. Itano. A periodic-like solution in channel flow. *Journal of Fluid Mechanics*, 481:67–76, Apr. 2003. doi: 10.1017/S0022112003003768. (cited on pages 31 and 174)

P. F. Tooby, G. L. Wick, and J. D. Isaacs. The motion of a small sphere in a rotating velocity field: A possible mechanism for suspending particles in turbulence. *Journal of Geophysical Research (1896-1977)*, 82 (15):2096–2100, 1977. doi: 10.1029/JC082i015p02096. (cited on page 40)

A. A. Townsend. Equilibrium layers and wall turbulence. *Journal of Fluid Mechanics*, 11(1):97–120, Aug. 1961. doi: 10.1017/S0022112061000883. (cited on page 25)

A. A. R. Townsend. *The Structure of Turbulent Shear Flow*. Cambridge University Press, 2nd edition, 1976. ISBN 0-521-29819-9. (cited on page 25)

M. Uhlmann. An immersed boundary method with direct forcing for the simulation of particulate flows. *Journal of Computational Physics*, 209(2):448–476, 2005. doi: 10.1016/j.jcp.2005.03.017. (cited on pages 49, 50, and 200)

M. Uhlmann. Interface-resolved direct numerical simulation of vertical particulate channel flow in the turbulent regime. *Physics of Fluids*, 20(5), 2008. doi: 10.1063/1.2912459. (cited on pages 62 and 70)

M. Uhlmann. DNS of scalar transport & solid particle motion: Numerical aspects and validation. Technical report, Karlsruhe Institute of Technology, Karlsruhe, 2019. (cited on page 53)

M. Uhlmann. Voronoï tessellation analysis of sets of randomly placed finite-size spheres. *Physica A: Statistical Mechanics and its Applications*, 555:124618, Oct. 2020. doi: 10.1016/j.physa.2020.124618. (cited on page 142)

M. Uhlmann and A. Chouippe. Clustering and preferential concentration of finite-size particles in forced homogeneous-isotropic turbulence. *Journal of Fluid Mechanics*, 812:991–1023, Feb. 2017. doi: 10.1017/jfm.2016.826. (cited on page 39)

M. Uhlmann and T. Doychev. Sedimentation of a dilute suspension of rigid spheres at intermediate Galileo numbers: The effect of clustering upon the particle motion. *Journal of Fluid Mechanics*, 752: 310–348, Aug. 2014. doi: 10.1017/jfm.2014.330. (cited on pages 42, 100, and 102)

M. Uhlmann and J. Dušek. The motion of a single heavy sphere in ambient fluid: A benchmark for interface-resolved particulate flow simulations with significant relative velocities. *International Journal of Multiphase Flow*, 59:221–243, 2014. doi: 10.1016/j.ijmultiphaseflow.2013.10.010. (cited on page 66)

L. Umlauf and H. Burchard. A generic length-scale equation for geophysical turbulence models. *Journal of Marine Research*, 61(2):235–265, Mar. 2003. doi: 10.1357/002224003322005087. (cited on page 9)

USEPA. Engineering Investigation of Sewer Overflow Problems: Roanoke, Virginia. Technical report, U.S. Environmental Protection Agency, 1970. (cited on page 2)

USEPA. Storm Water Problems and Control in Sanitary Sewers: Oakland and Berkeley, California. Technical report, U.S. Environmental Protection Agency, 1971. (cited on page 2)

USEPA. Combined Sewer Overflows Guidance for Screening and Ranking. Technical report, U.S. Environmental Protection Agency, 1995. (cited on page 2)

USPHS. Pollution effects of stormwater and overflows from combined sewer systems, a preliminary appraisal. Technical report, US Public Health Service, 1964. (cited on page 3)

L. van Rijn. Sediment Transport, Part I: Bed Load Transport. *Journal of Hydraulic Engineering*, 110(10): 1431–1456, Oct. 1984a. doi: 10.1061/(ASCE)0733-9429(1984)110:10(1431). (cited on page 35)

L. C. van Rijn. Sediment Transport, Part II: Suspended Load Transport. *Journal of Hydraulic Engineering*, 110(11):1613–1641, Nov. 1984b. doi: 10.1061/(ASCE)0733-9429(1984)110:11(1613). (cited on page 35)

L. C. van Rijn. Sediment Transport, Part III: Bed forms and Alluvial Roughness. *Journal of Hydraulic Engineering*, 110(12):1733–1754, Dec. 1984c. doi: 10.1061/(ASCE)0733-9429(1984)110:12(1733). (cited on page 35)

L. Vergeynst, B. Vallet, and P. A. Vanrolleghem. Modelling pathogen fate in stormwaters by a particle–pathogen interaction model using population balances. *Water Science and Technology*, 65(5): 823–832, Mar. 2012. doi: 10.2166/wst.2012.818. (cited on page 7)

R. Verzicco and P. Orlandi. A Finite-Difference Scheme for Three-Dimensional Incompressible Flows in Cylindrical Coordinates. *Journal of Computational Physics*, 123(2):402–414, Feb. 1996. doi: 10.1006/jcph.1996.0033. (cited on page 50)

T. Vicsek, A. Czirók, E. Ben-Jacob, I. Cohen, and O. Shochet. Novel Type of Phase Transition in a System of Self-Driven Particles. *Physical Review Letters*, 75(6):1226–1229, Aug. 1995. doi: 10.1103/PhysRevLett.75.1226. (cited on page 14)

I. Vinkovic, D. Doppler, J. Lelouvetel, and M. Buffat. Direct numerical simulation of particle interaction with ejections in turbulent channel flows. *International Journal of Multiphase Flow*, 37(2):187–197, Mar. 2011. doi: 10.1016/j.ijmultiphaseflow.2010.09.008. (cited on page 44)

D. Viswanath. Recurrent motions within plane Couette turbulence. *Journal of Fluid Mechanics*, 580: 339–358, June 2007. doi: 10.1017/S0022112007005459. (cited on pages 31 and 174)

Y. Wada, H. Miura, and M. Muraoka. Evaluation and Analysis of CSO Control with Storm Water Reservoir in Osaka, Japan. In *Global Solutions for Urban Drainage*, pages 1–8, Portland, Oregon, United States of America, 2002. American Society of Civil Engineers. doi: 10.1061/40644(2002)107. (cited on pages 2 and 3)

F. Waleffe. On the origin of the streak spacing in turbulent shear flows. *Annual Research Briefs-1990. Center for Turbulence Research, Stanford Univ./NASA-Ames*, pages 159–168, 1991. (cited on page 24)

F. Waleffe. Three-Dimensional Coherent States in Plane Shear Flows. *Physical Review Letters*, 81(19): 4140–4143, Nov. 1998. doi: 10.1103/PhysRevLett.81.4140. (cited on page 32)

F. Waleffe. Exact coherent structures in channel flow. *Journal of Fluid Mechanics*, 435:93–102, May 2001. doi: 10.1017/S0022112001004189. (cited on pages 30, 31, and 174)

F. Waleffe. Homotopy of exact coherent structures in plane shear flows. *Physics of Fluids*, 15(6):1517–1534, May 2003. doi: 10.1063/1.1566753. (cited on page 32)

F. Waleffe, J. Kim, and J. M. Hamilton. On the Origin of Streaks in Turbulent Shear Flows. In F. Durst, R. Friedrich, B. E. Launder, F. W. Schmidt, U. Schumann, and J. H. Whitelaw, editors, *Turbulent Shear Flows 8*, pages 37–49, Berlin, Heidelberg, 1993. Springer. ISBN 978-3-642-77674-8. doi: 10.1007/978-3-642-77674-8_4. (cited on page 24)

D. P. Wall and M. Nagata. Exact coherent states in channel flow. *Journal of Fluid Mechanics*, 788:444–468, Feb. 2016. doi: 10.1017/jfm.2015.685. (cited on page 174)

J. M. Wallace, H. Eckelmann, and R. S. Brodkey. The wall region in turbulent shear flow. *Journal of Fluid Mechanics*, 54(1):39–48, July 1972. doi: 10.1017/S0022112072000515. (cited on page 77)

E. Walters. *Fate and Transport of Fecal Indicator Bacteria in Flume Systems Mimicking an Oligotrophic River.* PhD thesis, TU München, 2013. (cited on pages 4, 5, and 7)

E. Walters, M. Graml, C. Behle, E. Müller, and H. Horn. Influence of Particle Association and Suspended Solids on UV Inactivation of Fecal Indicator Bacteria in an Urban River. *Water, Air, & Soil Pollution*, 225(1):1822, Jan. 2014. doi: 10.1007/s11270-013-1822-8. (cited on pages 3 and 4)

G. Wang and D. H. Richter. Modulation of the turbulence regeneration cycle by inertial particles in planar Couette flow. *Journal of Fluid Mechanics*, 861:901–929, Feb. 2019. doi: 10.1017/jfm.2018.936. (cited on page 41)

G. Wang, M. Abbas, and E. Climent. Modulation of large-scale structures by neutrally buoyant and inertial finite-size particles in turbulent Couette flow. *Physical Review Fluids*, 2(8):084302, Aug. 2017. doi: 10.1103/PhysRevFluids.2.084302. (cited on page 41)

G. Wang, M. Abbas, and E. Climent. Modulation of the regeneration cycle by neutrally buoyant finite-size particles. *Journal of Fluid Mechanics*, 852:257–282, Oct. 2018a. doi: 10.1017/jfm.2018.513. (cited on page 41)

L.-P. Wang and M. R. Maxey. Settling velocity and concentration distribution of heavy particles in homogeneous isotropic turbulence. *Journal of Fluid Mechanics*, 256:27–68, Nov. 1993. doi: 10.1017/S0022112093002708. (cited on page 40)

Y. Wang, K. M. Lam, and Y. Lu. Settling velocity of fine heavy particles in turbulent open channel flow. *Physics of Fluids*, 30(9):095106, Sept. 2018b. doi: 10.1063/1.5046333. (cited on page 44)

J. Wilkinson, A. Jenkins, M. Wyer, and D. Kay. Modelling faecal coliform dynamics in streams and rivers. *Water Research*, 29(3):847–855, Mar. 1995. doi: 10.1016/0043-1354(94)00211-O. (cited on pages 4, 6, 45, 62, 192, 199, and 204)

W. W. Willmarth and S. S. Lu. Structure of the Reynolds stress near the wall. *Journal of Fluid Mechanics*, 55(1):65–92, Sept. 1972. doi: 10.1017/S002211207200165X. (cited on page 77)

M. Wong and G. Parker. Reanalysis and Correction of Bed-Load Relation of Meyer-Peter and Müller Using Their Own Database. *Journal of Hydraulic Engineering*, 132(11):1159–1168, Nov. 2006. doi: 10.1061/(ASCE)0733-9429(2006)132:11(1159). (cited on pages 59 and 61)

D. A. Wright, K. Killham, L. A. Glover, and J. I. Prosser. Role of Pore Size Location in Determining Bacterial Activity during Predation by Protozoa in Soil. *Applied and Environmental Microbiology*, 61(10):3537–3543, Oct. 1995. doi: 10.1128/aem.61.10.3537-3543.1995. (cited on page 4)

J. Wu and R. Manasseh. Dynamics of dual-particles settling under gravity. *International Journal of Multiphase Flow*, 24(8):1343–1358, Dec. 1998. doi: 10.1016/S0301-9322(98)00029-9. (cited on page 42)

J.-S. Wu and G. M. Faeth. Sphere wakes in still surroundings at intermediate Reynolds numbers. *AIAA Journal*, 31(8):1448–1455, 1993. doi: 10.2514/3.11794. (cited on page 113)

J.-S. Wu and G. M. Faeth. Sphere wakes at moderate Reynolds numbers in a turbulent environment. *AIAA journal*, 32(3):535–541, 1994. (cited on pages 41 and 113)

J.-S. Wu and GM. Faeth. Effect of ambient turbulence intensity on sphere wakes at intermediate Reynolds number. *AIAA Journal*, 33(1):171–173, 1995. (cited on page 41)

X. Wu and P. Moin. Direct numerical simulation of turbulence in a nominally zero-pressure-gradient flat-plate boundary layer. *Journal of Fluid Mechanics*, 630:5–41, July 2009. doi: 10.1017/S0022112009006624. (cited on page 25)

I. J. Wygnanski and F. H. Champagne. On transition in a pipe. Part 1. The origin of puffs and slugs and the flow in a turbulent slug. *Journal of Fluid Mechanics*, 59(2):281–335, June 1973. doi: 10.1017/S0022112073001576. (cited on page 29)

C. Y. Yang and U. Lei. The role of the turbulent scales in the settling velocity of heavy particles in homogeneous isotropic turbulence. *Journal of Fluid Mechanics*, 371:179–205, Sept. 1998. doi: 10.1017/S0022112098002328. (cited on page 40)

Q. Yang, A. P. Willis, and Y. Hwang. Exact coherent states of attached eddies in channel flow. *Journal of Fluid Mechanics*, 862:1029–1059, Mar. 2019. doi: 10.1017/jfm.2018.1017. (cited on pages 30 and 174)

T. S. Yang and S. S. Shy. The settling velocity of heavy particles in an aqueous near-isotropic turbulence. *Physics of Fluids*, 15(4):868–880, Mar. 2003. doi: 10.1063/1.1557526. (cited on page 40)

X. Yin and D. L. Koch. Hindered settling velocity and microstructure in suspensions of solid spheres with moderate Reynolds numbers. *Physics of Fluids*, 19(9):093302, Sept. 2007. doi: 10.1063/1.2764109. (cited on page 43)

Z. Yu, Y. Xia, Y. Guo, and J. Lin. Modulation of turbulence intensity by heavy finite-size particles in upward channel flow. *Journal of Fluid Mechanics*, 913, Apr. 2021. doi: 10.1017/jfm.2020.1140. (cited on page 118)

L. I. Zaichik and V. M. Alipchenkov. Pair dispersion and preferential concentration of particles in isotropic turbulence. *Physics of Fluids*, 15(6):1776–1787, June 2003. doi: 10.1063/1.1569485. (cited on page 38)

L. I. Zaichik and V. M. Alipchenkov. Refinement of the probability density function model for preferential concentration of aerosol particles in isotropic turbulence. *Physics of Fluids*, 19(11):113308, Nov. 2007. doi: 10.1063/1.2813044. (cited on page 38)

L. I. Zaichik and V. M. Alipchenkov. Statistical models for predicting pair dispersion and particle clustering in isotropic turbulence and their applications. *New Journal of Physics*, 11(10):103018, Oct. 2009. doi: 10.1088/1367-2630/11/10/103018. (cited on page 38)

Y. Zhang and A. M. Baptista. SELFE: A semi-implicit Eulerian–Lagrangian finite-element model for cross-scale ocean circulation. *Ocean Modelling*, 21(3):71–96, Jan. 2008. doi: 10.1016/j.ocemod.2007.11.005. (cited on pages 8, 9, and 15)

L. Zhao, H. I. Andersson, and J. J. J. Gillissen. Interphasial energy transfer and particle dissipation in particle-laden wall turbulence. *Journal of Fluid Mechanics*, 715:32–59, Jan. 2013. doi: 10.1017/jfm.2012.492. (cited on page 41)

J. Zhou, R. J. Adrian, and S. Balachandar. Autogeneration of near-wall vortical structures in channel flow. *Physics of Fluids*, 8(1):288–290, Jan. 1996. doi: 10.1063/1.868838. (cited on page 25)

J. Zhou, R. J. Adrian, S. Balachandar, and T. M. Kendall. Mechanisms for generating coherent packets of hairpin vortices in channel flow. *Journal of Fluid Mechanics*, 387:353–396, May 1999. doi: 10.1017/S002211209900467X. (cited on pages 25 and 26)

C. Zhu, Z. Yu, D. Pan, and X. Shao. Interface-resolved direct numerical simulations of the interactions between spheroidal particles and upward vertical turbulent channel flows. *Journal of Fluid Mechanics*, 891, May 2020. doi: 10.1017/jfm.2020.159. (cited on page 118)

Dissertationsreihe am Institut für Hydromechanik
Karlsruher Institut für Technologie (KIT)

ISSN 1439-4111

Die Bände sind unter www.ksp.kit.edu als PDF frei verfügbar oder als Druckausgabe bestellbar.

2004/1 **Volker Weitbrecht**
Influence of dead-water zones on the dispersive mass-transport in rivers.
ISBN 3-937300-07-4

2004/2 **Christof Hinterberger**
Dreidimensionale und tiefengemittelte Large-Eddy-Simulation von Flachwasserströmungen.
ISBN 3-937300-18-X

2005/1 **Cornelia Frank**
Wirksamkeit von dünnen Windschutzstreifen auf Sockelwällen in luv- und leeseitiger Anordnung.
ISBN 3-937300-55-4

2005/2 **Carl Friedrich von Carmer**
Shallow turbulent wake flows: momentum and mass transfer due to large-scale coherent vortical structures.
ISBN 3-937300-63-5

2005/3 **Muhammad Ikhwan**
Investigation of flow and pressure characteristics around pyramidal buildings.
ISBN 3-937300-72-4

2005/4 **Herlina**
Gas transfer at the air-water interface in a turbulent flow environment.
ISBN 3-937300-74-0

2006/1 **Manual García-Villalba Navaridas**
Large eddy simulation of turbulent swirling jets.
ISBN 3-86644-015-4

2006/2 **Klaas E. Heinrich**
Hydraulische Steuerung einer Wasser-Alkohol-Zirkulation mittels Grundwasser-Zirkulations-Brunnen.
ISBN 3-86644-055-3

2006/3 **Tobias Bleninger**
Coupled 3D hydrodynamic models for submarine outfalls. Denvironmental hydraulic design and control of multiport diffusers.
ISBN 978-3-86644-160-6

2006/3 **Patrick Heneka**
Schäden durch Winterstürme – das Schadensrisiko von Wohngebäuden in Baden-Württemberg.
ISBN 978-3-86644-117-0

2007/1 **Maria Eletta Negretti**
Hydrodynamic instabilities and entrainment in two-layer stratified flows down a slope.
ISBN 978-3-86644-170-5

2007/2 **Gregor Kühn**
Untersuchungen zur Feinsedimentdynamik unter Turbulenzeinfluss.
ISBN 978-3-86644-205-4

2008/1 **Martin Detert**
Hydrodynamic processes at the water-sediment interface of streambeds.
ISBN 978-3-86644-244-3

2008/2 **Christof Gromke**
Einfluss von Bäumen auf die Durchlüftung von innerstädtischen Straßenschluchten.
ISBN 978-3-86644-339-6

2009/1 **Ekkehart Bethge**
Risikoberechnung zum Schadstoffeintrag aus Hochwasserretentionsräumen
in einen Grundwasserleiter.
ISBN 978-3-86644-426-3

2012/1 **Clemens Chan-Braun**
Turbulent open channel flow, sediment erosion and sediment transport.
ISBN 978-3-86644-900-8

2015/1 **Todor Doychev**
The dynamics of finite-size settling particles.
ISBN 978-3-7315-0307-1

2016/1 **Aman Ghebremichael Kidanemariam**
The formation of patterns in subaqueous sediment.
ISBN 978-3-7315-0532-7

2019/1 **Yoshiyuki Sakai**
Coherent structures and secondary motions in open duct flow.
ISBN 978-3-7315-0790-1

2023/1 **Michael Werner Tobias Krayer**
A hydrodynamical perspective on the turbulent transport of bacteria in rivers.
ISBN 978-3-7315-1310-0